The Structural Basis of Arrestin Functions

Vsevolod V. Gurevich
Editor

The Structural Basis of Arrestin Functions

Springer

Editor
Vsevolod V. Gurevich
Department of Pharmacology
Vanderbilt University
Nashville, TN
USA

ISBN 978-3-319-57552-0 ISBN 978-3-319-57553-7 (eBook)
DOI 10.1007/978-3-319-57553-7

Library of Congress Control Number: 2017938139

Printed on acid-free paper

This Springer imprint is published by Springer Nature
The registered company is Springer International Publishing AG
The registered company address is: Gewerbestrasse 11, 6330 Cham, Switzerland

Preface

Arrestin, like all key players in the G protein-coupled receptor (GPCR) signaling, was first discovered in the visual system, where it was shown to specifically bind active phosphorylated rhodopsin, "arresting" its coupling to a cognate G protein. Subsequently, three homologues of visual arrestin were cloned and functionally characterized. All members of the arrestin family bind active phosphorylated forms of their cognate receptors and stop G protein-mediated signaling of most GPCRs. Structurally ~45 kDa arrestin proteins are elongated two-domain molecules with the overall fold shared with (likely inherited from) arrestin domain-containing proteins (ARRDCs) involved in trafficking of membrane vesicles and proteins. Vertebrates (except bony fish that underwent an additional whole-genome duplication event) express only four arrestin subtypes, whereas other animals have even fewer arrestins. Recently, the focus has shifted to the non-visual arrestins, two subtypes of which are expressed in every vertebrate species. These two arrestins bind hundreds of different GPCRs and interact with numerous non-receptor partners, including various trafficking and signaling proteins. Interactions with some partners do not depend on receptor binding, whereas others are facilitated or suppressed by GPCR interactions. Thus, arrestins are at a crossroads of signaling pathways, participating in the integration of external and internal stimuli into coherent behavior of the cell. High biological importance stimulated structure-function studies of arrestins. This book summarizes the structural insights into arrestin functions gained in the last few years, focusing on the mechanisms of arrestin-mediated regulation of GPCRs and other signaling proteins in healthy cells, as well as in disease. The chapters, written by the people who made those discoveries, describe the molecular mechanisms of arrestin binding to GPCRs and other partners, emphasizing the therapeutic potential of modifying individual arrestin functions. By binding GPCRs and other partners, arrestins organize multi-protein signaling complexes and localize them to specific compartments in the cell. Importantly, arrestins act via protein–protein interactions, which underlie most vital cellular functions. On the basis of structure-function information, modified arrestins or their mono-functional elements can serve as prototypical next-generation tools for research and therapy that channel cell signaling to desired outcomes.

Nashville, TN, USA Vsevolod V. Gurevich

Contents

Part I
Introduction

Chapter 1
Arrestins: Discovery of the Family and Functional Role of Conformational Flexibility

Vsevolod V. Gurevich and Eugenia V. Gurevich

Abstract Arrestins were first discovered as negative regulators of G protein-mediated signaling by GPCRs: they bind GRK-phosphorylated active receptors and preclude their further coupling to cognate G proteins. Vertebrates have only four arrestin subtypes, two of which are specifically expressed in photoreceptors in the retina, whereas the two non-visual arrestins are ubiquitously expressed and apparently regulate hundreds of different GPCRs. Later studies showed that, in addition to receptors, non-visual arrestins interact with dozens of other signaling proteins. Current view is that arrestins are versatile regulators organizing multi-protein signaling complexes and localizing them to particular subcellular compartments. Arrestins can exist in several distinct conformations, the best-studied being basal and receptor-bound (often termed "active"), which differentially engage various partners. Identification of the arrestin elements engaged by each partner and construction of signaling-biased arrestins where individual functions are selectively disrupted or enhanced, helps us to elucidate their biological roles in the cell.

Keywords Arrestin · GPCR · Desensitization · Signaling proteins · Scaffolding · Protein conformation

Brief History of Arrestins

All key players in G protein-coupled receptor (GPCR) signalling were first discovered in photoreceptors, and arrestin is no exception. Unlike others, this protein was discovered before its biological function was established: it was first described

---

V.V. Gurevich (✉) · E.V. Gurevich (✉)
Department of Pharmacology, Vanderbilt University, 2200 Pierce Avenue, Preston Research Building, Rm. 454, Nashville, TN 37232, USA
e-mail: vsevolod.gurevich@vanderbilt.edu

E.V. Gurevich
e-mail: eugenia.gurevich@vanderbilt.edu

V.V. Gurevich (ed.), *The Structural Basis of Arrestin Functions*,
DOI 10.1007/978-3-319-57553-7_1

as soluble antigen (S-antigen for short) targeted by auto-antibodies in uveitis (Wacker et al. 1977; Wacker and Lipton 1965). Its gene is still called *Sag* in the HUGO database. Meanwhile, pioneering studies by Kuhn identified a 48 kDa protein (that was the name he gave it) that specifically binds light-activated rhodopsin (Kuhn 1978). He then showed that the binding of 48 kDa protein increases upon rhodopsin phosphorylation (Kuhn et al. 1984) and that it blocks light-induced activation of photoreceptor phosphodiesterase by rhodopsin (Wilden et al. 1986). Next, it was established that S-antigen and 48 kDa protein were the same protein (Pfister et al. 1984, 1985). As S-antigen/48 kDa protein stopped ("arrested") rhodopsin signaling, it received a new name, arrestin (Zuckerman and Cheasty 1986). Arrestin was reported to bind ATP (Zuckerman and Cheasty 1988) and Ca^{2+} (Huppertz et al. 1990), but subsequent experiments debunked these claims (Palczewski and Hargrave 1991). Within a few years, arrestin was cloned (Yamaki et al. 1987; Shinohara et al. 1987). Its primary sequence suggested possible structural homology with the α-subunit of transducin (Yamaki et al. 1987; Tsuda et al. 1988), but the crystal structures of these proteins disproved this hypothesis (Noel et al. 1993; Sondek et al. 1994; Hirsch et al. 1999; Granzin et al. 1998).

Seminal discovery that β2-adrenergic receptor (β2AR) (Dixon et al. 1986) structurally resembles rhodopsin (Ovchinnikov 1982) revealed the existence of a family of rhodopsin-like GPCRs (now called class A). Ever since the idea that key signaling and desensitization mechanisms are shared by all GPCRs is widely accepted in the field. The kinase that preferentially phosphorylates agonist-activated β2AR, similar to activation-dependent phosphorylation of rhodopsin by rhodopsin kinase (Shichi and Somers 1978), was cloned in 1989 (Benovic et al. 1989). By analogy with the rhodopsin kinase [which was cloned later (Lorenz et al. 1991); modern systematic name GRK1], this new kinase was named βARK (β2AR kinase) (Benovic et al. 1989) (modern systematic name GRK2). βARK was purified and found not to reduce β2AR-mediated activation of cognate G protein much, which suggested that there must be an equivalent of arrestin that binds phosphorylated β2AR (Benovic et al. 1987). Then the first non-visual arrestin was cloned by homology (Lohse et al. 1990). It showed clear preference for β2AR over rhodopsin, and therefore was termed β-arrestin, while the original arrestin was renamed visual or rod arrestin (Lohse et al. 1990, 1992). All three names turned out to be not quite correct. The cloning of cone-specific arrestin [termed cone (Craft et al. 1994) or X-arrestin (Murakami et al. 1993)] made the term "visual" equivocal. Subsequent finding that the expression of this specific subtype in cone photoreceptors was relatively low (Chan et al. 2007), whereas cones contain ~50-fold higher level of "rod" subtype (Nikonov et al. 2008) showed that the terms "cone" and "rod" arrestin are inaccurate. After cloning of another non-visual subtype, which was called β-arrestin2 (Attramadal et al. 1992) [independently discovered under the name hTHY-ARRX (Rapoport et al. 1992)], the original non-visual arrestin was renamed β-arrestin1. Finally, the demonstration that β-arrestin1 binds not only β2AR, but numerous GPCRs (Barak et al. 1997; Gurevich et al. 1993, 1995), showed that the term "β-arrestin" is also inaccurate. Arrestin nomenclature remains confusing to this day. Here we use systematic names of arrestin proteins, numbered in the order of cloning: arrestin-1 (historic names S-antigen, 48 kDa protein, visual or rod arrestin), arrestin-2 (β-arrestin or β-arrestin1), arrestin-3 (β-arrestin2 or hTHY-ARRX), and arrestin-4 (cone or X-arrestin).

The first function of arrestins discovered was selective binding to active phosphorylated GPCRs, which blocked receptor interactions with G proteins by simple competition (Wilden 1995; Krupnick et al. 1997a). This aspect was recently confirmed by crystallography: Gs in the complex with β2AR (Rasmussen et al. 2007) and arrestin-1 in complex with rhodopsin (Kang et al. 2015) engage the same cavity that opens on the cytoplasmic side between the helices upon GPCR activation (Farrens et al. 1996). The binding of non-visual arrestins to non-receptor partners was only discovered in 1996, when clathrin, the key component of the internalization machinery of coated pits, was shown to bind both non-visual arrestins directly (Goodman et al. 1996). Many additional non-GPCR partners of arrestin proteins were described since: clathrin adaptor AP2 (Laporte et al. 1999), protein kinases c-Src (Luttrell et al. 1999; DeFea et al. 2000a), JNK3 (McDonald et al. 2000), and ERK1/2 (DeFea et al. 2000b; Luttrell et al. 2001) and their upstream activators were shown to bind non-visual arrestins. The number of arrestin binding partners now exceeds 100 (Xiao et al. 2007), and new ones keep being discovered, including ubiquitin ligases AIP4 (Bhandari et al. 2007) and parkin (Ahmed et al. 2011), protein deubiquitinase USP33 (Shenoy et al. 2009), etc.

The Functional Cycle of Arrestins

Arrestins bind many non-receptor signaling proteins, but the concept of arrestin functional cycle is still dominated by receptor binding: the first recognized functional (and likely structural) states of arrestins were "free" and GPCR-bound (Gurevich and Gurevich 2003, 2004, 2006a). A very convincing early study demonstrated that receptor binding involves a global conformational change in arrestin-1 (Schleicher et al. 1989), which was subsequently shown to include the release of the arrestin C-terminus (Palczewski et al. 1991a, b). Thus, the finding that both clathrin and AP2 binding sites are localized in the C-termini of non-visual arrestins (Goodman et al. 1996; Laporte et al. 1999; Kim and Benovic 2002) made perfect sense: only receptor-bound arrestin needs to help recruit its partner to the coated pit (Gurevich and Gurevich 2003). Indeed, the expression of separated C-terminus of arrestin-2 inhibited GPCR internalization, apparently via successful competition with the arrestin-receptor complexes for clathrin and AP2 (Krupnick et al. 1997b). The first studies of arrestin-mediated activation of protein kinases c-Src, JNK3, and ERK1/2 showed that the signaling in these pathways was triggered by GPCR activation, i.e., was mediated by receptor-bound arrestins (Luttrell et al. 1999, 2001; McDonald et al. 2000; DeFea et al. 2000a, b). These findings established a new paradigm of the G protein-independent arrestin-mediated GPCR signalling (DeWire et al. 2007), while also implying that free arrestin molecules do not do anything of importance. Based on this assumption, the basal conformation of arrestins was termed "inactive". The findings that at least two binding partners, ubiquitin ligases Mdm2 (Song et al. 2006) and parkin (Ahmed et al. 2011), preferentially bind arrestins in their basal conformation, and that "pre-activated" arrestin-3 mutant with enhanced receptor-binding ability fails to facilitate the activation of JNK3 (Breitman et al. 2012), suggest that this notion is unlikely to be correct. In addition, most reviews of the subject also implied that only non-visual arrestins do anything interesting, whereas visual arrestin-1 and -4 only bind active phosphorylated photopigments, terminating their signaling via cognate G proteins. However, recent findings that arrestin-1 and arrestin-4 bind Mdm2 and JNK3 (Song et al. 2006, 2007), arrestin-1 binds NSF and regulates its activity in rods (Huang et al. 2010), and that arrestin-1 binds AP2 in photoreceptors with high enough affinity to deplete it and cause cell death (Moaven et al. 2013) suggest otherwise. Finally, arrestins bind microtubules (Hanson et al. 2006a, 2007), and this interaction also induces a conformational change, distinct from that induced by receptor binding (Hanson et al. 2006a), which suggests that arrestins can assume yet another conformation, microtubule-bound. Since phenomenology of arrestin-mediated signalling has been extensively reviewed (DeWire et al. 2007; Gurevich and Gurevich 2006b, 2014), here we focus on conformational selectivity of arrestin-binding proteins. Existing evidence suggests that arrestin binding partners have different preferences: some interact with receptor-bound arrestins with the highest affinity, some with free (basal) forms, some with microtubule-associated, whereas there are partners that do not show a clear preference, apparently engaging arrestin elements that are exposed in all these states and do not change much.

The Structural Evidence for Distinct Arrestin Conformations

The first structures of the four vertebrate subtypes, arrestin-1 (Hirsch et al. 1999; Granzin et al. 1998), arrestin-2 (Han et al. 2001; Milano et al. 2002), arrestin-3 (Zhan et al. 2011a), and arrestin-4 (Sutton et al. 2005), represented the basal state. They turned out to be very similar overall, with a few subtle differences (Fig. 1.1). All arrestins are elongated two-domain molecules with relatively few contacts between the domains, perfectly suited for a large conformational change suggested by earlier biophysical (Schleicher et al. 1989) and mutagenesis (Gurevich and Benovic 1993) studies. Three main inter-domain interactions were identified (Hirsch et al. 1999): (1) the "polar core", an arrangement of five charged residues from both N- and C-domains and the C-tail, which are largely solvent-excluded; (2) three-element interaction of the C-tail with β-strand I and α-helix I, which anchors the C-tail to the N-domain (corresponding to predicted on the basis of mutagenesis data (Gurevich and Benovic 1993; Gurevich et al. 1994) interaction of the N- and C-termini of arrestin); and a number of hydrophobic interactions between the residues in both domains at their interface. Interestingly, the polar core

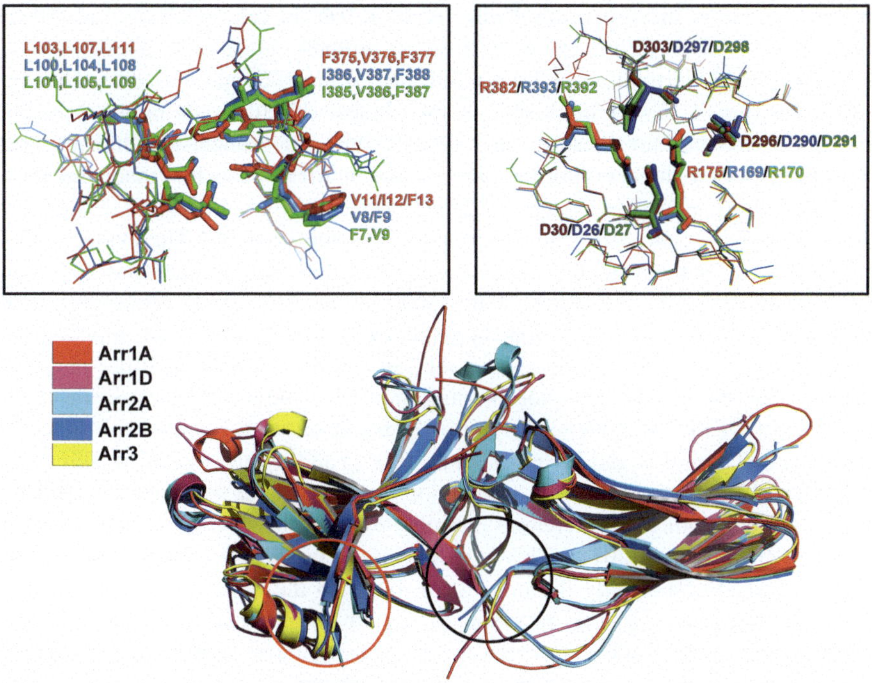

Fig. 1.1 Basal conformation of different arrestin subtypes. Superimposition of the crystal structures of the two monomers in arrestin-1 tetramer [*Arr1A red*, and *Arr1D pink*; PDB ID 1CF1 (Hirsch et al. 1999)], two monomers of arrestin-2 dimer [*Arr2A light blue*, and *Arr2B dark blue*; PDB ID 1G4M (Han et al. 2001)], and arrestin-3 [*Arr3 yellow*; PDB ID 3P2D (Zhan et al. 2011a)] shows remarkably similar cores of both domains and variable structure of the loops. Importantly, the variability of each loop in different monomers of the same arrestin (compare *Arr1A* and *Arr1D*, as well as *Arr2A* and *Arr2B*) is essentially as great as between arrestin subtypes, suggesting that it reflects the flexibility of these loops, rather than their subtype-specific conformations. *Black* and *red circles* show the location of the two key intra-molecular interactions that hold arrestins in their basal state, the polar core and the three-element interaction between β-strand I, α-helix I, and the β-strand XX in the C-tail, respectively. The panels above show detailed structure of these elements. *Right* the polar core, main phosphate sensor; *left* the three-element interaction. In both panels residue numbers are indicated, as follows: arrestin-1 *red*, arrestin-2 *blue*, arrestin-3 *green*

of arrestin-1 included Arg175, earlier predicted to be the key part of the phosphate sensor ensuring preferential binding to phosphorylated receptors (Gurevich and Benovic 1995, 1997). Indeed, the disruption by receptor-attached phosphates of the salt bridge between Arg175 and Asp296 in the polar core was shown to be the mechanism of arrestin "activation" by the receptor-attached phosphates (Hirsch et al. 1999; Vishnivetskiy et al. 1999). Importantly, the disruption of this salt bridge by charge reversal mutations of either residue yields "pre-activated" arrestins that do not require phosphates, binding both phosphorylated and unphosphorylated active GPCRs with high affinity (Gurevich and Benovic 1995, 1997). This phosphorylation-independent interaction was shown to be strong enough to block G protein activation in vitro (Gray-Keller et al. 1997), in cells (Celver et al. 2001, 2002; Kovoor et al. 1999), and even in living animals (Song et al. 2009a).

Parts of the C-terminus that are not attached to the N-domain via three-element interaction (Fig. 1.1) were not visible in any of these structures, strongly suggesting that these elements are "disordered", i.e., flexible. The conclusion that receptor binding induces a global conformational change in arrestin was first made based on high Arrhenius activation energy of arrestin-1 binding to rhodopsin (Schleicher et al. 1989) before any structural information became available. The evidence that this conformational change involves the release of the arrestin C-terminus was also reported long before structures (Palczewski et al. 1991a). Interestingly, both phosphorhodopsin and polyanion heparin were found to induce this release (Palczewski et al. 1991a), suggesting that any molecule with multiple negative charges can do it. The receptor binding-induced C-terminus release was later confirmed by EPR, as evidenced by an increase in distance between spin label in the C-tail and both β-strand I (Hanson et al. 2006b) and α-helix (Vishnivetskiy et al. 2010) involved in three-element interaction with it (Hirsch et al. 1999), and by NMR, demonstrating an increased mobility of the distal C-tail upon receptor binding (Zhuang et al. 2013). This agrees with the prediction of a previous mutagenesis study that not only the polar core, but also the three-element interaction must be disrupted by the receptor to ensure high-affinity arrestin binding (Vishnivetskiy et al. 2000). An extensive study using pulse EPR method DEER to measure intra-molecular distances in free and rhodopsin-bound arrestin-1 (Kim et al. 2012) revealed several additional receptor binding-induced conformational changes: the outward movement of the "finger loop" in the central crest of the receptor-binding surface, large movement of the adjacent "139-loop" to the side and towards the N-domain, as well as the movements of distal tips of both domains. Small deletions in the 139-loop facilitated both the release of the C-tail (Kim et al. 2012) and arrestin-1 binding to rhodopsin (Vishnivetskiy et al. 2013). Importantly, very similar (although not identical) changes were detected in receptor-bound arrestin-2 and -3 by DEER-based measurements of intra-molecular distances (Zhuo et al. 2014), supporting the idea that the molecular mechanism of receptor binding-induced activation is shared by all members of the arrestin family (Gurevich and Gurevich 2004).

Several crystal structures of arrestins in "active" receptor-bound-like conformation were solved recently (Fig. 1.2): "pre-activated" form of arrestin-1, short splice variant p44 (Kim et al. 2013), C-terminally truncated arrestin-2 in complex with multi-phosphorylated peptide corresponding to the C-terminus of vasopressin V2 receptor (Shukla et al. 2013), and more recently pre-activated arrestin-1 mutant in complex with constitutively active unphosphorylated rhodopsin (Kang et al. 2015). Several features that are evident in all three structures are likely the hallmarks of arrestin "activation" by a GPCR. The most prominent one is the rotation of the two domains relative to each other by ~17°–20°, which was predicted to become possible upon the disruption of the polar core and three-element interaction based on "slippery" nature of purely hydrophobic interactions between the domains (Gurevich and Gurevich 2006a). Domain rotation of this magnitude in

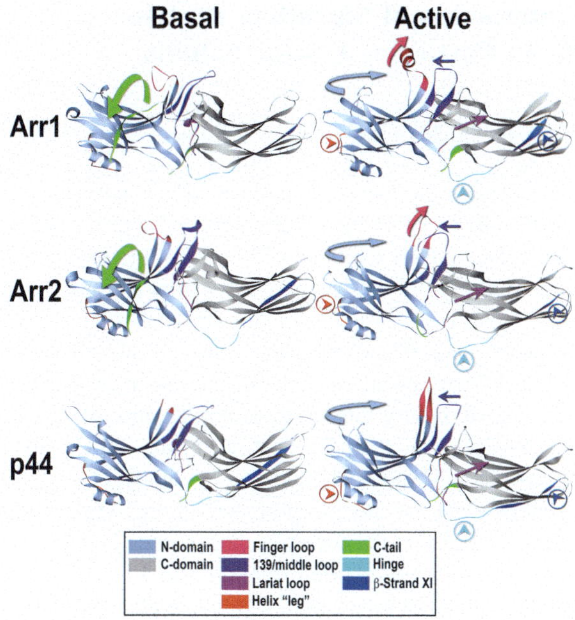

Fig. 1.2 Comparison of the basal and "active" conformations of arrestin proteins. Crystal structures of arrestin-1 (Arr1) [PDB ID 1CF1 (*left*) (Hirsch et al. 1999) and 4ZWJ (*right*) (Kang et al. 2015)], arrestin-2 (Arr2) [PDB ID 1G4M (*left*) (Han et al. 2001) and 4JQI (*right*) (Shukla et al. 2013)], and short splice variant of arrestin-1, p44 [PDB ID 3UGU (*left*) (Granzin et al. 2012) and 4J2Q (*right*) (Kim et al. 2013)] in their basal (*left*) and presumed active (*right*) conformations are shown. The elements undergoing conformational changes upon activation are shown, as follows: the C-tail *green* (it is pushed out of its basal position by receptor binding; absent in p44); inter-domain hinge *light blue* (assumes the same conformation in all activated arrestins, in contrast to a variety of conformations in the basal state); β-strand XI *dark blue* (undergoes register shift by one residue in arrestin-2, which "flips" it by 180°, and by two residues in arrestin-1 and p44); finger loop *red* (moves towards the receptor and assumes helical conformation); 139/middle loop *violet* (moves towards the N-domain and changes shape); lariat loop *magenta* (moves out of its basal position, removing two out of three negative charges from the polar core); the "leg" of the α-helix *pink* (changes conformation). The twist of the two domains (N-domain *teal*; C-domain *gray*) relative to each other by 17°–20° is shown by an *arrow*

receptor-bound arrestin was also predicted by molecular modeling (Modzelewska et al. 2006), based on the requirement of a long inter-domain "hinge" in arrestin for receptor binding (Vishnivetskiy et al. 2002). The other was a disruption of the three-element interaction and consequent release of the arrestin C-tail, which became totally invisible in these structures. The movement of the 139-loop [termed middle loop in the arrestin-2 (Shukla et al. 2013)] was also invariably observed. The tip of the finger loop in the arrestin-1-rhodopsin complex assumed α-helical conformation (Kang et al. 2015) earlier predicted based on arrestin peptide co-structure with rhodopsin (Szczepek et al. 2014).

Preferential Interaction of Signalling Proteins with Arrestins in Particular Conformations

Whereas arrestin-mediated activation of protein kinases c-Src (Luttrell et al. 1999), JNK3 (McDonald et al. 2000), ERK1/2 (Luttrell et al. 2001), and p38 (Bruchas et al. 2006) was originally reported to depend on GPCR activation, suggesting that this must be a function of receptor-bound arrestins, subsequent studies showed that this is not always the case. So far every study supports the notion that ERK1/2 bind with reasonably high affinity only to receptor-associated arrestins, even though their upstream activators MEK1 and c-Raf1 do not show this clear preference (Coffa et al. 2011). In contrast, the ability of arrestin-3 [but not other isoforms, even highly homologous arrestin-2 (Attramadal et al. 1992; Sterne-Marr et al. 1993)] to facilitate the activation of JNK family kinases does not depend on GPCRs (Miller et al. 2001; Song et al. 2009b; Seo et al. 2011; Zhan et al. 2011b, 2013; Kook et al. 2014). When the activity of both ERK and JNK is analyzed in the same cell, the former is strictly dependent on the receptor activation, whereas the latter is not (Breitman et al. 2012). There is no general rule for arrestin non-receptor partners: some prefer "active" receptor bound conformation, like ERK1/2 (Coffa et al. 2011), clathrin and AP2 (Kim and Benovic 2002), some prefer basal state, like Mdm2 (Song et al. 2006, 2007) or parkin (Ahmed et al. 2011), whereas others appear to bind both, like JNK3, MEK1, c-Raf1, etc. (Song et al. 2006, 2007, 2009b; Coffa et al. 2011).

These distinct conformational preferences of arrestin binding partners likely have functional significance, which is obvious in some cases, and remains to be elucidated in others (Gurevich and Gurevich 2003). Biologically, preferential binding of key components of the internalization machinery clathrin and AP2 to the active arrestins ensure that receptor-bound arrestins promote GPCR endocytosis, whereas free arrestins in the cytoplasm do not get in the way (Gurevich and Gurevich 2003). Structurally, this is ensured by the anchoring of the C-terminus to the body of the arrestin molecule (Fig. 1.1) and its release upon receptor binding (Palczewski et al. 1991a; Hanson et al. 2006b; Vishnivetskiy et al. 2010; Zhuang et al. 2013; Kim et al. 2012). Indeed, separated arrestin C-terminus expressed in cells binds clathrin and AP2, outcompeting the arrestin-receptor complexes, thereby suppressing arrestin-dependent GPCR internalization via coated pits (Orsini and Benovic 1998). Apparently, preferential interaction of ERK1/2 with receptor-bound arrestins (Coffa et al. 2011) ensures GPCR control of ERK activity, although the biological importance of this phenomenon remains unclear. Mdm2 preference for free arrestins (Song et al. 2006), along with the finding that Mdm2 recruited via arrestins ubiquitinates GPCRs (Ahmed et al. 2011; Shenoy et al. 2001, 2009), ensures only limited ubiquitination of active receptor due to release of bound Mdm2 from receptor-associated arrestins. This is consistent with the role of receptor ubiquitination in its trafficking, rather than in degradation (Shenoy et al. 2009).

Functional role of comparable binding of signaling proteins, such as JNK3 (Song et al. 2006), MEK1, c-Raf1, MKK4, and MKK7 (Coffa et al. 2011;

Song et al. 2009b; Zhan et al. 2011b, 2013) to free and receptor-bound arrestins is unclear. In fact, JNK3 and upstream kinases MKK4, MKK7, and ASK1 bind a small 25-residue peptide of arrestin-3 (Zhan et al. 2016), which is unlikely to have a preferred conformation as a separate entity. In general, an enormous number of proteins were reported to bind arrestins (Xiao et al. 2007). We are only beginning to elucidate conformational preferences of arrestin non-receptor partners, let alone the biological significance of these preferences. It appears that quite a few proteins readily bind microtubule-associated arrestins (Hanson et al. 2007; Coffa et al. 2011), and the role of these interactions also remains obscure.

Thus, while in recent years a lot of novel information about arrestin conformations appeared and the ability of arrestins to change shape is now widely recognized (Gurevich and Gurevich 2014), comprehensive elucidation of the biological role of the conformational changes in arrestins requires further investigation. Accumulating data suggesting that distinct arrestin conformations promote different branches of the arrestin-mediated signaling (Nuber et al. 2016; Lee et al. 2016) are intriguing, but are not comprehensive enough to draw general conclusions. While it is tempting to speculate that different patterns of GPCR phosphorylation by distinct GRKs are decoded by arrestins that assume different conformation upon binding [barcode hypothesis (Tobin et al. 2008; Nobles et al. 2011; Lau et al. 2011)], and there is some structural evidence supporting this notion (Yang et al. 2015), the jury is still out. Possible biological role of distinct conformations of the receptor-bound arrestins is discussed in Chaps. 11, 13 and 18.

To obtain unambiguous answers we need co-structures of arrestins with differentially phosphorylated GPCRs and downstream signaling proteins. However, the complexes of three or more proteins including GPCRs are unlikely to be crystallized any time soon. Considering that the first arrestin-receptor structure has been solved (Kang et al. 2015), it would be helpful to co-crystallize the same non-visual arrestin with the same GPCR phosphorylated at different positions. Hopefully, these structures would reveal conformational differences in bound arrestin induced by distinct patterns of receptor phosphorylation. In addition, we need structures of arrestin complexes with non-receptor signaling proteins, which might reveal subtle conformational preferences of different binding partners. Collectively these structures have a chance to prove the barcode hypothesis, which appears very attractive because it makes perfect sense biologically.

Another aspect that requires further investigation is the biological role of different arrestin-mediated signaling pathways in the cell. Virtually every signaling protein reported to be activated via arrestins, can be activated via other arrestin-independent mechanisms. This is true for Src, ERK1/2, JNK1/2/3, p38, Akt, PI3K, etc. In most cases arrestin-mediated activity constitutes a relatively small proportion of the total. Thus, to be biologically meaningful, arrestin-mediated activation of these signaling proteins must have some unique characteristics. One possibility is the subcellular localization of kinases activated with the help of arrestin, as opposed to other activators or scaffolds. There is evidence that this might be the case for MAP kinases: arrestin-activated ERK1/2 (Luttrell et al. 1999, 2001) and JNK3 (McDonald et al. 2000; Breitman et al. 2012) were reported to remain in the cytoplasm, where these kinases likely phosphorylate different targets than in the nucleus. Arrestins often organize multi-protein signaling complexes after binding to GPCRs, which localizes signaling proteins activated with the help of arrestins to the plasma membrane and/or endosomes, directing their activity to the targets co-localized in the same compartment. In the same vein, Mdm2 was reported to be recruited to the microtubules via arrestins, and that appears to direct its activity towards microtubule-associated proteins (Hanson et al. 2007). In general, arrestins likely play a role in compartment-specific signaling, but in most cases this needs to be demonstrated, and the biological role of this compartmentalization needs to be elucidated.

To summarize, unexpected discoveries of the multi-faceted role of arrestins as ubiquitous signaling regulators yielded many new questions that need to be answered experimentally.

Acknowledgements Supported by NIH Grants EY011500, GM077561, GM109955 (VVG), NS065868, and DA030103 (EVG).

References

Ahmed MR, Zhan X, Song X, Kook S, Gurevich VV, Gurevich EV (2011) Ubiquitin ligase parkin promotes Mdm2-arrestin interaction but inhibits arrestin ubiquitination. Biochemistry 50:3749–3763

Attramadal H, Arriza JL, Aoki C, Dawson TM, Codina J, Kwatra MM, Snyder SH, Caron MG, Lefkowitz RJ (1992) Beta-arrestin2, a novel member of the arrestin/beta-arrestin gene family. J Biol Chem 267:17882–17890

Barak LS, Ferguson SS, Zhang J, Caron MG (1997) A beta-arrestin/green fluorescent protein biosensor for detecting G protein-coupled receptor activation. J Biol Chem 272:27497–27500

Benovic JL, Kühn H, Weyand I, Codina J, Caron MG, Lefkowitz RJ (1987) Functional desensitization of the isolated β-adrenergic receptor by the β-adrenergic receptor kinase: potential role of an analog of the retinal protein arrestin (48 kDa protein). Proc Natl Acad Sci 84:8879–8882

Benovic JL, DeBlasi A, Stone WC, Caron MG, Lefkowitz RJ (1989) Beta-adrenergic receptor kinase: primary structure delineates a multigene family. Science 246:235–240

Bhandari D, Trejo J, Benovic JL, Marchese A (2007) Arrestin-2 interacts with the ubiquitin-protein isopeptide ligase atrophin-interacting protein 4 and mediates endosomal sorting of the chemokine receptor CXCR4. J Biol Chem 282:36971–36979

Breitman M, Kook S, Gimenez LE, Lizama BN, Palazzo MC, Gurevich EV, Gurevich VV (2012) Silent scaffolds: inhibition of c-Jun N-terminal kinase 3 activity in the cell by a dominant-negative arrestin-3 mutant. J Biol Chem 287:19653–19664

Bruchas MR, Macey TA, Lowe JD, Chavkin C (2006) Kappa opioid receptor activation of p38 MAPK is GRK3- and arrestin-dependent in neurons and astrocytes. J Biol Chem 281:18081–18089

Celver J, Lowe J, Kovoor A, Gurevich VV, Chavkin C (2001) Threonine 180 is required for G protein-coupled receptor kinase 3 and b-arrestin mediated desensitization of the m-opioid receptor in Xenopus oocytes. J Biol Chem 276:4894–4900

Celver J, Vishnivetskiy SA, Chavkin C, Gurevich VV (2002) Conservation of the phosphate-sensitive elements in the arrestin family of proteins. J Biol Chem 277:9043–9048

Chan S, Rubin WW, Mendez A, Liu X, Song X, Hanson SM, Craft CM, Gurevich VV, Burns ME, Chen J (2007) Functional comparisons of visual arrestins in rod photoreceptors of transgenic mice. Invest Ophthalmol Vis Sci 48:1968–1975

Coffa S, Breitman M, Hanson SM, Callaway K, Kook S, Dalby KN, Gurevich VV (2011) The effect of arrestin conformation on the recruitment of c-Raf1, MEK1, and ERK1/2 activation. PLoS ONE 6:e28723

Craft CM, Whitmore DH, Wiechmann AF (1994) Cone arrestin identified by targeting expression of a functional family. J Biol Chem 269:4613–4619

DeFea KA, Vaughn ZD, O'Bryan EM, Nishijima D, Déry O, Bunnett NW (2000a) The proliferative and antiapoptotic effects of substance P are facilitated by formation of a beta-arrestin-dependent scaffolding complex. Proc Natl Acad Sci USA 97:11086–11091

DeFea KA, Zalevsky J, Thoma MS, Déry O, Mullins RD, Bunnett NW (2000b) Beta-arrestin-dependent endocytosis of proteinase-activated receptor 2 is required for intracellular targeting of activated ERK1/2. J Cell Biol 148:1267–1281

DeWire SM, Ahn S, Lefkowitz RJ, Shenoy SK (2007) Beta-arrestins and cell signaling. Ann Rev Physiol 69:483–510

Dixon RA, Kobilka BK, Strader DJ, Benovic JL, Dohlman HG, Frielle T, Bolanowski MA, Bennett CD, Rands E, Diehl RE, Mumford RA, Slater EE, Sigal IS, Caron MG, Lefkowitz RJ, Strader CD (1986) Cloning of the gene and cDNA for mammalian beta-adrenergic receptor and homology with rhodopsin. Nature 321:75–79

Farrens DL, Altenbach C, Yang K, Hubbell WL, Khorana HG (1996) Requirement of rigid-body motion of transmembrane helices for light activation of rhodopsin. Science 274:768–770

Goodman OB Jr, Krupnick JG, Santini F, Gurevich VV, Penn RB, Gagnon AW, Keen JH, Benovic JL (1996) Beta-arrestin acts as a clathrin adaptor in endocytosis of the beta2-adrenergic receptor. Nature 383:447–450

Granzin J, Wilden U, Choe HW, Labahn J, Krafft B, Buldt G (1998) X-ray crystal structure of arrestin from bovine rod outer segments. Nature 391:918–921

Granzin J, Cousin A, Weirauch M, Schlesinger R, Büldt G, Batra-Safferling R (2012) Crystal structure of p44, a constitutively active splice variant of visual arrestin. J Mol Biol 416:611–618

Gray-Keller MP, Detwiler PB, Benovic JL, Gurevich VV (1997) Arrestin with a single amino acid substitution quenches light-activated rhodopsin in a phosphorylation-independent fasion. Biochemistry 36:7058–7063

Gurevich VV, Benovic JL (1993) Visual arrestin interaction with rhodopsin: sequential multisite binding ensures strict selectivity towards light-activated phosphorylated rhodopsin. J Biol Chem 268:11628–11638

Gurevich VV, Benovic JL (1995) Visual arrestin binding to rhodopsin: diverse functional roles of positively charged residues within the phosphorylation-recognition region of arrestin. J Biol Chem 270:6010–6016

Gurevich VV, Benovic JL (1997) Mechanism of phosphorylation-recognition by visual arrestin and the transition of arrestin into a high affinity binding state. Mol Pharmacol 51:161–169

Gurevich VV, Gurevich EV (2003) The new face of active receptor bound arrestin attracts new partners. Structure 11:1037–1042

Gurevich VV, Gurevich EV (2004) The molecular acrobatics of arrestin activation. Trends Pharmacol Sci 25:105–111

Gurevich VV, Gurevich EV (2006a) The structural basis of arrestin-mediated regulation of G protein-coupled receptors. Pharm Ther 110:465–502

Gurevich EV, Gurevich VV (2006b) Arrestins are ubiquitous regulators of cellular signaling pathways. Genome Biol 7:236

Gurevich VV, Gurevich EV (2014) Extensive shape shifting underlies functional versatility of arrestins. Curr Opin Cell Biol 27:1–9

Gurevich VV, Richardson RM, Kim CM, Hosey MM, Benovic JL (1993) Binding of wild type and chimeric arrestins to the m2 muscarinic cholinergic receptor. J Biol Chem 268:16879–16882

Nuber S, Zabel U, Lorenz K, Nuber A, Milligan G, Tobin AB, Lohse MJ, Hoffmann C (2016) β-Arrestin biosensors reveal a rapid, receptor-dependent activation/deactivation cycle. Nature 531:661–664

Orsini MJ, Benovic JL (1998) Characterization of dominant negative arrestins that inhibit beta2-adrenergic receptor internalization by distinct mechanisms. J Biol Chem 273:34616–34622

Ovchinnikov YA (1982) Rhodopsin and bacteriorhodopsin: structure-function relationship. FEBS Lett 148:179–191

Palczewski K, Hargrave PA (1991) Studies of ligand binding to arrestin. J Biol Chem 266:4201–4206

Palczewski K, Pulvermuller A, Buczylko J, Hofmann KP (1991a) Phosphorylated rhodopsin and heparin induce similar conformational changes in arrestin. J Biol Chem 266:18649–18654

Palczewski K, Buczyłko J, Imami NR, McDowell JH, Hargrave PA (1991b) Role of the carboxyl-terminal region of arrestin in binding to phosphorylated rhodopsin. J Biol Chem 266:15334–15339

Pfister C, Dorey C, Vadot E, Mirshahi M, Deterre P, Chabre M, Faure JP (1984) Identification of the so-called 48 K protein that interacts with illuminated rhodopsin in retinal rods, and the retinal S antigen, inductor of experimental autoimmune uveoretinitis. C R Acad Sci III 299:261–265

Pfister C, Chabre M, Plouet J, Tuyen VV, De Kozak Y, Faure JP, Kühn H (1985) Retinal S antigen identified as the 48 K protein regulating light-dependent phosphodiesterase in rods. Science 228:891–893

Rapoport B, Kaufman KD, Chazenbalk GD (1992) Cloning of a member of the arrestin family from a human thyroid cDNA library. Mol Cell Endocrinol 84:R39–R43

Rasmussen SG, Choi HJ, Rosenbaum DM, Kobilka TS, Thian FS, Edwards PC, Burghammer M, Ratnala VR, Sanishvili R, Fischetti RF, Schertler GF, Weis WI, Kobilka BK (2007) Crystal structure of the human beta2 adrenergic G-protein-coupled receptor. Nature 450:383–387

Schleicher A, Kuhn H, Hofmann KP (1989) Kinetics, binding constant, and activation energy of the 48-kDa protein-rhodopsin complex by extra-metarhodopsin II. Biochemistry 28:1770–1775

Seo J, Tsakem EL, Breitman M, Gurevich VV (2011) Identification of arrestin-3-specific residues necessary for JNK3 activation. J Biol Chem 286:27894–27901

Shenoy SK, McDonald PH, Kohout TA, Lefkowitz RJ (2001) Regulation of receptor fate by ubiquitination of activated beta 2-adrenergic receptor and beta-arrestin. Science 294:1307–1313

Shenoy SK, Modi AS, Shukla AK, Xiao K, Berthouze M, Ahn S, Wilkinson KD, Miller WE, Lefkowitz RJ (2009) Beta-arrestin-dependent signaling and trafficking of 7-transmembrane receptors is reciprocally regulated by the deubiquitinase USP33 and the E3 ligase Mdm2. Proc Nat Acad Sci USA 106:6650–6655

Shichi H, Somers RL (1978) Light-dependent phosphorylation of rhodopsin. Purification and properties of rhodopsin kinase. J Biol Chem 253:7040–7046

Shinohara T, Dietzschold B, Craft CM, Wistow G, Early JJ, Donoso LA, Horwitz J, Tao R (1987) Primary and secondary structure of bovine retinal S antigen (48 kDa protein). Proc Natl Acad Sci 84:6975–6979

Shukla AK, Manglik A, Kruse AC, Xiao K, Reis RI, Tseng WC, Staus DP, Hilger D, Uysal S, Huang LY, Paduch M, Tripathi-Shukla P, Koide A, Koide S, Weis WI, Kossiakoff AA, Kobilka BK, Lefkowitz RJ (2013) Structure of active beta-arrestin-1 bound to a G-protein-coupled receptor phosphopeptide. Nature 497:137–141

Sondek J, Lambright DG, Noel JP, Hamm HE, Sigler PB (1994) GTPase mechanism of Gproteins from the 1.7-A crystal structure of transducin alpha-GDP-AIF-4. Nature 372:276–279

Song X, Raman D, Gurevich EV, Vishnivetskiy SA, Gurevich VV (2006) Visual and both non-visual arrestins in their “inactive” conformation bind JNK3 and Mdm2 and relocalize them from the nucleus to the cytoplasm. J Biol Chem 281:21491–21499

Song X, Gurevich EV, Gurevich VV (2007) Cone arrestin binding to JNK3 and Mdm2: conformational preference and localization of interaction sites. J Neurochem 103:1053–1062
Song X, Vishnivetskiy SA, Gross OP, Emelianoff K, Mendez A, Chen J, Gurevich EV, Burns ME, Gurevich VV (2009a) Enhanced arrestin facilitates recovery and protects rods lacking rhodopsin phosphorylation. Curr Biol 19:700–705
Song X, Coffa S, Fu H, Gurevich VV (2009b) How does arrestin assemble MAP kinases into a signaling complex? J Biol Chem 284:685–695
Sterne-Marr R, Gurevich VV, Goldsmith P, Bodine RC, Sanders C, Donoso LA, Benovic JL (1993) Polypeptide variants of beta-arrestin and arrestin3. J Biol Chem 268:15640–15648
Sutton RB, Vishnivetskiy SA, Robert J, Hanson SM, Raman D, Knox BE, Kono M, Navarro J, Gurevich VV (2005) Crystal structure of cone arrestin at 2.3 Å: evolution of receptor specificity. J Mol Biol 354:1069–1080
Szczepek M, Beyriere F, Hofmann KP, Elgeti M, Kazmin R, Rose A, Bartl FJ, von Stetten D, Heck M, Sommer ME, Hildebrand PW, Scheerer P (2014) Crystal structure of a common GPCR-binding interface for G protein and arrestin. Nat Commun 5:4801
Tobin AB, Butcher AJ, Kong KC (2008) Location, location, location...site-specific GPCR phosphorylation offers a mechanism for cell-type-specific signalling. Trends Pharmacol Sci 29:413–420
Tsuda M, Syed M, Bugra K, Whelan JP, McGinnis JF, Shinohara T (1988) Structural analysis of mouse S-antigen. Gene 73:11–20
Vishnivetskiy SA, Paz CL, Schubert C, Hirsch JA, Sigler PB, Gurevich VV (1999) How does arrestin respond to the phosphorylated state of rhodopsin? J Biol Chem 274:11451–11454
Vishnivetskiy SA, Schubert C, Climaco GC, Gurevich YV, Velez M-G, Gurevich VV (2000) An additional phosphate-binding element in arrestin molecule: implications for the mechanism of arrestin activation. J Biol Chem 275:41049–41057
Vishnivetskiy SA, Hirsch JA, Velez M-G, Gurevich YV, Gurevich VV (2002) Transition of arrestin in the active receptor-binding state requires an extended interdomain hinge. J Biol Chem 277:43961–43968
Vishnivetskiy SA, Francis DJ, Van Eps N, Kim M, Hanson SM, Klug CS, Hubbell WL, Gurevich VV (2010) The role of arrestin alpha-helix I in receptor binding. J Mol Biol 395:42–54
Vishnivetskiy SA, Baameur F, Findley KR, Gurevich VV (2013) Critical role of the central 139-loop in stability and binding selectivity of arrestin-1. J Biol Chem 288:11741–11750
Wacker WB, Lipton MM (1965) Experimental allergic uveitis: homologous retina as uveitogenic antigen. Nature 206:253–254
Wacker WB, Donoso LA, Kalsow CM, Yankeelov JA Jr, Organisciak DT (1977) Experimental allergic uveitis. Isolation, characterization, and localization of a soluble uveitopathogenic antigen from bovine retina. J Immunol 119:1949–1958
Wilden U (1995) Duration and amplitude of the light-induced cGMP hydrolysis in vertebrate photoreceptors are regulated by multiple phosphorylation of rhodopsin and by arrestin binding. Biochemistry 34:1446–1454
Wilden U, Hall SW, Kühn H (1986) Phosphodiesterase activation by photoexcited rhodopsin is quenched when rhodopsin is phosphorylated and binds the intrinsic 48-kDa protein of rod outer segments. Proc Natl Acad Sci 83:1174–1178
Xiao K, McClatchy DB, Shukla AK, Zhao Y, Chen M, Shenoy SK, Yates JR, Lefkowitz RJ (2007) Functional specialization of beta-arrestin interactions revealed by proteomic analysis. Proc Natl Acad Sci USA 104:12011–12016
Yamaki K, Takahashi Y, Sakuragi S, Matsubara K (1987) Molecular cloning of the S-antigen cDNA from bovine retina. Biochem Biophys Res Commun 142:904–910
Yang F, Yu X, Liu C, Qu CX, Gong Z, Liu HD, Li FH, Wang HM, He DF, Yi F, Song C, Tian CL, Xiao KH, Wang JY, Sun JP (2015) Phospho-selective mechanisms of arrestin conformations and functions revealed by unnatural amino acid incorporation and (19)F-NMR. Nat Commun 6:8202

exclusively expressed in rod and cone receptor cells at high levels, demonstrating specificity for their cognate receptors, rhodopsin and cone opsins. The other category is non-visual arrestins or β-arrestins, which consists of arrestin-2 (β-arrestin-1) and arrestin-3 (β-arrestin-2). The non-visual arrestins were first discovered in the late 1980s, and reported to be homologues of visual arrestin in non-retinal tissues (Benovic et al. 1987). Non-visual arrestins are expressed ubiquitously in all cell types with the highest expression levels observed in the brain and spleen (Attramadal et al. 1992; Lohse et al. 1990; Gurevich et al. 2002, 2004). The classical function of non-visual arrestins is to desensitize most GPCR signaling and initiate receptor internalization. In order to turn off the G protein-mediated signaling to prevent persistent activation, the activated receptor is first phosphorylated by a G protein coupled receptor kinase (GRK) (Gurevich et al. 2012). Receptor phosphorylation by GRK specifically prepares the activated receptor for arrestin binding (Carman and Benovic 1998). Once arrestin binds the activated and phosphorylated receptor, it physically blocks further G protein-mediated signaling and targets the receptor for internalization, whereupon the receptor can be either degraded or recycled back to the surface for another round of signaling (Gurevich and Gurevich 2004). For example, in rod photoreceptors, it is the visual arrestin-1 that specifically stops the signaling of rhodopsin. Once a photon of light is absorbed, rhodopsin undergoes a conformational change that activates the associated visual G protein transducin (Noma et al. 2007) and triggers a second messenger cascade. Upon activation of the receptor, rhodopsin kinase (systematic name GRK1) phosphorylates multiple sites on rhodospin's C-tail, preparing the phosphorylated and activated receptor to bind visual arrestin-1 (Maeda et al. 2003). Arrestin binding prevents rhodopsin from interacting with any more transducin molecules, ending the G protein-dependent signaling (McBee et al. 2001; Gurevich et al. 2011). This arrestin-mediated desensitization mechanism is essential and universal to almost all GPCRs.

GPCRs constitute the largest, most versatile and most ubiquitous class of membrane receptors, with more than 800 members identified in the human genome (Lagerstrom and Schioth 2008). They bind to a diverse category of ligands which include hormones, peptides, neurotransmitters, chemokines and lipids (Bockaert and Philippe Pin 1999). Upon activation, they regulate a variety of intracellular signaling pathways to produce appropriate cellular responses, such as cell growth, differentiation, metabolism and also vision and taste (Pierce et al. 2002). Since GPCR signaling critically regulates a wide range of physiological and pathophysiological processes, these receptors are among the most important drug targets, accounting for approximately one third of currently marketed drugs (Ma and Zemmel 2002).

Although different GPCRs have various primary structures, they do share a conserved seven-transmembrane domain architecture, which consists of a single polypeptide chain that spans the lipid membrane seven times (Baldwin et al. 1997). At the transmembrane region, there is a seven α-helical bundle with hydrophobic helices linked by three extracellular and three intracellular loops. Several crystal structures have confirmed that the GPCRs have a tertiary structure resembling a barrel, with the seven transmembrane helices forming a cavity within the plasma membrane that serves as a ligand-binding domain that is often covered by extracellular loop-2 (Cherezov et al. 2007; Palczewski et al. 2000; Jaakola et al. 2008; Wu et al. 2010; Huang et al. 2015; Hua et al. 2016; Yin et al. 2015; Dore et al. 2014). GPCRs in vertebrates are commonly divided into five families based on their sequence and structural similarity: rhodopsin (family A), secretin (family B), glutamate (family C), adhesion and Frizzled/Taste2 (Fredriksson et al. 2003). The rhodopsin family is by far the largest and most diverse class and is characterized by several conserved sequence motifs (Fredriksson et al. 2003). The rhodopsin-like receptors include many important drug targets, including chemokine receptors, adrenergic receptors, angiotensin receptors, dopamine receptors, serotonin receptors, histamine receptors, neuropeptide Y receptors and glycoprotein receptors. The crystal structure of dark-state bovine rhodopsin reported in 2000 was considered an important milestone for understanding the structure and function of GPCRs (Palczewski et al. 2000).

The interaction of arrestins with GPCRs generally requires two consecutive structural changes in the receptor: the ligand-induced conformational changes associated with the activation and phosphorylation by GRK at intracellular loops and/or the C-terminal tail of the receptor. Once bound to arrestin, the receptor is linked to the clathrin-dependent endocytic machinery and the complex persists on a time scale of minutes to hours (Charest et al. 2005). The traditional GPCR function is that these receptors catalyze the activation of heterotrimeric G proteins, whose dissociated subunits interact with second messenger-generating or -degrading enzymes to amplify the signal (Hepler and Gilman 1992). Interestingly, recently it has also been demonstrated that the binding of arrestins to GPCRs initiates diverse signaling pathways that are independent of G proteins (Reiter et al. 2012). Preferential activation of one of a number of possible downstream pathways of a receptor by a particular ligand is referred to as biased agonism, which means a certain ligand can activate either G protein-dependent or arrestin-dependent signaling pathways (DeWire et al. 2007). The molecular mechanism of GPCR biased agonism suggested the ability of the receptor to adopt distinct conformations. A large number of studies using different full and partial agonist ligands, site-directed mutagenesis and probes placed in different places on the receptors (Hoffmann et al. 2008; Lohse et al. 2008; Seifert and Dove 2009) have confirmed the existence of distinct active conformations of GPCRs in response to various ligands. Moreover, accumulated evidence has also revealed that GPCRs are dynamic proteins and their conformational plasticity contributes to their interactions with multiple signaling partners including G proteins, GRKs and arrestins (Mahoney and Sunahara 2016; Kenakin 2013).

In recent years, the investigation of arrestin-dependent signaling via GPCRs has identified that both non-visual arrestins not only mediate receptor desensitization and internalization, but also regulate GPCR signal transduction in a G protein-independent manner (Lefkowitz and Shenoy 2005; Shenoy et al. 2001; Shenoy and Lefkowitz 2003; DeFea 2007; Abraham et al. 2016; Mancini et al. 2015; Yang et al. 2016). There is an ever increasing list of kinases and regulatory proteins that bind specifically to one or both non-visual arrestin isoforms (Lefkowitz and Shenoy 2005). For example, non-visual arrestin scaffolding of intracellular signaling molecules was first demonstrated for the non-receptor tyrosine kinase c-Src (Miller et al. 2000; DeFea et al. 2000a; Luttrell et al. 1999). In these studies, Src was shown to assemble on the agonist-stimulated β_2-adrenergic receptor (β_2AR) in a non-visual arrestin-dependent manner. This arrestin-dependent c-Src recruitment to the receptors results in the activation of extracellular signal-regulated kinases (ERK1/2) (Miller et al. 2000; DeFea et al. 2000a; Luttrell et al. 1999). A mutational analysis indicated that some mutant receptors do not couple to their cognate G proteins but still recruit non-visual arrestins in response to agonist stimulation (Gáborik et al. 2003; Wei et al. 2003a). Another group, using the angiotensin II type 1a receptor (AT1aR) as a model, found that exogenous arrestin-2 or -3 expression resulted in decreased agonist-stimulated phosphoinositide hydrolysis, yet increased ERK activation (Tohgo et al. 2002). This suggests that non-visual arrestins might inhibit G protein signaling through increased desensitization but increase ERK phosphorylation through an arrestin-dependent mechanism. By recruiting various protein partners including trafficking proteins and signaling molecules directly to the GPCR, non-visual arrestins can connect activated GPCRs to diverse signaling pathways, which leads to the phosphorylation of numerous intracellular targets (Xiao et al. 2007, 2010). Accumulated evidence has shown that this non-visual arrestin-mediated signaling regulates important cellular responses such as cell growth, differentiation, cytoskeleton rearrangement, chemotaxis, and apoptosis (DeWire et al. 2007). For example, one study found that the directed movement of lymphocytes towards chemokines is regulated specifically by GRK6-mediated phosphorylation and arrestin-3 binding to CXCR4 (Fong et al. 2002). Moreover, in the case of the apoptotic pathway, researchers found that in mouse embryonic fibroblasts arrestin-2 was necessary for the insulin-like growth factor-1 receptor (IGF1-R) stimulated pathway through phosphatidylinositol 3 kinase (PI3K) activation, which inhibits apoptosis (Povsic et al. 2003; Miller and Lefkowitz 2001).

Despite these efforts, the field is just beginning to understand the molecular mechanism of arrestin-mediated signaling. These pathways are regulated via non-visual arrestins by bringing signaling proteins within spatial proximity of each other, therefore facilitating protein-protein interactions. Non-visual arrestins often function as signaling scaffolds. Multiprotein complexes organized by arrestins were termed signalosomes (Shenoy and Lefkowitz 2003). Some of the best-characterized non-visual arrestin mediated signalosomes upon stimulation of different GPCRs induce RhoA-dependent stress fiber formation, inhibit nuclear factor κB (NF-κB) targeted gene expression through IκB stabilization, induce ERK-dependent protein translation and antiapoptotic effects, and so on (Reiter et al. 2012).

Several types of signalosomes organized by non-visual arrestins promote the activation of MAP kinases of the three main subfamilies: ERK (Luttrell et al. 2001), JNK (McDonald et al. 2000), and p38 (Bruchas et al. 2006). Accumulated evidence suggests that both arrestin-2 and arrestin-3 can form a signaling complex with different GPCRs and MAPK cascade components, though a clear model of this signaling complex has not been established (Burack and Shaw 2000; Pearson et al. 2001; Luttrell et al. 2001; DeFea et al. 2000b; Christopoulos et al. 2003; McDonald et al. 2000). In mammals, there are three major MAPK pathways: the extracellular signal-regulated kinase (ERK) signaling pathway, the c-Jun NH2-terminal kinase (JNK) pathway and the p38 pathway. Each signaling cascade is composed of three kinase modules, where the kinases sequentially activate downstream component by phosphorylation. MAPKs regulate a diverse range of cellular responses induced by many different activators (Roux and Blenis 2004). Generally, there are two mechanisms that control the efficiency and specificity of MAPK signaling: recognition motifs and scaffolds that organize them into multiprotein complexes (Zeke et al. 2009; Brown and Sacks 2009). Several scaffold proteins that specifically organize the MAPK cascade components have been discovered, the classical examples of which are the Ste5 protein in yeast and the KSR (kinase suppressor of Ras) protein in mammals (Zeke et al. 2009; Brown and Sacks 2009). Ste5 specifically serves as a scaffold in yeast mating pathway by directly interacting with the kinase components of a particular MAPK cascade (Chol et al. 1994; Printen and Sprague 1994). Similarly, KSR proteins scaffold signaling modules of the ERK cascade and modulate signaling in the Ras-dependent signaling pathway (Therrien et al. 1996). The functions of these kinds of scaffold proteins are similar. They physically assemble the individual kinases and upstream regulators and control MAPK pathway localization within the cell. In addition, they can prevent MAPK signaling proteins from competing inputs. Most importantly, they are required for efficient signaling (Therrien et al. 1996; Garrenton et al. 2006; Morrison and Davis 2003; Good et al. 2011). Despite no sequence and size similarity to KSR and Ste5 protein, non-visual arrestins carry out a similar scaffolding function in the regulation of MAPKs (Shenoy and Lefkowitz 2003).

Recently, a number of studies have attempted to explore the mechanism of arrestin-dependent assembly of MAPKs and also the downstream consequences of arrestin-dependent signaling (Tohgo et al. 2002; Xiao et al. 2010; Luttrell and Gesty-Palmer 2010; Breitman et al. 2012; Seo et al. 2011; Coffa et al. 2011; Walters et al. 2009; Zhan et al. 2016; Wisler et al. 2015; Khoury et al. 2014). DeFea et al. (2000b) showed that agonist stimulation of protease-activated receptor (PAR2) results in the formation of a complex containing the activated receptor, arrestin-2, Raf-1, and phosphorylated ERK. Arrestin-3 has been found to scaffold an entire signaling module for JNK3 in response to GPCR signaling, such as AT1aR (McDonald et al. 2000). Another study using purified proteins has shown that ERK2 directly binds free arrestin-2 and arrestin-3, as well as receptor-associated arrestin-2 and arrestin-3 (Coffa et al. 2011). Moreover, they also found that in COS-7 cells arrestin-2 and -3 associated with receptors significantly enhanced ERK2 binding (Coffa et al. 2011). Another study based on immunoprecipitation assays found that both arrestin domains interact with all components of the two MAPK cascades (ASK1-MKK4-JNK3 and Raf-1-MEK1-ERK2), which suggests a model of arrestin-dependent assembly of the MAPK signaling module: arrestin binds all three kinases along its short axis with each kinase directly interacting with both domains of arrestin (Song et al. 2009). However, studies using the AT1aR and co-immunoprecipitation described a different model of arrestin-3 mediated assembly of MAPKs: MEK1 indirectly binds arrestin-3 through contacts with Raf and ERK, whereas the latter two kinases directly bind arrestin-3 (DeWire et al. 2007). In addition to the investigation of the physical interactions within arrestin-dependent signalosome, there are many studies focusing on the physiological responses resulting from arrestin-dependent ERK activation. It has been found that arrestin-mediated ERK activation results in retention of active ERK in cytosolic endocytic vesicles, instead of trafficking into the nucleus (Lefkowitz and Shenoy 2005; Luttrell et al. 2001; Tohgo et al. 2003). In the cytoplasm, ERK in this arrestin-mediated signalosome can phosphorylate non-nuclear substrates that regulate possible physiological effects on cell motility, chemotaxis, and apoptosis (Wei et al. 2003a; Tohgo et al. 2002; Luttrell et al. 2001).

Rhodopsin-like GPCRs were functionally divided into broad classes based on the stability of the arrestin-receptor complex (Oakley et al. 2000). Class A receptors such as β_2AR form weak and transient complexes with non-visual arrestins, so that they do not constrain ERK activity within endosomes. Instead, active ERK immediately dissociates from the signalosome and translocates into the nucleus, while the receptors are rapidly recycled to the plasma membrane (Christopoulos et al. 2003). In contrast, Class B receptors such as AT1aR form stable receptor-arrestin complexes so that active ERK1/2 remains associated with the receptor in endocytic vesicles and likely phosphorylates cytosolic substrates (Lefkowitz and Shenoy 2005; Luttrell et al. 2001; Oakley et al. 1999, 2000; Kendall and Luttrell 2009; Wei et al. 2003b). It appears that arrestin dependent ERK activation is mainly generated by Class B receptors and does not have the typical nuclear ERK functions, however, the precise downstream targets largely remain unknown. Furthermore, the key that determines receptor affinity for the two non-visual arrestins and the lifetime of the receptor-arrestin complex is the distinct phosphorylation sites, most notably in the C-termini of the receptors (Tohgo et al. 2003; Oakley et al. 2000; Rh et al. 2001). Interestingly, one study showed that swapping the C-termini between class A and class B receptors changes their non-visual arrestin binding behavior (Lohse and Hoffmann 2014).

There are many human disorders associated with excessive signaling by various GPCRs. Mutations in GPCRs can cause acquired and inherited diseases such as retinitis pigmentosa (RP), nephrogenic diabetes insipidus, severe fertility disorders, and even carcinomas (Lagerstrom and Schioth 2008; Ma and Zemmel 2002). It is hypothesized that the excessive signaling of GPCRs can be dampened by enhanced arrestins, which were designed to bind active unphosphorylated GPCRs with high affinity (Kovoor et al. 1999; Celver et al. 2002; Gurevich and Gurevich 2013). It has been widely found that excessive MAPK signaling results in many severe pathological processes such as cancer. Non-visual arrestins play an important role in regulating different MAPK pathways; a clear understanding of the mechanisms of arrestin-dependent assembly of MAPKs into a signaling complex is very important and promising for the development of therapies that selectively manipulate MAPK cascades. Although there have been models and various studies (DeWire et al. 2007; Coffa et al. 2011; Song et al. 2009) focusing on arrestin-dependent assembly of MAPKs, mechanisms of the signaling complex still remain largely unexplored. It is very likely that scaffold proteins adopt more sophisticated mechanisms in regulating the signaling network, such as using cooperative or allosteric assembly of components. Moreover, one of the most important features of signaling complexes is their dynamic nature. Dynamic conformational changes are always involved in the formation and dissociation of multiple protein complexes. Advanced biophysical and biochemical techniques have allowed us to obtain both static structural information and monitor the signaling events in a time-resolved fashion to probe the dynamics of interactions between components of signaling complexes (Shukla and Wodak 2016).

It is well known that all cellular behaviors, such as proliferation, differentiation, apoptosis, migration, etc., are mediated and regulated by signaling events that are driven by protein-protein interactions. Targeting individual protein-protein interactions has been proposed to have great therapeutic potential (Gurevich and Gurevich 2012). For example, if we could selectively disrupt or enhance individual protein-protein interactions, we may be able to force cancer cells to stop proliferating or "tell" dying neurons to stay alive by manipulating corresponding signaling pathways. Therefore, the ability to modulate protein-protein interactions in a desired manner may help us to cure some of the severe diseases instead of just managing the symptoms, although there are a lot of challenges that need to be addressed. A comprehensive understanding of structural and functional properties of signaling proteins is essential to develop signaling biased proteins with modified structure and functions. In fact, arrestin proteins are likely to be a perfect target to test this idea because non-visual arrestins function as important signaling scaffolds in the cell through interactions with numerous signaling proteins.

References

Abraham DM, Davis RT, Warren CM, Mao L, Wolska BM, Solaro RJ, Rockman HA (2016) β-Arrestin mediates the Frank-Starling mechanism of cardiac contractility. Proc Natl Acad Sci 113:14426–14431

Attramadal H, Arriza JL, Aoki C, Dawson TM, Codina J, Kwatra MM, Snyder SH, Caron MG, Lefkowitz RJ (1992) Beta-arrestin2, a novel member of the arrestin/beta-arrestin gene family. J Biol Chem 267:17882–17890

Baldwin JM, Schertler GFX, Unger VM (1997) An alpha-carbon template for the transmembrane helices in the rhodopsin family of G-protein-coupled receptors. J Mol Biol 272:144–164

Benovic JL, Kühn H, Weyand I, Codina J, Caron MG, Lefkowitz RJ (1987) Functional desensitization of the isolated beta-adrenergic receptor by the beta-adrenergic receptor kinase: potential role of an analog of the retinal protein arrestin (48-kDa protein). Proc Natl Acad Sci 84:8879–8882

Bockaert J, Philippe Pin J (1999) Molecular tinkering of G protein-coupled receptors: an evolutionary success. EMBO J 18:1723–1729

Breitman M, Kook S, Gimenez LE, Lizama BN, Palazzo MC, Gurevich EV, Gurevich VV (2012) Silent scaffolds. J Biol Chem 287:19653–19664

Brown MD, Sacks DB (2009) Protein scaffolds in MAP kinase signalling. Cell Signal 21:462–469

Bruchas MR, Macey TA, Lowe JD, Chavkin C (2006) Kappa opioid receptor activation of p38 MAPK is GRK3- and arrestin-dependent in neurons and astrocytes. J Biol Chem 281:18081–18089

Burack WR, Shaw AS (2000) Signal transduction: hanging on a scaffold. Curr Opin Cell Biol 12:211–216

Carman CV, Benovic JL (1998) G-protein-coupled receptors: turn-ons and turn-offs. Curr Opin Neurobiol 8:335–344

Celver J, Vishnivetskiy SA, Chavkin C, Gurevich VV (2002) Conservation of the phosphate-sensitive elements in the arrestin family of proteins. J Biol Chem 277:9043–9048

Charest PG, Terrillon S, Bouvier M (2005) Monitoring agonist-promoted conformational changes of β-arrestin in living cells by intramolecular BRET. EMBO Rep 6:334–340

Cherezov V, Rosenbaum DM, Hanson MA, Rasmussen SGF, Thian FS, Kobilka TS, Choi H-J, Kuhn P, Weis WI, Kobilka BK, Stevens RC (2007) High-resolution crystal structure of an engineered human β2-adrenergic G protein-coupled receptor. Science 318:1258–1265

Chol K-Y, Satterberg B, Lyons DM, Elion EA (1994) Ste5 tethers multiple protein kinases in the MAP kinase cascade required for mating in *S. cerevisiae*. Cell 78:499–512

Christopoulos A, Christopoulos G, Morfis M, Udawela M, Laburthe M, Couvineau A, Kuwasako K, Tilakaratne N, Sexton PM (2003) Novel receptor partners and function of receptor activity-modifying proteins. J Biol Chem 278:3293–3297

Coffa S, Breitman M, Hanson SM, Callaway K, Kook S, Dalby KN, Gurevich VV (2011) The effect of arrestin conformation on the recruitment of c-Raf1, MEK1, and ERK1/2 activation. PLoS ONE 6:e28723

DeFea KA (2007) Stop that cell! β-arrestin-dependent chemotaxis: a tale of localized actin assembly and receptor desensitization. Ann Rev Physiol 69:535–560

DeFea KA, Vaughn ZD, O'Bryan EM, Nishijima D, Déry O, Bunnett NW (2000a) The proliferative and antiapoptotic effects of substance P are facilitated by formation of a β-arrestin-dependent scaffolding complex. Proc Natl Acad Sci 97:11086–11091

DeFea KA, Zalevsky J, Thoma MS, Déry O, Mullins RD, Bunnett NW (2000b) β-Arrestin-dependent endocytosis of proteinase-activated receptor 2 is required for intracellular targeting of activated Erk1/2. J Cell Biol 148:1267–1282

DeWire SM, Ahn S, Lefkowitz RJ, Shenoy SK (2007) β-arrestins and cell signaling. Ann Rev Physiol 69:483–510

Dore AS, Okrasa K, Patel JC, Serrano-Vega M, Bennett K, Cooke RM, Errey JC, Jazayeri A, Khan S, Tehan B, Weir M, Wiggin GR, Marshall FH (2014) Structure of class C GPCR metabotropic glutamate receptor 5 transmembrane domain. Nature 511:557–562

Fong AM, Premont RT, Richardson RM, Yu Y-RA, Lefkowitz RJ, Patel DD (2002) Defective lymphocyte chemotaxis in β-arrestin2- and GRK6-deficient mice. Proc Natl Acad Sci 99:7478–7483

Fredriksson R, Lagerström MC, Lundin L-G, Schiöth HB (2003) The G-protein-coupled receptors in the human genome form five main families. Phylogenetic analysis, paralogon groups, and fingerprints. Mol Pharmacol 63:1256–1272

Gáborik Z, Jagadeesh G, Zhang M, Spät A, Catt KJ, Hunyady L (2003) The role of a conserved region of the second intracellular loop in AT1 angiotensin receptor activation and signaling. Endocrinology 144:2220–2228

Garrenton LS, Young SL, Thorner J (2006) Function of the MAPK scaffold protein, Ste5, requires a cryptic PH domain. Genes Dev 20:1946–1958

Good MC, Zalatan JG, Lim WA (2011) Scaffold proteins: hubs for controlling the flow of cellular information. Science 332:680–686

Gurevich VV, Gurevich EV (2004) The molecular acrobatics of arrestin activation. Trends Pharmacol Sci 25:105–111

Gurevich VV, Gurevich EV (2012) Synthetic biology with surgical precision: targeted reengineering of signaling proteins. Cell Signal 24:1899–1908

Gurevich VV, Gurevich EV (2013) Structural determinants of arrestin functions. In: Luttrell LM (ed) The Molecular Biology of Arrestins. Progress in molecular biology and translational science, vol 118. Elsevier, Oxford, pp 57–92

Gurevich EV, Benovic JL, Gurevich VV (2002) Arrestin2 and arrestin3 are differentially expressed in the rat brain during postnatal development. Neuroscience 109:421–436

Gurevich EV, Benovic JL, Gurevich VV (2004) Arrestin2 expression selectively increases during neural differentiation. J Neurochem 91:1404–1416

Gurevich VV, Hanson SM, Song X, Vishnivetskiy SA, Gurevich EV (2011) The functional cycle of visual arrestins in photoreceptor cells. Prog Retin Eye Res 30:405–430

Gurevich EV, Tesmer JJ, Mushegian A, Gurevich VV (2012) G protein-coupled receptor kinases: more than just kinases and not only for GPCRs. Pharmacol Ther 133:40–69

Hepler JR, Gilman AG (1992) G proteins. Trends Biochem Sci 17:383–387

Hoffmann C, Zürn A, Bünemann M, Lohse MJ (2008) Conformational changes in G-protein-coupled receptors—the quest for functionally selective conformations is open. Br J Pharmacol 153:S358–S366

Hua T, Vemuri K, Pu M, Qu L, Han GW, Wu Y, Zhao S, Shui W, Li S, Korde A, Laprairie RB, Stahl EL, Ho JH, Zvonok N, Zhou H, Kufareva I, Wu B, Zhao Q, Hanson MA, Bohn LM, Makriyannis A, Stevens RC, Liu ZJ (2016) Crystal structure of the human cannabinoid receptor CB1. Cell 167(750–762):e714

Huang W, Manglik A, Venkatakrishnan AJ, Laeremans T, Feinberg EN, Sanborn AL, Kato HE, Livingston KE, Thorsen TS, Kling RC, Granier S, Gmeiner P, Husbands SM, Traynor JR, Weis WI, Steyaert J, Dror RO, Kobilka BK (2015) Structural insights into micro-opioid receptor activation. Nature 524:315–321

Jaakola V-P, Griffith MT, Hanson MA, Cherezov V, Chien EYT, Lane JR, Ijzerman AP, Stevens RC (2008) The 2.6 angstrom crystal structure of a human A2A adenosine receptor bound to an antagonist. Science 322:1211–1217

Kenakin T (2013) New concepts in pharmacological efficacy at 7TM receptors: IUPHAR review 2. Br J Pharmacol 168:554–575

Kendall RT, Luttrell LM (2009) Diversity in arrestin function. Cell Mol Life Sci 66:2953–2973

Khoury E, Nikolajev L, Simaan M, Namkung Y, Laporte SA (2014) Differential regulation of endosomal GPCR/beta-arrestin complexes and trafficking by MAPK. J Biol Chem 289: 23302–23317

Kovoor A, Celver J, Abdryashitov RI, Chavkin C, Gurevich VV (1999) Targeted construction of phosphorylation-independent β-arrestin mutants with constitutive activity in cells. J Biol Chem 274:6831–6834

Lagerstrom MC, Schioth HB (2008) Structural diversity of G protein-coupled receptors and significance for drug discovery. Nat Rev Drug Discov 7:339–357

Lefkowitz RJ, Shenoy SK (2005) Transduction of receptor signals by ß-arrestins. Science 308:512–517

Lohse MJ, Hoffmann C (2014) Arrestin interactions with G protein-coupled receptors. Handb Exp Pharmacol 219:15–56

Lohse M, Benovic J, Codina J, Caron M, Lefkowitz R (1990) Beta-arrestin: a protein that regulates beta-adrenergic receptor function. Science 248:1547–1550

Lohse MJ, Nikolaev VO, Hein P, Hoffmann C, Vilardaga J-P, Bünemann M (2008) Optical techniques to analyze real-time activation and signaling of G-protein-coupled receptors. Trends Pharmacol Sci 29:159–165

Luttrell LM, Gesty-Palmer D (2010) Beyond desensitization: physiological relevance of arrestin-dependent signaling. Pharmacol Rev 62:305–330

Luttrell LM, Ferguson SSG, Daaka Y, Miller WE, Maudsley S, Della Rocca GJ, Lin F-T, Kawakatsu H, Owada K, Luttrell DK, Caron MG, Lefkowitz RJ (1999) β-Arrestin-dependent formation of β2 adrenergic receptor-Src protein kinase complexes. Science 283:655–661

Luttrell LM, Roudabush FL, Choy EW, Miller WE, Field ME, Pierce KL, Lefkowitz RJ (2001) Activation and targeting of extracellular signal-regulated kinases by β-arrestin scaffolds. Proc Natl Acad Sci 98:2449–2454

Ma P, Zemmel R (2002) Value of novelty? Nat Rev Drug Discov 1:571–572
Maeda T, Imanishi Y, Palczewski K (2003) Rhodopsin phosphorylation: 30 years later. Prog Retin Eye Res 22:417–434
Mahoney JP, Sunahara RK (2016) Mechanistic insights into GPCR-G protein interactions. Curr Opin Struct Biol 41:247–254
Mancini AD, Bertrand G, Vivot K, Carpentier É, Tremblay C, Ghislain J, Bouvier M, Poitout V (2015) β-Arrestin recruitment and biased agonism at free fatty acid receptor 1. J Biol Chem 290:21131–21140
McBee JK, Palczewski K, Baehr W, Pepperberg DR (2001) Confronting complexity: the interlink of phototransduction and retinoid metabolism in the vertebrate retina. Prog Retin Eye Res 20:469–529
McDonald PH, Chow CW, Miller WE, Laporte SA, Field ME, Lin FT, Davis RJ, Lefkowitz RJ (2000) β-arrestin 2: a receptor-regulated MAPK scaffold for the activation of JNK3. Science 290:1574–1577
Miller WE, Lefkowitz RJ (2001) Arrestins as signaling molecules involved in apoptotic pathways: a real eye opener. Sci STKE 2001:pe1
Miller WE, Maudsley S, Ahn S, Khan KD, Luttrell LM, Lefkowitz RJ (2000) β-arrestin1 interacts with the catalytic domain of the tyrosine kinase c-SRC. J Biol Chem 275:11312–11319
Morrison DK, Davis RJ (2003) Regulation of MAP kinase signaling modules by scaffold proteins in mammals. Ann Rev Cell Dev Biol 19:91–118
Noma T, Lemaire A, Naga Prasad SV, Barki-Harrington L, Tilley DG, Chen J, Le Corvoisier P, Violin JD, Wei H, Lefkowitz RJ, Rockman HA (2007) β-arrestin-mediated β1-adrenergic receptor transactivation of the EGFR confers cardioprotection. J Clin Investig 117:2445–2458
Oakley RH, Laporte SA, Holt JA, Barak LS, Caron MG (1999) Association of β-arrestin with G protein-coupled receptors during clathrin-mediated endocytosis dictates the profile of receptor resensitization. J Biol Chem 274:32248–32257
Oakley RH, Laporte SA, Holt JA, Caron MG, Barak LS (2000) Differential affinities of visual arrestin, βarrestin1, and βarrestin2 for G protein-coupled receptors delineate two major classes of receptors. J Biol Chem 275:17201–17210
Oakley RF, Laporte SA, Holt JA, Barak LS, Caron MG (2001) Molecular determinants underlying the formation of stable intracellular G protein-coupled receptor-β-arrestin complexes after receptor endocytosis. J Biol Chem 276:19452–19460
Palczewski K, Kumasaka T, Hori T, Behnke CA, Motoshima H, Fox BA, Trong IL, Teller DC, Okada T, Stenkamp RE, Yamamoto M, Miyano M (2000) Crystal structure of rhodopsin: a G protein-coupled receptor. Science 289:739–745
Pearson G, Robinson F, Beers Gibson T, Xu B-E, Karandikar M, Berman K, Cobb MH (2001) Mitogen-activated protein (MAP) kinase pathways: regulation and physiological functions. Endocr Rev 22:153–183
Pierce KL, Premont RT, Lefkowitz RJ (2002) Seven-transmembrane receptors. Nat Rev Mol Cell Biol 3:639–650
Povsic TJ, Kohout TA, Lefkowitz RJ (2003) β-arrestin1 mediates insulin-like growth factor 1 (IGF-1) activation of phosphatidylinositol 3-kinase (PI3K) and anti-apoptosis. J Biol Chem 278:51334–51339
Printen JA, Sprague GF (1994) Protein-protein interactions in the yeast pheromone response pathway: Ste5p interacts with all members of the MAP kinase cascade. Genetics 138:609–619
Reiter E, Ahn S, Shukla AK, Lefkowitz RJ (2012) Molecular mechanism of beta-arrestin-biased agonism at seven-transmembrane receptors. Annu Rev Pharmacol Toxicol 52:179–197
Roux PP, Blenis J (2004) ERK and p38 MAPK-activated protein kinases: a family of protein kinases with diverse biological functions. Microbiol Mol Biol Rev 68:320–344
Seifert R, Dove S (2009) Functional selectivity of GPCR ligand stereoisomers: new pharmacological opportunities. Mol Pharmacol 75:13–18
Seo J, Tsakem EL, Breitman M, Gurevich VV (2011) Identification of arrestin-3-specific residues necessary for JNK3 kinase activation. J Biol Chem 286:27894–27901

Shenoy SK, Lefkowitz RJ (2003) Multifaceted roles of beta-arrestins in the regulation of seven-membrane-spanning receptor trafficking and signalling. Biochem J 375:503–515

Shenoy SK, McDonald PH, Kohout TA, Lefkowitz RJ (2001) Regulation of receptor fate by ubiquitination of activated β2-adrenergic receptor and β-arrestin. Science 294:1307–1313

Shukla AK, Wodak SJ (2016) Editorial overview: multi-protein assemblies in signaling. Curr Opin Struct Biol 41:v–vii

Song X, Coffa S, Fu H, Gurevich VV (2009) How does arrestin assemble MAPKs into a signaling complex? J Biol Chem 284:685–695

Therrien M, Michaud NR, Rubin GM, Morrison DK (1996) KSR modulates signal propagation within the MAPK cascade. Genes Dev 10:2684–2695

Tohgo A, Pierce KL, Choy EW, Lefkowitz RJ, Luttrell LM (2002) β-Arrestin scaffolding of the ERK cascade enhances cytosolic ERK activity but inhibits ERK-mediated transcription following angiotensin AT1a receptor stimulation. J Biol Chem 277:9429–9436

Tohgo A, Choy EW, Gesty-Palmer D, Pierce KL, Laporte S, Oakley RH, Caron MG, Lefkowitz RJ, Luttrell LM (2003) The stability of the G protein-coupled receptor-β-arrestin interaction determines the mechanism and functional consequence of ERK activation. J Biol Chem 278:6258–6267

Walters RW, Shukla AK, Kovacs JJ, Violon JD, DeWire SM, Lam CM, Chen JR, Muehlbauer MJ, Whalen EJ, Levkowitz RJ (2009) β-arrestin1 mediates nicotinic acid-induced flushing, but not its antilipolytic effect, in mice. J Clin Investig 119:1312–1321

Wei H, Ahn S, Shenoy SK, Karnik SS, Hunyady L, Luttrell LM, Lefkowitz RJ (2003a) Independent β-arrestin 2 and G protein-mediated pathways for angiotensin II activation of extracellular signal-regulated kinases 1 and 2. Proc Natl Acad Sci 100:10782–10787

Wei H, Ahn S, Shenoy SK, Karnik SS, Hunyady L, Luttrell LM, Lefkowitz RJ (2003b) Independent β-arrestin 2 and G protein-mediated pathways for angiotensin II activation of extracellular signal-regulated kinases 1 and 2. Proc Natl Acad Sci USA 100:10782–10787

Wisler JW, Harris EM, Raisch M, Mao L, Kim J, Rockman HA, Lefkowitz RJ (2015) The role of beta-arrestin2-dependent signaling in thoracic aortic aneurysm formation in a murine model of Marfan syndrome. Am J Physiol Heart Circ Physiol 309:H1516–H1527

Wu B, Chien EYT, Mol CD, Fenalti G, Liu W, Katritch V, Abagyan R, Brooun A, Wells P, Bi FC, Hamel DJ, Kuhn P, Handel TM, Cherezov V, Stevens RC (2010) Structures of the CXCR4 chemokine GPCR with small-molecule and cyclic peptide antagonists. Science 330:1066–1071

Xiao K, McClatchy DB, Shukla AK, Zhao Y, Chen M, Shenoy SK, Yates JR, Lefkowitz RJ (2007) Functional specialization of β-arrestin interactions revealed by proteomic analysis. Proc Natl Acad Sci 104:12011–12016

Xiao K, Sun J, Kim J, Rajagopal S, Zhai B, Villén J, Haas W, Kovacs JJ, Shukla AK, Hara MR, Hernandez M, Lachmann A, Zhao S, Lin Y, Cheng Y, Mizuno K, Ma'ayan A, Gygi SP, Lefkowitz RJ (2010) Global phosphorylation analysis of β-arrestin-mediated signaling downstream of a seven transmembrane receptor (7TMR). Proc Natl Acad Sci 107: 15299–15304

Yang S, Ben-Shalom R, Ahn M, Liptak AT, van Rijn RM, Whistler JL, Bender KJ (2016) Beta-arrestin-dependent dopaminergic regulation of calcium channel activity in the axon initial segment. Cell Rep 16:1518–1526

Yin J, Mobarec JC, Kolb P, Rosenbaum DM (2015) Crystal structure of the human OX2 orexin receptor bound to the insomnia drug suvorexant. Nature 519:247–250

Zeke A, Lukács M, Lim WA, Reményi A (2009) Scaffolds: interaction platforms for cellular signalling circuits. Trends Cell Biol 19:364–374

Zhan X, Stoy H, Kaoud TS, Perry NA, Chen Q, Perez A, Els-Heindl S, Slagis JV, Iverson TM, Beck-Sickinger AG, Gurevich EV, Dalby KN, Gurevich VV (2016) Peptide mini-scaffold facilitates JNK3 activation in cells. Sci Rep 6:21025

Part II
Localization of Functional Elements on Arrestins

Chapter 3
Initial Crystallographic Studies of Visual Arrestin: Insights and Perspectives

Joel A. Hirsch

Abstract Crystallographic studies of visual arrestin in the late nineties built upon fundamental biochemical work that had identified arrestins as proteins central to desensitization of GPCR signaling. The structural findings revealed the arrestin fold that had two related domains and a C-tail, which folded onto the molecule's surface. The structures had a curious "polar core" that sat at the fulcrum of the two domains and which acts as an active site for conformational activation. The structures also served as the basis for elaborating the arrestin fold, found in all the kingdoms of life. Finally, the results suggested that quaternary structure i.e. the oligomeric state played a role in self-regulation.

Keywords Arrestin · Crystal structure · Arrestin fold · Phosphate sensor · Activation · Conformational change

Early Biochemical Studies

The G protein signaling cycle, with the receptor as the focus of signal initiation, was largely worked out in the seventies. The receptors that became paradigmatic for this canonical G protein-mediated signaling were rhodopsin (Rho), the receptor for photons in the retina, and β2-adrenergic receptor (βAR), the receptor for the hormone epinephrine. These receptors were cloned in 1980s and the first glimpses of their molecular structure were obtained (Ovchinnikov 1982; Dixon et al. 1986). The full elucidation of their 3D structure took many decades, as they are integral

This chapter is dedicated to the memory of our colleagues Drs. Serge Pares, Carsten Schubert, and Paul Sigler, who played essential roles in the elucidation of arrestin structures and their implications.

J.A. Hirsch (✉)
Department of Biochemistry and Molecular Biology, Sagol School of Neuroscience, George S. Wise Faculty of Life Sciences, Tel Aviv University, 6997810 Tel Aviv, Israel
e-mail: jhirsch@post.tau.ac.il

V.V. Gurevich (ed.), *The Structural Basis of Arrestin Functions*,
DOI 10.1007/978-3-319-57553-7_3

membrane proteins, which in general have often been refractory to straightforward structure determination. However, the structural biology of the transducing, or in the words of Rodbell, transferring molecules advanced more rapidly with the details of hetero-trimeric G proteins published in the nineties. Subsequent to elucidation of the G-protein cycle in the early eighties, the question of non-agonist specific receptor desensitization was addressed, again in the two aforementioned paradigms. The molecular players for the desensitization system were discovered by the German group of Herman Kuhn (Wilden et al. 1986a), working on Rho and in parallel, the Lefkowitz group in the USA, working on βAR (Benovic et al. 1986). The groups identified a two-step cascade whereby the receptor was phosphorylated by a specific kinase, named G-protein receptor kinase (GRK), on its intracellular C-terminus at multiple sites. The phosphorylation itself slowed coupling with the cognate G protein. A soluble protein with approximate mass of 48 kDa (hence, named by Kuhn p48, also known as S-antigen) terminated the signal by association with Rho (Kuhn et al. 1984; Wilden et al. 1986a). Likewise, a short time thereafter, a protein was identified biochemically by the Lefkowitz group with analogous function in the βAR system (Benovic et al. 1987). These protein "terminators" were dubbed arrestins. Visual arrestin was cloned and published in 1987 by two groups (Shinohara et al. 1987; Yamaki et al. 1987), while β-arrestin, cognate for βAR, was cloned by Lefkowitz and co-workers (Lohse et al. 1990) in 1990. Hoffman's group demonstrated that visual arrestin had a strong affinity for light-activated and phosphorylated Rho (Schleicher et al. 1989). This specific receptor form proved to be the preferred one by a significant ratio, implying that arrestin exquisitely discriminated for the appropriate receptor state when quenching the signal.

What the Structural Biologists Knew

With the cloning of arrestin's cDNA, achieved by biochemical purification of the intact protein, peptide sequencing, and isolation of matching nucleotide sequences from libraries, a primary structure was in hand (Shinohara et al. 1987; Yamaki et al. 1987). Sequence analysis suggested some homology between arrestin and Gα transducin (Gt), a cognate heterotrimeric G protein of rhodopsin. Secondary structure prediction based on the sequence and circular dichroism spectroscopy indicated that the protein was almost complete beta-strand/sheet with quite a small percentage of α-helical structure (Shinohara et al. 1987). Classical biochemical structural probing by limited proteolysis performed by Palczewski and coworkers showed that the protein's N- and C-termini were easily cleaved but a resistant core, spanning residues 6 through 365, was likely to present a stable folded molecule (Palczewski et al. 1991a). An important conclusion was that the protein contained a flexible C-tail of about 40 residues. Concomitant with these biochemical studies, Gurevich and Benovic developed an in vitro translation system for preparation of radiolabeled recombinant arrestin that could be engineered for use in a functional truncation analysis (Gurevich and Benovic 1992). Their salient findings were that a mini-arrestin, spanning residues 1 through 191, retained significant binding activity to Rho. In addition, they observed that the C-tail, had important impact on arrestin's specificity of receptor state binding (Gurevich and Benovic 1992, 1993). In parallel, Palczewski reported the isolation of a natural splice variant, purified from rod outer segments, called p44 that lacked the C-tail described earlier (Palczewski et al. 1994). Furthermore, p44 binds constitutively to activated but non-phosphorylated Rho. Finally, the polyanionic sugar heparin and phosphorylated Rho C-terminus were shown to bind arrestin (Schleicher et al. 1989; Palczewski et al. 1991b). The conclusion from these studies indicated that the arrestin C-tail was an important self-regulatory element, which held arrestin in a basal, inactive state until it was activated by the activated and phosphorylated receptor.

Crystallographic Studies

Due to the biological and biomedical significance of the GPCR signaling systems, elucidation of the components' 3D structures became an immediate goal in several laboratories. Ongoing successful efforts to understand the mechanism of the hetero-trimeric G protein cycle elicited the crystallographic structures of the visual and inhibitory G protein complexes in various nucleotide states. The prerequisite for these studies was abundant and homogeneous protein preparations. Due to the nature of the visual system, it was possible to purify from bovine rod outer segments (ROS) copious amounts of Rho, Gt, and arrestin. In parallel, recombinant expression was pursued, using both bacterial and baculovirus/insect cell culture systems. In the end, both approaches proved to be productive in the hands of the competing laboratories, which included the Sigler group at Yale, the Sprang and Gilman groups at Southwestern, and the German groups, which worked on arrestin.

For the arrestin work, ROS were the starting material and the membranes themselves were used to purify arrestin by adsorption and subsequent chromatography (Wilden et al. 1986b). Wilden and colleagues provided this preparation for crystallization, which was achieved at room temperature (Wilden et al. 1997). In contrast, our group pursued recombinant bacterial expression (Schubert et al. 1999), employing a modified preparation protocol based on Gurevich and coworkers' work, published in 1997 (Gurevich and Benovic 1997). The system did not employ any real protein engineering so that no affinity tags were used. Hence, we used column chromatography, in particular heparin Sepharose to obtain the requisite protein (Hirsch et al. 1999). In our hands, crystals were only obtained at 4 °C, although there was considerable polymorphism. Some of these forms did exhibit reasonable diffraction. Many did not. The crystal forms that were the basis of structure determination for both research groups were related and had in common an asymmetric unit comprised of a dimer of dimers with the same configuration, discussed below. These forms had high solvent content ($\sim$70%), confounding efforts to measure quality data sets. In our hands, crystals were generally non-isomorphic, further challenging phasing efforts. With synchrotron radiation, diffraction limits were pushed to 3.3 Å for the Julich group led by Granzin et al. (1998) and to 2.8 Å for the Yale group (Hirsch et al. 1999). Experimental phasing was obtained by isomorphous replacement (Julich) or anomalous diffraction and isomorphous replacement (Yale).

The tertiary structure of arrestin from these studies revealed a protein with two clear domains, labeled the N- and C-domains, connected by a linker lacking secondary structure. Both domains are β-sandwiches i.e. each comprises two stacked β-sheets, each bearing four strands. Each domain is autonomous, lacking any interleaving elements from the other domain. Indeed, the N-domain has the protein's single α-helix, formed from a linker connecting two strands. The β-sandwiches have significant curvature, so that both appear as cupped palms (Fig. 3.1). These β-sandwiches have been classified as members in the immunoglobulin superfamily by projects such as SCOP. The third structural element found in the structures of the Yale group was the C-tail. This element spans residue 363 through the C-terminus. Poorly defined electron density characterized residues 363–371 and 395–402, likely due to high mobility of these polypeptide sections. Importantly, the segment from 372 to 395 was clearly modeled. Indeed, it forms, in part, a β-strand that interacts with the β-sheet of the N-domain. This C-tail has been observed in several arrestin family crystal forms and its location and conformation provide essential functional insights, in addition to being consistent with a large body of biochemical findings.

The structure revealed what had been discerned early on with publication of the primary sequence, namely that the two primary domains were organized in a tandem fashion and had some level of sequence similarity, albeit rather low (about 19%). The tertiary structure clearly exhibits an intra-molecular dyad, relating the two β-sandwiches (Fig. 3.1). Perhaps even more striking is the topology of the sandwiches, in particular the connecting turns and loops, which is conserved between domains. The sequence and topology similarities are exhibited at the level of superpositions such that a majority of the Ca atoms have 1.8 Å RMSD with a pseudo two-fold axis (~160°) (Hirsch et al. 1999). And while the fold classifiers

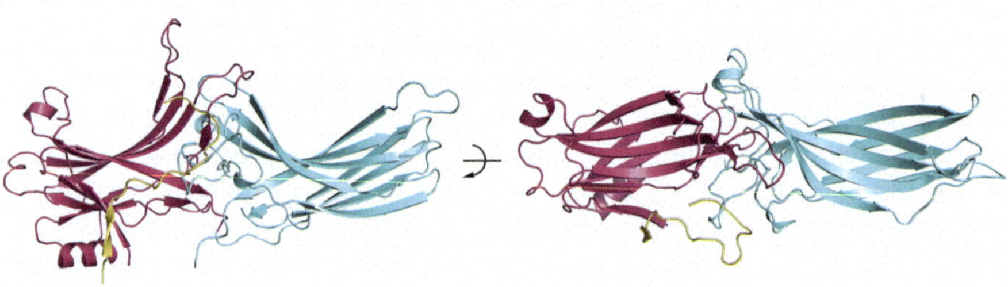

Fig. 3.1 The crystallographic 3D structure of bovine visual arrestin at 2.8 Å resolution. A protomer is shown, depicting the secondary structure elements. A *side* perspective is shown at *left* and a *top* perspective is displayed at *right*. Note the pseudo twofold axis perpendicular to the plane in the *right panel* that relates the two domains. The N-domain is depicted in magenta while the C-domain is *cyan*. The C-tail is *colored yellow*. The illustration is based on coordinates of PDB entry 1CF1 and was prepared using PyMol

have created unique motifs for each domain (N and C), most known arrestin family sequences have the tandem domain architecture (Alvarez 2008). This empirical observation can be readily justified by inspection of the structure. The two domains abut each other at the respective ends and that interface is largely hydrophobic in nature with a caveat described below. So while each domain is topologically independent, the domains are unlikely to be stable or soluble individually in solution.

Close examination of the structure, directed by earlier mutagenesis and functional characterization (Gurevich and Benovic 1995), brought focus to an unusual and concentrated constellation of charged residues situated in the fulcrum of the molecule, i.e. between the N- and C-domains. These residues were largely buried, with the attendant energetic penalties involved, posing a riddle to any thoughtful biophysical chemist. Indeed, Mayo and coworkers have outlined three categories for the underlying rationale of buried polar residues: active sites, thermodynamic stability, and folding/structure specificity (Bolon and Mayo 2001). We named this constellation of buried salt bridges the polar core. The earlier structure-function studies had revealed that a single point mutant, Arg175Gln or Arg175Glu, could render visual arrestin constitutively active i.e. it was able to bind with avidity to activated Rho without need for phosphorylation of the receptor (Gurevich and Benovic 1995, 1997). This behavior was akin to the p44 splice variant described above. These previous structure-function studies had sought to map the phosphate sensor in arrestin with the assumption that the Rho phospho-C-tail activated arrestin into a receptor binding-competent state. Hence, the positively charged residues buried in the polar core could attract the phosphopeptide, triggering and inducing a conformational change in arrestin. Importantly, the crystallographic studies elegantly provided the atomic details for arrestin's charged residues that acted as an "active site" for catalyzing the protein's conformational changes. Specifically, Asp296 served as the counter-ion for Arg175, holding the two arrestin domains in the basal orientation. Other charged residues contributed to the polar core, in particular Arg382 from the C-tail. Structure-guided mutagenesis of this polar core and functional characterization validated and reinforced the crystallographic insights, and taken together resulted in a model of arrestin activation by disruption of the polar core upon association with light-activated phospho-Rho (Hirsch et al. 1999; Vishnivetskiy et al. 1999). The model included the idea that the C-tail maintained the basal state by its interactions with the polar core and neighboring surfaces. At activation, the C-tail is released, the polar core is exposed to the receptor's phosphopeptide and arrestin undergoes a conformational change. Strikingly, the polar core could be remodeled by reciprocally swapping charged residues. In addition, the protein's electrostatic properties were consistent with the notion that basic surfaces of arrestin attract the flexible receptor phosphopeptide and guide it to the polar core. The centrality of the polar core has been borne out in later structural work which includes the structure determination of p44, Arg175Glu, and the most recent Rho*-Arr complex (Kim et al. 2013; Granzin et al. 2015; Kang et al. 2015).

One of the curious observations we made about these initial structures was the finding that several regions encoded chameleon sequences. Chameleon sequences are polypeptide stretches, which can adopt multiple conformations including switches of secondary structure [for a recent review see Li et al. (2015)]. We noted in our original study [see Fig. 5 in Hirsch et al. (1999)] that residues 68–79, later called the finger loop (Hanson et al. 2006), undergo a transition between loop and β-strand/sheet conformations in the two different protomer copies of arrestin. In the recently solved Rho-arrestin complex crystal structure, the finger loop is one of three important patches of the protein-protein interface. Remarkably, in the receptor complex the loop forms an α-helix. It seems likely that chameleon sequences are well suited for serving as molecular recognition features and may be related to some of the general properties of intrinsically disordered proteins (Cumberworth et al. 2013).

Subsequent to the initial structures of visual arrestin, other arrestin family structures were determined (Han et al. 2001; Milano et al. 2002; Sutton et al. 2005), demonstrating the conserved arrestin fold fold. Other structural work revealed that Vps26, a protein important for cellular trafficking and part of the retromer, a protein complex that directs transmembrane cargo from endosomes to the Golgi, surprisingly had an arrestin fold, despite no apparent sequence similarity (Shi et al. 2006). Moreover, Vps26 had something resembling a polar core, a constellation of charged residues holding the N- and C-domains together, albeit in a different configuration than that found in the arrestins. The function of this polar core, though, may not be analogous to that of arrestin as based on recent structural work of the retromer complex (Lucas et al. 2016). Alvarez performed a comprehensive bioinformatic analysis of the arrestin super-family, incorporating molecular phylogenetics on the basis of expanded genomic data (Alvarez 2008). These structural findings and sequence observations have led to the conclusion that there is an arrestin fold, found throughout the eukaryotic kingdom, extant in fungi and metazoa. Indeed, fungi have a wide variety of arrestins, containing homologs to both the N- and C-domain, but including other protein-protein interaction or protein phospholipid motifs such as PY or FYVE (Aubry et al. 2009). Using phylogenetic analysis, Alvarez delineated two subfamilies, α and β. The former (traditionally called arrestin domain-containing proteins, ARRDCs) has many representatives in the fungal and metazoan world while the latter is found only in metazoa, with *Hydra* being the most primal species. The β subfamily includes arrestin1-4, i.e. visual/β-arrestins that have been characterized so extensively in vertebrates. Notably, no species has exclusively single domain arrestins, indicating that the gene duplication responsible for the tandem N- and C-domains is very ancient and consistent with our understanding of the fold structure. Vps26 is more closely related to the α arrestins but is clearly ancient also, as it is found in all eukaryotes including plants, which lack the β arrestin subfamily. Interestingly, SpsoOM proteins, related to the α arrestins have been found in archaea and bacteria (Alvarez 2008).

Quaternary Structure

As mentioned earlier, the asymmetric unit for both crystal structures comprised a dimer of dimers, with two distinct protomer conformations. These tetrameric complexes had the same exact quaternary structure despite having been crystallized under vastly different conditions and forming different crystallographic habits. The Yale group noted this conserved molecular symmetry and asked if such a particle was found in solution or restricted to the crystalline form and therefore irrelevant to molecular physiology. In the first reports of biochemical isolation, an oligomeric state had been detected (Wacker et al. 1977). To address this question rigorously, we used a sedimentation equilibrium (Schubert et al. 1999). Our conclusions were that indeed tetrameric arrestin is found in solution at high concentrations but that there is a monomer-dimer-tetramer equilibrium and this mixture may be a relevant quaternary state in photoreceptor cells as based on the very high concentration of arrestin in these cells (Strissel et al. 2006; Song et al. 2011; Hanson et al. 2007a). At the time, we proposed that self-association engenders an inactive storage form, rapidly converted to the monomer for proper and rapid quenching of signal. In elegant subsequent studies reviewed in Chen et al. (2014), this proposal was modified but the basic idea remains valid; that arrestin self-association plays an important role in arrestin function by enabling the correct concentration for the monomeric species capable of receptor binding (Hanson et al. 2007b) while at the same time maintaining a large dynamic range of signal amplitude. Further structural and biochemical work demonstrated that the solution tetramer was configured differently than the crystalline form, although certain molecular interfaces were the same (Hanson et al. 2008). The solution tetramer model utilizes all potential molecular interfaces in a closed ring, explaining why it is stable, whereas the crystalline tetramer (asymmetric unit) does not, enabling lattice formation. It remains possible that the solution tetramer structure was actually crystallized by the Yale group in the gamut of polymorphic crystals obtained, but due to poor diffraction of these crystals, no structure was determined.

Thus, the first crystal structures of visual arrestin and subsequent structure-based mutagenesis and biophysical studies yielded the first glimpse into molecular mechanism of arrestin activation and arrestin regulation by self-association. The structures showed that arrestin is a metastable protein, perfectly designed for a global conformational change. Moreover, the structures revealed a novel arrestin fold, which was subsequently shown to be shared by all receptor-binding arrestins and at least some arrestin-domain containing proteins.

References

Alvarez CE (2008) On the origins of arrestin and rhodopsin. BMC Evol Biol 8:222

Aubry L, Guetta D, Klein G (2009) The arrestin fold: variations on a theme. Curr Genomics 10:133–142

Benovic JL, Strasser RH, Caron MG, Lefkowitz RJ (1986) Beta-adrenergic receptor kinase: identification of a novel protein kinase that phosphorylates the agonist-occupied form of the receptor. Proc Natl Acad Sci USA 83:2797–2801

Benovic JL, Kuhn H, Weyand I, Codina J, Caron MG, Lefkowitz RJ (1987) Functional desensitization of the isolated beta-adrenergic receptor by the beta-adrenergic receptor kinase: potential role of an analog of the retinal protein arrestin (48-kDa protein). Proc Natl Acad Sci USA 84:8879–8882

Bolon DN, Mayo SL (2001) Polar residues in the protein core of *Escherichia coli* thioredoxin are important for fold specificity. Biochemistry 40:10047–10053

Chen Q, Zhuo Y, Kim M, Hanson SM, Francis DJ, Vishnivetskiy SA, Altenbach C, Klug CS, Hubbell WL, Gurevich VV (2014) Self-association of arrestin family members. Handb Exp Pharmacol 219:205–223

Cumberworth A, Lamour G, Babu MM, Gsponer J (2013) Promiscuity as a functional trait: intrinsically disordered regions as central players of interactomes. Biochem J 454:361–369

Dixon RA, Kobilka BK, Strader DJ, Benovic JL, Dohlman HG, Frielle T, Bolanowski MA, Bennett CD, Rands E, Diehl RE, Mumford RA, Slater EE, Sigal IS, Caron MG, Lefkowitz RJ, Strader CD (1986) Cloning of the gene and cDNA for mammalian beta-adrenergic receptor and homology with rhodopsin. Nature 321:75–79

Granzin J, Wilden U, Choe HW, Labahn J, Krafft B, Buldt G (1998) X-ray crystal structure of arrestin from bovine rod outer segments. Nature 391:918–921

Granzin J, Stadler A, Cousin A, Schlesinger R, Batra-Safferling R (2015) Structural evidence for the role of polar core residue Arg175 in arrestin activation. Sci Rep 5:15808

Gurevich VV, Benovic JL (1992) Cell-free expression of visual arrestin. Truncation mutagenesis identifies multiple domains involved in rhodopsin interaction. J Biol Chem 267:21919–21923

Gurevich VV, Benovic JL (1993) Visual arrestin interaction with rhodopsin: sequential multisite binding ensures strict selectivity towards light-activated phosphorylated rhodopsin. J Biol Chem 268:11628–11638

Gurevich VV, Benovic JL (1995) Visual arrestin binding to rhodopsin: diverse functional roles of positively charged residues within the phosphorylation-recignition region of arrestin. J Biol Chem 270:6010–6016

Gurevich VV, Benovic JL (1997) Mechanism of phosphorylation-recognition by visual arrestin and the transition of arrestin into a high affinity binding state. Mol Pharmacol 51:161–169

Han M, Gurevich VV, Vishnivetskiy SA, Sigler PB, Schubert C (2001) Crystal structure of beta-arrestin at 1.9 Å: possible mechanism of receptor binding and membrane translocation. Structure 9:869–880

Hanson SM, Francis DJ, Vishnivetskiy SA, Kolobova EA, Hubbell WL, Klug CS, Gurevich VV (2006) Differential interaction of spin-labeled arrestin with inactive and active phosphorhodopsin. Proc Natl Acad Sci USA 103:4900–4905

Hanson SM, Gurevich EV, Vishnivetskiy SA, Ahmed MR, Song X, Gurevich VV (2007a) Each rhodopsin molecule binds its own arrestin. Proc Nat Acad Sci USA 104:3125–3128

Hanson SM, Van Eps N, Francis DJ, Altenbach C, Vishnivetskiy SA, Arshavsky VY, Klug CS, Hubbell WL, Gurevich VV (2007b) Structure and function of the visual arrestin oligomer. EMBO J 26:1726–1736

Hanson SM, Dawson ES, Francis DJ, Van Eps N, Klug CS, Hubbell WL, Meiler J, Gurevich VV (2008) A model for the solution structure of the rod arrestin tetramer. Structure 16:924–934

Hirsch JA, Schubert C, Gurevich VV, Sigler PB (1999) The 2.8 Å crystal structure of visual arrestin: a model for arrestin's regulation. Cell 97:257–269

Kang Y, Zhou XE, Gao X, He Y, Liu W, Ishchenko A, Barty A, White TA, Yefanov O, Han GW, Xu Q, de Waal PW, Ke J, Tan MH, Zhang C, Moeller A, West GM, Pascal BD, Van Eps N, Caro LN, Vishnivetskiy SA, Lee RJ, Suino-Powell KM, Gu X, Pal K, Ma J, Zhi X, Boutet S, Williams GJ, Messerschmidt M, Gati C, Zatsepin NA, Wang D, James D, Basu S, Roy-Chowdhury S, Conrad CE, Coe J, Liu H, Lisova S, Kupitz C, Grotjohann I, Fromme R, Jiang Y, Tan M, Yang H, Li J, Wang M, Zheng Z, Li D, Howe N, Zhao Y, Standfuss J, Diederichs K, Dong Y, Potter CS, Carragher B, Caffrey M, Jiang H, Chapman HN, Spence JC,

Fromme P, Weierstall U, Ernst OP, Katritch V, Gurevich VV, Griffin PR, Hubbell WL, Stevens RC, Cherezov V, Melcher K, Xu HE (2015) Crystal structure of rhodopsin bound to arrestin by femtosecond X-ray laser. Nature 523:561–567

Kim YJ, Hofmann KP, Ernst OP, Scheerer P, Choe HW, Sommer ME (2013) Crystal structure of pre-activated arrestin p44. Nature 497:142–146

Kuhn H, Hall SW, Wilden U (1984) Light-induced binding of 48-kDa protein to photoreceptor membranes is highly enhanced by phosphorylation of rhodopsin. FEBS Lett 176:473–478

Li W, Kinch LN, Karplus PA, Grishin NV (2015) ChSeq: a database of chameleon sequences. Protein Sci 24:1075–1086

Lohse MJ, Benovic JL, Codina J, Caron MG, Lefkowitz RJ (1990) Beta-arrestin: a protein that regulates beta-adrenergic receptor function. Science 248:1547–1550

Lucas M, Gershlick DC, Vidaurrazaga A, Rojas AL, Bonifacino JS, Hierro A (2016) Structural mechanism for cargo recognition by the retromer complex. Cell 167(1623–1635):e1614

Milano SK, Pace HC, Kim YM, Brenner C, Benovic JL (2002) Scaffolding functions of arrestin-2 revealed by crystal structure and mutagenesis. Biochemistry 41:3321–3328

Ovchinnikov YA (1982) Rhodopsin and bacteriorhodopsin: structure-function relationship. FEBS Lett 148:179–191

Palczewski K, Buczylko J, Imami NR, McDowell JH, Hargrave PA (1991a) Role of the carboxyl-terminal region of arrestin in binding to phosphorylated rhodopsin. J Biol Chem 266:15334–15339

Palczewski K, Pulvermuller A, Buczylko J, Hofmann KP (1991b) Phosphorylated rhodopsin and heparin induce similar conformational changes in arrestin. J Biol Chem 266:18649–18654

Palczewski K, Buczylko J, Ohguro H, Annan RS, Carr SA, Crabb JW, Kaplan MW, Johnson RS, Walsh KA (1994) Characterization of a truncated form of arrestin isolated from bovine rod outer segments. Protein Sci 3:314–324

Schleicher A, Kuhn H, Hofmann KP (1989) Kinetics, binding constant, and activation energy of the 48-kDa protein-rhodopsin complex by extra-metarhodopsin II. Biochemistry 28:1770–1775

Schubert C, Hirsch JA, Gurevich VV, Engelman DM, Sigler PB, Fleming KG (1999) Visual arrestin activity may be regulated by self-association. J Biol Chem 274:21186–21190

Shi H, Rojas R, Bonifacino JS, Hurley JH (2006) The retromer subunit Vps26 has an arrestin fold and binds Vps35 through its C-terminal domain. Nat Struct Mol Biol 13:540–548

Shinohara T, Dietzschold B, Craft CM, Wistow G, Early JJ, Donoso LA, Horwitz J, Tao R (1987) Primary and secondary structure of bovine retinal S antigen (48-kDa protein). Proc Natl Acad Sci USA 84:6975–6979

Song X, Vishnivetskiy SA, Seo J, Chen J, Gurevich EV, Gurevich VV (2011) Arrestin-1 expression in rods: balancing functional performance and photoreceptor health. Neuroscience 174:37–49

Strissel KJ, Sokolov M, Trieu LH, Arshavsky VY (2006) Arrestin translocation is induced at a critical threshold of visual signaling and is superstoichiometric to bleached rhodopsin. J Neurosci 26:1146–1153

Sutton RB, Vishnivetskiy SA, Robert J, Hanson SM, Raman D, Knox BE, Kono M, Navarro J, Gurevich VV (2005) Crystal structure of cone arrestin at 2.3 Å: evolution of receptor specificity. J Mol Biol 354:1069–1080

Vishnivetskiy SA, Paz CL, Schubert C, Hirsch JA, Sigler PB, Gurevich VV (1999) How does arrestin respond to the phosphorylated state of rhodopsin? J Biol Chem 274:11451–11454

Wacker WB, Donoso LA, Kalsow CM, Yankeelov JA Jr, Organisciak DT (1977) Experimental allergic uveitis. Isolation, characterization, and localization of a soluble uveitopathogenic antigen from bovine retina. J Immunol 119:1949–1958

Wilden U, Hall SW, Kuhn H (1986a) Phosphodiesterase activation by photoexcited rhodopsin is quenched when rhodopsin is phosphorylated and binds the intrinsic 48-kDa protein of rod outer segments. Proc Natl Acad Sci 83:1174–1178

Wilden U, Wust E, Weyand I, Kuhn H (1986b) Rapid affinity purification of retinal arrestin (48 kDa protein) via its light-dependent binding to phosphorylated rhodopsin. FEBS Lett 207:292–295

Wilden U, Choe HW, Krafft B, Granzin J (1997) Crystallization and preliminary X-ray analysis of arrestin from bovine rod outer segment. FEBS Lett 415:268–270

Yamaki K, Takahashi Y, Sakuragi S, Matsubara K (1987) Molecular cloning of the S-antigen cDNA from bovine retina. Biochem Biophys Res Commun 142:904–910

Chapter 4
Structural Basis for β-Arrestins in GPCR Trafficking

Sungsoo M. Yoo, Dong Soo Kang and Jeffrey L. Benovic

Abstract β-arrestins are adaptor proteins that function to regulate G protein-coupled receptor (GPCR) desensitization, trafficking and signaling. β-arrestin involvement in receptor trafficking is mediated by GPCR-promoted conformational changes that drive β-arrestin interaction with adaptor protein 2 (AP2), clathrin and phosphoinositides, with each of these interactions being essential for effective β-arrestin-mediated trafficking. Structural information on β-arrestin interaction with the endocytic machinery provides mechanistic insight into how these interactions are coordinated. Here we provide an overview of the structural insight on β-arrestin-mediated trafficking of GPCRs.

Keywords Arrestin · Endocytosis · G protein-coupled receptor · Signaling · Trafficking · X-ray crystallography

Introduction

G protein-coupled receptors (GPCRs) mediate many physiological responses via their ability to respond to a diverse array of extracellular stimuli (Moore et al. 2007). Since GPCR signaling leads to activation of numerous downstream effectors, GPCR activity needs to be tightly regulated so as to not result in pathological signaling. There are three major modes of regulation of GPCR activity: desensitization, endocytosis, and down-regulation. Desensitization uncouples heterotrimeric

S.M. Yoo · J.L. Benovic (✉)
Department of Biochemistry and Molecular Biology,
Thomas Jefferson University, Philadelphia, PA, USA
e-mail: jeffrey.benovic@jefferson.edu

S.M. Yoo
e-mail: sungsoo.yoo@jefferson.edu

D.S. Kang
WuXi AppTec, Suzhou, Jiangsu, China
e-mail: dongsoo.kang@wuxiapptec.com

V.V. Gurevich (ed.), *The Structural Basis of Arrestin Functions*,
DOI 10.1007/978-3-319-57553-7_4

G protein signaling from the activated GPCR, endocytosis sequesters the receptor from the cell surface, and down-regulation controls the expression level of the GPCR. Many aspects involved in regulating GPCR signaling depend on the endocytosis of the receptor. For example, various signaling events, such as ERK signaling through certain GPCRs, involve receptor endocytosis (Daaka et al. 1998). In addition, GPCR-mediated G-protein signaling can occur at the plasma membrane as well as at the endosome after endocytosis (Irannejad et al. 2013). The endocytosis of many GPCRs is mediated by β-arrestins, in clathrin- and AP2-dependent manner (Moore et al. 2007). Although it has been 20 years since the initial discovery of β-arrestin's role in mediating GPCR endocytosis (Ferguson et al. 1996; Goodman et al. 1996), much remains to be discovered about the molecular mechanisms of this process. What is the structural basis of β-arrestin mediating the different stages of endocytosis? Is the structure of β-arrestin during endocytosis different from its receptor bound structure? When does β-arrestin dissociate from the receptor during endocytosis and what is its conformation?

Part of the reason for the difficulty in delineating the mechanisms involved in β-arrestin-mediated endocytosis of GPCRs is the scarcity of structural data that captures β-arrestin in complex with its endocytic binding partners. None of the 'active' arrestin crystal structures are in a complex with endocytic proteins nor do they contain structural information on the C-terminal tail of β-arrestin where the endocytic proteins bind. In fact, the only crystal structure of arrestin in a complex with an endocytic molecule is β-arrestin1 bound to the N-terminal domain of clathrin (Kang et al. 2009). Curiously, even in this structure, β-arrestin was found to be in its inactive state, and the implication of this particular structure for the molecular mechanism behind arrestin-mediated endocytosis remains to be elucidated. Nevertheless, given that endocytosis is a unique function of the non-visual β-arrestins, there are reasons to believe that novel insight into the molecular mechanisms behind an exclusive function for β-arrestins can be learned through studying the structure of arrestins during endocytosis. Here we provide an overview of current knowledge on the role of β-arrestins in clathrin-mediated endocytosis of GPCRs.

Overview of Clathrin Mediated Endocytosis

GPCR endocytosis mainly occurs via a clathrin and dynamin-dependent process that involves interaction with β-arrestin, although many exceptions exist (Gurevich and Gurevich 2015). Clathrin-mediated endocytosis occurs in several stages (McMahon and Boucrot 2011; Schmid and McMahon 2007; Traub and Bonifacino 2013). During the nucleation/cargo-selection stage, cargo-recognizing adaptor proteins such as AP2 and 'accessory proteins' such as epsins localize to the site where endocytosis will occur. Traditionally, cargo adaptor proteins bind cargo and clathrin, serving to link the two proteins. The term accessory proteins was commonly used to categorize proteins that were neither adaptors nor clathrin. However, many of the 'accessory proteins' were found to interact with specific cargos, such as ubiquitin interaction with epsins (Traub and Bonifacino 2013). Thus, the majority of proteins known as accessory proteins can simply be called adaptors or Clathrin Associated Sorting Proteins (CLASP), the term also often used to refer to any adaptor proteins besides clathrin that are involved in clathrin-mediated endocytosis (Traub and Bonifacino 2013). Although it is generally thought that some adaptor protein binding to cargo or PIP_2 may prompt the nucleation of clathrin, defining the identities of such proteins necessary or sufficient to initiate endocytosis is an active area of research (Traub 2009). Clathrin coat assembly and the budding stage follows, where clathrin molecules recruited by AP2 and other accessory proteins polymerize. It is thought that the polymerization of clathrin may contribute to the formation or stabilization of the curvature in the budded membrane (Kirchhausen et al. 2014). The vesicle scission stage ensues where dynamin mediates the scission of the clathrin-coated vesicle from the plasma membrane in a GTP hydrolysis-dependent manner. Once detached from the plasma membrane, shedding of the clathrin coat ensues via a mechanism involving HSC70 and its cofactor auxilin (McMahon and Boucrot 2011).

Investigation of the 'connectivity' amongst the various endocytic proteins showed how the endocytic process is designed to accommodate a diverse array of cargos via their connections to AP2 and clathrin (Schmid and McMahon 2007; Schmid et al. 2006; Praefcke et al. 2004). Each cargo is recognized by designated adaptors, such as β-arrestins, and these adaptors are able to connect to AP2 by binding to the AP2 appendage domains (Schmid and McMahon 2007). Thus, AP2 initially integrates the various cargos, via appendage domain interactions, into a concentrated hub. These AP2 connections with other adaptor proteins are thought to make way for new connections with clathrin as endocytosis progresses (Schmid and McMahon 2007; Schmid et al. 2006; Praefcke et al. 2004). That is, the previous connections that adaptor proteins were making with AP2 are replaced by new interactions with clathrin. This is in agreement with the fact that many of the AP2 binding sites on these accessory and adaptor proteins are very close to their binding site for clathrin making it unlikely that simultaneous interaction with AP2 and clathrin can occur, due to steric inhibition (Schmid and McMahon 2007). β-arrestins also have binding sites for both AP2 and clathrin in tandem.

Another interesting feature of protein interactions involved in endocytosis is the prevalence of low affinity interactions amongst the various molecules, where multiple occurrences are required for accumulation of such interactions and subsequent endocytosis (Schmid and McMahon 2007; Traub and Bonifacino 2013; Di Paolo and De Camilli 2006). It is necessary that the interactions are low affinity to be able to abort the endocytic process when the conditions are not fully met; such as enough cargo and the presence of PIP_2. Indeed, the interactions involved in β-arrestin-mediated endocytosis are in the μM range similar to many other interactions involved in endocytosis (Kang et al. 2009; Edeling et al. 2006a, b).

Overview of β-Arrestin-Mediated Endocytosis

Two seminal papers showed that β-arrestins, which were already known to function in the desensitization of GPCRs, also mediated the endocytosis of the β_2-adrenergic receptor (Ferguson et al. 1996) via their ability to bind to clathrin (Goodman et al. 1996). Since this finding, many GPCRs have been shown to undergo β-arrestin-mediated clathrin-dependent endocytosis. It was later shown that β-arrestins also bound to the clathrin adaptor protein AP2 (Laporte et al. 1999) at its β2 appendage domain (Laporte et al. 2002) and this interaction was also shown to be important for GPCR endocytosis (Laporte et al. 1999, 2002).

With the development of GFP-tagged arrestins, it became possible to monitor real-time movement of arrestins in response to receptor activation (Barak et al. 1997). This led to discoveries that gave further insight into the behavior of β-arrestins during endocytosis in a cellular context. In response to receptor activation, β-arrestins are recruited to the plasma membrane within seconds and within minutes localize to clathrin-coated pits (Barak et al. 1999; Zhang et al. 1999). Instead of β-arrestins driving the assembly of clathrin coated pits, it was shown that β-arrestin interaction with AP2 localizes the β-arrestin/receptor complex to already-formed pits (Santini et al. 2002; Scott et al. 2002). In addition, the use of GFP-tagged proteins revealed that there are two classes of GPCRs with distinct patterns of arrestin binding during endocytosis (Oakley et al. 2000). When the phosphorylation pattern on the C-terminal tail of a GPCR contains three or more tandem phosphorylation sites, then β-arrestins appear to remain bound to the receptor after endocytosis, co-localizing at endosomes (Oakley et al. 2001). This type of receptor is referred to as a Class B receptor. This is in contrast to Class A receptors, which contain no more than two tandem phosphorylation sites, where arrestins dissociate at or near the plasma membrane during endocytosis. Another characteristic of Class A receptor endocytosis is that β-arrestin2 is more efficient in being recruited to the receptor and promoting receptor endocytosis, compared to β-arrestin1 (Oakley et al. 2000).

A role for phosphoinositides in assisting β-arrestin-mediated endocytosis was also discovered in the late 1990s (Gaidarov et al. 1999). The initial rationale for such a role was considered due to the apparent functional similarities between adaptor protein AP2 and β-arrestin in GPCR endocytosis. Both were acting as adaptors that can recognize and bind cargos and bind to clathrin at their terminal domains to mediate endocytosis. Indeed, PIP_2 and PIP_3 were able to bind to β-arrestins and promote interaction between receptor and β-arrestin2 in vitro (Gaidarov et al. 1999). Also, β-arrestin2 mutants that could not bind phosphoinositides were largely unable to mediate β_2-adrenergic receptor endocytosis, apparently due to an inability to promote localization of the receptor/β-arrestin complex to clathrin-coated pits (Gaidarov et al. 1999).

Structural View of β-Arrestin-Mediated Endocytosis

From a structural point of view, there are several structures of arrestins in a basal state (Granzin et al. 1998; Hirsch et al. 1999; Han et al. 2001; Milano et al. 2002) and in an 'activated' state (Granzin et al. 2012, 2015; Shukla et al. 2013; Kim et al. 2013; Kang et al. 2015; Zhou et al. 2016), the activated state being either in complex with a GPCR or phosphorylated GPCR peptide or containing mutations to disrupt the basal state. However, arrestin-mediated endocytosis not only involves active arrestins and phosphorylated receptor, but also AP2, clathrin, PIP_2 and possibly ubiquitination. It is reasonable to speculate that the structure of β-arrestin engaged in GPCR endocytosis might have a distinct conformation from the 'active' arrestin that is solely interacting with the receptor. However, there is very little structural information on β-arrestin in this state. This is partly due to the fact that the binding sites for endocytic proteins on β-arrestin are located on the C terminal tail, a region that is not observed in crystal structures due to its flexibility. In addition, it should also be noted that there is currently no structure of other similar adaptor proteins with an extended flexible linker in complex with either AP2 or clathrin. The crystal structures that are available are structures that contain a short segment of the linker in complex with a domain of AP2 or clathrin. Given that it is by design that the long linkers are flexible, it may be difficult to obtain full length β-arrestin complexed with either of these endocytic proteins unless, perhaps, in a clathrin coat. We will overview what is known about the structure of activated arrestin as well as the structural basis for β-arrestins role in GPCR endocytosis.

Structure of Basal and Active β-Arrestin in Endocytosis

The key interactions that keep arrestin in its basal state are the polar core and the three-element region (Hirsch et al. 1999; Han et al. 2001) (Fig. 4.1a). Interestingly, both sets of interactions are dependent on the C-terminal tail that stretches from the C-terminal domain and wraps the N-terminal domain over the alpha helix region (three element interaction) and the lariat loop (polar core interaction) (Milano et al. 2002). When arrestin binds to an activated phosphorylated GPCR, it is thought that interactions between the phospho-groups on the C-terminal tail of the receptor and arrestin leads to the disruption of both the polar core and the three-element interactions (Fig. 4.1b). Indeed, the structure of a constitutively active visual arrestin mutant, R175E (the phosphate sensor residue at the center of the polar core in visual arrestins), contains a disrupted three-element interaction (Granzin et al. 2015) (see Chap. 11). p44 splice variant of visual arrestin, which is truncated at its C-terminus so as to not have the three-element interaction sites available, also had some change in the polar core interactions (Granzin et al. 2012). When p44 arrestin was crystallized in the presence of opsin, the polar core was fully disrupted (Kim et al. 2013), even though opsin was not present in the solved structure. Thus, there is

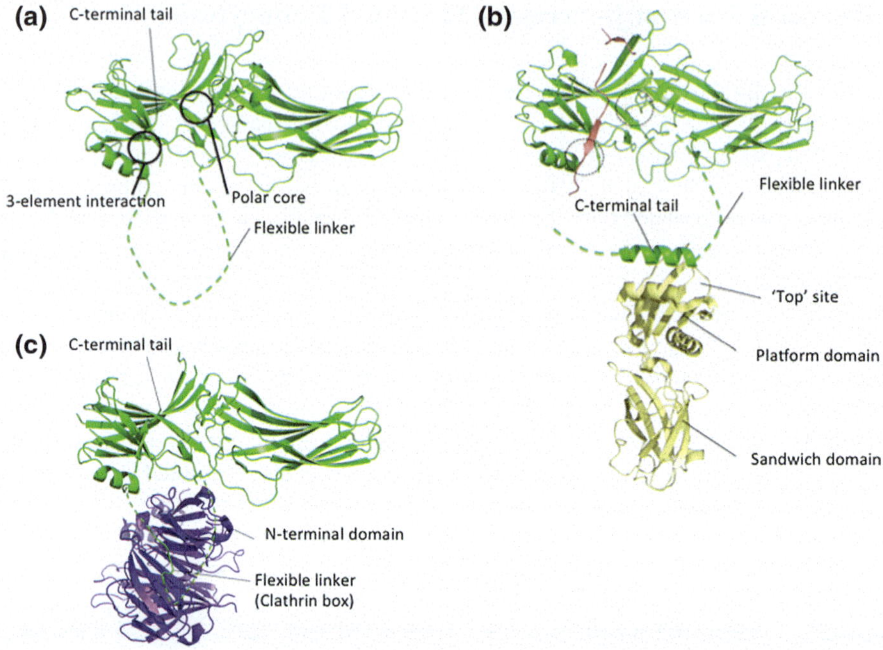

Fig. 4.1 Structure of β-arrestin and β-arrestin endocytic complexes. **a** β-Arrestin1 (*green*) in its basal state. The *black circles* highlight the 3-element and polar core interactions that keep arrestin in its basal state. Flexible linker region of β-arrestin where clathrin binds and the C-terminal tail of arrestin where AP2 binds are also *highlighted*. It should be noted that the 3-element and polar core interactions are mediated in part by residues in the C-terminal tail of arrestin (adapted from PDB: 1JSY). **b** Activated state of β-arrestin1. The image depicts the structure of β-arrestin1 bound to a phosphorylated C-terminal tail peptide from the V2R vasopressin receptor (*red* peptide) with a hypothetical structure of the released C-terminal tail interacting with the β2-appendage domain of AP2 (*yellow*). The crystal structure of β-arrestin1 in complex with phosphorylated V2R vasopressin receptor peptide shows how the phosphorylated peptide disrupts both the 3-element and polar core interactions to promote the activated state of arrestin. The *black circles* in *dashed lines* denote the sites where the 3-element interaction and polar core interaction are disrupted. The C-terminal tail of arrestin is released from the N-terminal domain of arrestin when arrestin is interacting with a phosphorylated receptor. The residues on the C-terminal tail that participate in the 3-element interaction and polar core interactions now form an alpha helix during interaction with the 'top' site of β-appendage's platform domain (adapted from PDB: 4JQI and 2IV8). **c** β-arrestin1 bound to clathrin (*purple*). The crystal structure of β-arrestin1S in complex with the N-terminal domain of the clathrin heavy chain. The clathrin N-terminal domain interacts with the clathrin box motif within the flexible linker region. Arrestin was found to be in its basal state (adapted from PDB: 3GC3)

some level of inter-dependence between the polar core and the three-element interactions. As a result of the disruption in either of these interactions, several common changes are observed in the active arrestin conformation: upswing of the finger loop, movement of the middle loop, and rotation between the N- and C-domains (Kim et al. 2013; Shukla et al. 2013; Kang et al. 2015). Many of these changes have been shown to promote interaction with the active conformation of rhodopsin (Kang et al. 2015). Moreover, the biggest change that results from the activation of arrestin is the release of the ~60 amino acid C-terminal tail. The tail can be divided into two regions, the C terminal linker between Pro^{356} and the three-element interaction region and the three-element interacting region to the far C-terminus (in β-arrestin1). The tail harbors the two primary sites that are important for endocytosis: the LIELD clathrin box motif in the C-terminal linker (Krupnick et al. 1997), and the DDDIV**F**ED**F**ARQ**R**LKG motif for AP2 binding where the residues in bold are essential for AP2 interaction (Edeling et al. 2006a, b; Schmid et al. 2006; Laporte et al. 2000; Kim and Benovic 2002). It should be noted that IV**F** is involved in the three-element interaction and the R in ARQ is involved in the polar core interaction.

An interesting structural characteristic of arrestins is that the C-terminal tail, which contains the binding sites for clathrin and AP2, is released only after engaging the phosphorylated active receptor. Given the distinct endocytic pattern of β-arrestin1 versus β-arrestin2 when mediating Class A receptor endocytosis, and how this distinction disappears in Class B receptor endocytosis (Oakley et al. 2000), it is reasonable to suspect a critical role for the phosphorylation pattern of a GPCR to induce distinct conformations in β-arrestin1 versus β-arrestin2. Given the fact that such distinct endocytic behavior between β-arrestin1 and 2 with regards to the endocytosis of Class A receptors is largely dependent on the C-terminal tail of β-arrestin (Oakley et al. 2000), it is likely that the conformational change that leads to such differences may be traced to how the phosphorylation pattern affects the conformation of the C-terminal tail. Indeed, there have been reports where AP2 and clathrin were shown to bind more efficiently to β-arrestin2 compared to β-arrestin1 in response to agonist stimulation of the β_2-adrenergic receptor (Laporte et al. 1999; Goodman et al. 1996). It may be that the C-terminal tail of β-arrestin2 has an inherently higher affinity towards the endocytic proteins. Indeed, the clathrin binding box sequence is distinct in the two proteins (LIEFE for β-arrestin2 vs. LIELD for β-arrestin1) (Krupnick et al. 1997) and also there are additional amino acids between the clathrin box and adaptin binding motif in β-arrestin2. Importantly, β-arrestin2 has significantly fewer acidic residues in the C-terminus beyond the residues shown to positively affect clathrin interaction (Kern et al. 2009). However, given that the distinct endocytic behavior between β-arrestin1 and 2 largely disappears when interacting with Class B receptors (Oakley et al. 2000), this suggests that the differences between β-arrestin1 and 2 in response to Class A receptors may be how accessible the C-terminal tail becomes in response to the phosphorylated receptor. That is, it may be that when β-arrestin1 and 2 are interacting with heavily phosphorylated Class B receptors, the C-terminal tail of arrestin may be equally accessible to endocytic proteins, whereas when arrestins are interacting with less phosphorylated Class A receptors, β-arrestin2 may allow more access to endocytic proteins than β-arrestin1. There are several reports that monitored how the phosphorylation patterns of the receptor affected the conformation of arrestin in general (Nobles et al. 2011; Yang et al. 2015; Nuber et al. 2016; Lee et al. 2016). What is less clear is the differential effect of the phosphorylation pattern on the accessibility of the β-arrestin C-terminal tail to endocytic proteins, especially in the context of comparing β-arrestin1 versus β-arrestin2. Interestingly, a recent report suggests that a distinct conformation of the C-terminal tail of β-arrestin2 exists, depending on whether it is interacting with Class A or Class B receptors (Lee et al. 2016).

Structure of β-Arrestin as an Endocytic Adaptor

β-Arrestins share a modular design with many other adaptors and accessory proteins involved in clathrin mediated endocytosis (Owen et al. 2004); it has a discretely folded core domain that recognizes the cargo and phospholipid and a long flexible linker that contains motifs for interacting with clathrin and other adaptors. The N-terminal domain of arrestin's 'core' detects the phosphorylated tail of the receptor while its C-terminal domain binds PIP_2. The C-terminal tail of β-arrestin1/2 has tandem binding sites for the appendage domain of AP2 and the clathrin heavy chain N-terminal domain.

Arrestin interaction with 'cargo' (i.e., GPCRs) is discussed in Chaps. 5–9. Here we will discuss arrestin interaction with phospholipid and its conformational implications. Despite its critical role in assisting β-arrestin-mediated endocytosis (Gaidarov et al. 1999), the molecular mechanism behind how PIPs affect arrestin structure and function is not well characterized. Part of the difficulty lies in the fact that there are at least two sites on non-visual arrestins that bind to PIPs, and their effect on receptor recruitment and arrestin oligomerization differs, depending on the identity of the arrestin and the phosphoinositide.

The first observation of the effect of phosphoinositides on arrestin function was the finding that IP_6 inhibits visual arrestin binding to rhodopsin (Palczewski et al. 1991). The affinity of IP_6 for visual arrestin was found to be around 5 μM and its site of interaction was localized to the N-domain of arrestin (Palczewski et al. 1991; Gaidarov et al. 1999). For β-arrestins, however, PIPs were observed to promote the interaction between β-arrestin and GPCRs, and be necessary for GPCR localization to clathrin coated pits and endocytosis (Gaidarov et al. 1999). The primary binding site for PIP interaction with β-arrestins was found in the C-domain and was distinct from where IP_6 binds to visual arrestin (Gaidarov et al. 1999). This site includes residues Lys^{233}, Arg^{237} and Lys^{251} (Gaidarov et al. 1999; Milano et al. 2002) and had 100-fold higher affinity for PIPs and IP_6 compared to a site in the N-domain. These particular residues are largely absent from visual arrestin. As a corollary, it is reasonable to suspect a unique effect for PIPs on the conformation of β-arrestins. In fact, there are reasons to suspect that the way visual and non-visual arrestins interact with receptors might not be exactly the same. For example, visual arrestin is specific for binding to rhodopsin while β-arrestins appear to have a broader binding profile for GPCRs (Shenoy and Lefkowitz 2011). Moreover, incorporation of specific visual arrestin residues into β-arrestin1 promoted more efficient binding of β-arrestin1 to rhodopsin (Vishnivetskiy et al. 2011). In addition, while wild type β-arrestin2 and mutant β-arrestin2 that cannot interact with phosphoinositides bind to rhodopsin with equal efficiency, PIP_2 addition selectively enhances wild type β-arrestin2 binding to receptor while having no effect on mutant β-arrestin2 (Gaidarov et al. 1999). Again, given that the mutated residues are specific for non-visual arrestins, the PIP_2 effect is expected in non-visual arrestins only.

Another functional consequence of arrestin interaction with PIPs or IPs is the effect on the oligomerization state. Both visual and non-visual arrestins share an IP_6 binding site on the N-domain, and IP_6 interaction with visual arrestin inhibits self-association (Hanson et al. 2008) and blocks interaction with the receptor. In contrast, IP_6 interaction with β-arrestins promotes oligomerization (Milano et al. 2002; Hanson et al. 2008). It is thought that IP_6 can interact with both N- and C-terminal binding sites on β-arrestin to promote oligomerization. Curiously, despite IP_6 having opposite effects on oligomerization of visual and β-arrestins, it has the same inhibitory effect on β-arrestin recruitment to GPCRs (Gaidarov et al. 1999). While PIPs promote recruitment of β-arrestin2 to GPCRs, their effect on arrestin oligomerization has not yet been reported.

Like many endocytic adaptor proteins, β-arrestins have a binding site for the β2-appendage domain of AP2 (Laporte et al. 1999, 2002). AP2 is a central player in integrating a diverse array of cargos into clathrin coated pits by its ability to bind various adaptor and accessory proteins through its appendage domain (Praefcke et al. 2004; Schmid et al. 2006; Schmid and McMahon 2007; Traub and Bonifacino 2013). However, the mechanisms involved in the coordination of these interactions and their role in clathrin coat assembly and endocytosis remain under active investigation. AP2 has a core domain through which it recognizes its cargos and PIP_2, and long unstructured linkers that are connected to alpha and beta appendage domains (Edeling et al. 2006b). Inhibition of β-arrestin interaction with AP2 blocks arrestin/GPCR recruitment to clathrin coated pits and inhibits receptor endocytosis (Laporte et al. 1999, 2000). Interestingly, this is different from blocking β-arrestin interaction with clathrin where arrestin/receptor complexes still localize to clathrin-coated pits, but nevertheless fail to undergo endocytosis (Laporte et al. 2000). The remarkable trait that is somewhat unique to arrestin compared to many adaptor proteins is how the binding motifs for endocytic proteins are only available in response to cargo activation, allowing for swift temporal regulation of cargo endocytosis (Traub 2009). The binding motif for AP2 on the C-terminal tail of β-arrestin includes residues that are involved in the three-element and polar core interactions (Han et al. 2001; Milano et al. 2002). Thus, it is necessary for arrestin to become activated and the three-element interaction to be disrupted, in order for β-arrestin to effectively interact with AP2. Another attribute that is unique to β-arrestin amongst other endocytic proteins is that the affinity between the cargo and β-arrestin is very high, in the low nM range (Gurevich et al. 1995; Sohlemann et al. 1995) compared to other adaptors that primarily rely on short motifs for interaction (Owen et al. 2004). Additionally, the adaptin-binding motif DDDIV**F**EDF**A**RQ**R**LKG of β-arrestin is one of the highest affinity amongst other adaptin binding motifs found in other adaptors (Edeling et al. 2006a, b). Thus, once the receptor is activated, β-arrestin will quickly bind the receptor with high affinity and will be able to compete with other adaptors in binding AP2. This mode of interaction is somewhat quantitatively different from many adaptors that rely on lower affinity, avidity-based interactions that propel clathrin budding by accumulation. It's worth noting that β-arrestins do not drive clathrin coat assembly but rather promote recruitment of arrestin/receptor complexes to pre-formed clathrin coated pits (Santini et al. 2002; Scott et al. 2002). The detailed mechanism of how β-arrestins and GPCRs get recruited to clathrin coated pits and how β-arrestins contribute to endocytosis of GPCRs needs further investigation.

Once the C-terminal tail containing the AP2 binding motif is released from activated β-arrestin, it appears to undergo a conformational change from a short beta-strand to an alpha-helix to mediate interaction with the β-appendage domain of AP2 (Fig. 4.1b) (Edeling et al. 2006a; Schmid et al. 2006) with an ~2–2.5 μM affinity (Edeling et al. 2006a, b; Schmid et al. 2006; Burtey et al. 2007). The β-appendage domain is special in that it is a site for clathrin coat assembly by binding to clathrin via its hinge region via the two sites within the appendage domain; the 'top' site of the 'platform' domain and/or the 'side' site of the 'sandwich' domain (Schmid et al. 2006; Edeling et al. 2006a, b). Concomitantly, the β-appendage domain's platform and sandwich binding sites are used for interaction with various accessory/adaptor proteins (Schmid et al. 2006). β-arrestins, meanwhile, bind to the 'top' platform site, which is at the 'top' side of the appendage domain (Fig. 4.1b) (Laporte et al. 2002; Milano et al. 2002; Kim and Benovic 2002; Edeling et al. 2006a, b). Thus, β-arrestins, like other accessory proteins, are thought to first bind to AP2, and as the clathrin coat assembles, β-arrestin interaction with AP2 is lost due to the competing interaction of AP2 with the clathrin coat (Schmid et al. 2006; Schmid and McMahon 2007). While the loss of interaction with AP2 forces some adaptor/accessory proteins, such as eps15, to be excluded from the maturing clathrin-coated bud, β-arrestin interaction with clathrin and PIP_2 is thought to keep arrestin within the coated vesicles (Schmid et al. 2006; Edeling et al. 2006a). Presumably, β-arrestin either dissociates from the vesicle as the coat disassembles post-endocytosis, or stays bound to the GPCR depending on its affinity, which, in turn, is dependent on whether the receptor is Class A (low affinity) or Class B (high affinity). Indeed, as will be discussed later, the affinity of β-arrestin for clathrin seems to depend on the assembly state of the clathrin (Kang et al. 2009; Goodman et al. 1996), and, at least for β-arrestin1, its interaction with AP2 seems to compete with the interaction with clathrin (Schmid et al. 2006).

Clathrin was the first endocytic protein that was shown to bind β-arrestin (Goodman et al. 1996). An individual clathrin triskelion structure comprises three long 'legs', each of which has 'feet', i.e., the N-terminal domain, which are used to bind to various adaptors (Kirchhausen et al. 2014). Clearly, β-arrestin binding to clathrin is a crucial step that is necessary for arrestin-mediated endocytosis, as disruption of the interaction effectively blocks GPCR endocytosis (Goodman et al. 1996; Krupnick et al. 1997; Kim and Benovic 2002; Laporte et al. 1999). The primary binding site for clathrin shared by β-arrestins is the canonical clathrin binding box motif (LΦ × Φ[DE]) (Krupnick et al. 1997; Shih et al. 1995; ter Haar et al. 2000) located in the extended flexible C-terminal linker region. This is between the three-element interaction site and residue 356, the last two regions of β-arrestin that are visible in the crystal structure in its basal state. The release of the C-terminal tail from the three-element interaction site in response to interaction with the receptor promotes efficient binding between β-arrestin and clathrin as well as AP2 (Goodman et al. 1996; Laporte et al. 1999). Indeed, a constitutively active mutant of β-arrestin binds both clathrin and AP2 more efficiently than wild-type β-arrestin in vitro (Kim and Benovic 2002; Kern et al. 2009).

Surprisingly, however, β-arrestin was found in its basal state in the only available structure of a β-arrestin complex with the clathrin N-terminal domain (Fig. 4.1c) (Kang et al. 2009). Presumably, although β-arrestin binding to clathrin is greatly facilitated by arrestin activation, it is not required. This is plausible since the location of the clathrin box motif is in the middle of an extended flexible region. Given the preference for an active conformation of arrestin in binding clathrin, and considering the cellular context of how β-arrestin binds clathrin after it has formed a complex with a receptor, the crystal structure probably does not reflect the complete picture of clathrin and β-arrestin interaction during endocytosis. However, it should be noted that we really don't know what the structural requirement for β-arrestin-mediated endocytosis is. Additionally, β-arrestin must undergo several stages during endocytosis: (i) interaction with the phosphorylated active receptor; (ii) interaction with AP2, PIP_2 and clathrin during clathrin coat assembly; and (iii) dissociation from the receptor at or near the plasma membrane for Class A GPCRs. Thus, there may be several conformations of β-arrestin that correlate with the different stages of endocytosis. Moreover, compared to β-arrestin2, β-arrestin1 was relatively less dependent on AP2-binding for mediating the endocytosis of the β_2AR, especially in HEK293 cells (Kim and Benovic 2002). Although understanding the meaning of such an observation requires more investigation, one cannot rule out the possibility that AP2 binding might not be extensively utilized for β-arrestin1 during endocytosis compared to β-arrestin2, which could mean, at least during the endocytosis of Class A receptors, β-arrestin1 may not be in its fully activated state where β-arrestin1's C-terminal tail is fully accessible to AP2, but in a conformation where it is still interacting with clathrin, in agreement with the structure. Indeed, β-arrestin1 was recruited to clathrin-coated pits to a lesser degree compared to β-arrestin2 upon β_2AR stimulation (Santini et al. 2000). Thus, more investigation is needed to understand the cellular context of the current structure of β-arrestin in complex with clathrin.

The binding affinity of β-arrestin1 for clathrin's N-terminal domain was ~1–2 μM (Kang et al. 2009) which is in agreement with affinities of other adaptor proteins involved in dynamic interactions (Owen et al. 2004) or 'dynamic instability' of endocytosis (Schmid and McMahon 2007). During clathrin coat assembly, however, it is thought that the coat itself can drive the clathrin interaction with adaptors, despite the affinity of the interaction being low. This is presumably due to the decreased off-rate for clathrin monomers in a cage as it becomes incorporated into a solid-like matrix in the clathrin cage (Schmid et al. 2006; Schmid and McMahon 2007). Interestingly, β-arrestin was reported previously to interact with clathrin cages with a much higher affinity (60 nM for β-arrestin1) compared to how it interacts with just the N-terminal domain (1–2 μM) (Goodman et al. 1996; Kang et al. 2009).

There is a second binding site for clathrin on β-arrestin1L, the longer splice variant of β-arrestin1, specifically within the 8 extra amino acids that form a loop between beta-strands 18 and 19 (Kang et al. 2009). This novel clathrin binding motif, $(L/I)_2$GXL, was identified in the crystal structure of clathrin with β-arrestin1L, where one of the two clathrins was found to be binding to this motif via its cleft between blades 4 and 5. Its functionality in endocytosis had been verified in cells, and has been shown to bind the clathrin N-terminal domain in vitro (Kang et al. 2009). Interestingly, these 8 extra amino acids are shared only with visual arrestins (Kang et al. 2009), and, in the recent rhodopsin-arrestin complex crystal structure (Kang et al. 2015), the 8 residues were thought to act as a hydrophobic anchor that protrudes into the plasma membrane when arrestin is bound to rhodopsin (see also Chap. 7). In the β-arrestin1L-clathrin complex crystal structure, the second clathrin was bound to the known clathrin binding box motif at the C-terminal tail via its traditional binding site between blades 1 and 2. Thus far, three binding motifs for clathrin have been identified, that is, the clathrin box, the W-box, and the $(L/I)_2$GXL motif. Each of these motifs has a corresponding binding site with one of the four different sites within the clathrin N-terminal domain that are used to bind adaptor proteins: groove between blades 1 and 2, groove between blades 4 and 5, blade 7 and the 'top' side of the beta propeller of the N-terminal domain (Lemmon and Traub 2012). Thus, it would be theoretically possible for β-arrestin1L to bind to two distinct clathrin N-terminal domains simultaneously (Kang et al. 2009). Recent reports suggest that these binding sites on clathrin may be functionally redundant, and mutation of all four sites was necessary to disrupt endocytosis (Willox and Royle 2012). This may suggest that the interactions between clathrin binding motifs and the N-terminal domain of clathrin may be more flexible than commonly thought, and the mechanism of β-arrestin binding to clathrin might be more complicated than currently envisioned.

References

Barak LS, Ferguson SSG, Zhang J, Caron MG (1997) A β-arrestin/green fluorescent protein biosensor for detecting G protein-coupled receptor activation. J Biol Chem 272(44): 27497–27500

Barak LS, Warabi K, Feng X, Caron MG, Kwatra MM (1999) Real-time visualization of the cellular redistribution of G protein-coupled receptor kinase 2 and beta-arrestin 2 during homologous desensitization of the substance P receptor. J Biol Chem 274(11):7565–7569

Burtey A, Schmid EM, Ford MGJ, Rappoport JZ, Scott MGH, Marullo S, Simon SM, McMahon HT, Benmerah A (2007) The conserved isoleucine-valine-phenylalanine motif couples activation state and endocytic functions of β-arrestins. Traffic 8(7):914–931

Daaka Y, Luttrell LM, Ahn S, Della Rocca GJ, Ferguson SS, Caron MG, Lefkowitz RJ (1998) Essential role for G protein-coupled receptor endocytosis in the activation of mitogen-activated protein kinase. J Biol Chem 273(2):685–688

Di Paolo G, De Camilli P (2006) Phosphoinositides in cell regulation and membrane dynamics. Nature 443(7112):651–657

Edeling MA, Mishra SK, Keyel PA, Steinhauser AL, Collins BM, Roth R, Heuser JE, Traub LM (2006a) Molecular switches involving the AP-2 β2 appendage regulate endocytic cargo selection and clathrin coat assembly. Dev Cell 10(3):329–342

Edeling MA, Smith C, Owen D (2006b) Life of a clathrin coat: insights from clathrin and AP structures. Nat Rev Mol Cell Biol 7(1):32–44

Ferguson SS, Downey WE, Colapietro AM, Barak LS, Ménard L, Caron MG (1996) Role of beta-arrestin in mediating agonist-promoted G protein-coupled receptor internalization. Science 271(5247):363–366

Gaidarov I, Krupnick JG, Falck JR, Benovic JL, Keen JH (1999) Arrestin function in G protein-coupled receptor endocytosis requires phosphoinositide binding. EMBO J 18(4): 871–881

Goodman OB, Krupnick JG, Santini F, Gurevich VV, Penn RB, Gagnon AW, Keen JH, Benovic JL (1996) Beta-arrestin acts as a clathrin adaptor in endocytosis of the beta2-adrenergic receptor. Nature 383(6599):447–450

Granzin J, Wilden U, Choe H, Labahn J, Krafft B, Buldt G (1998) X-ray crystal structure of arrestin from bovine rod outer segments. Nature 391(6670):918–921

Granzin J, Cousin A, Weirauch M, Schlesinger R, Buldt G, Batra-Safferling R (2012) Crystal structure of p44, a constitutively active splice variant of visual arrestin. J Mol Biol 416(5): 611–618

Granzin J, Stadler A, Cousin A, Schlesinger R, Batra-Safferling R (2015) Structural evidence for the role of polar core residue Arg175 in arrestin activation. Sci Rep 5:15808

Gurevich VV, Gurevich EV (2015) Arrestins: critical players in trafficking of many GPCRs. Prog Mol Biol Transl Sci 132:1–14

Gurevich VV, Dion SB, Onorato JJ, Ptasienski J, Kim CM, Sterne-Marr R, Howey MM, Benovic JL (1995) Arrestin interactions with G protein-coupled receptors. Direct binding studies of wild type and mutant arrestins with rhodopsin, beta 2-adrenergic, and m2 muscarinic cholinergic receptors. J Biol Chem 270(2):720–731

Han M, Gurevich VV, Vishnivetskiy SA, Sigler PB, Schubert C (2001) Crystal structure of beta-arrestin at 1.9 A: possible mechanism of receptor binding and membrane translocation. Structure 9(9):869–880

Hanson SM, Vishnivetskiy SA, Hubbell WL, Gurevich VV (2008) Opposing effects of inositol hexakisphosphate on rod arrestin and arrestin2 self-association. Biochemistry 47(3):1070–1075

Hirsch J, Schubert C, Gurevich VV, Sigler PB (1999) The 2.8 A crystal structure of visual arrestin: a model for arrestin's regulation. Cell 97(2):257–269

Irannejad R, Tomshine JC, Tomshine JR, Chevalier M, Mahoney JP, Steyaert J, Rasmussen SG, Sunahara RK, El-Samad H, Huang B, von Zastrow M (2013) Conformational biosensors reveal GPCR signalling from endosomes. Nature 495(7442):534–538

Kang DS, Kern RC, Puthenveedu MA, von Zastrow M, Williams JC, Benovic JL (2009) Structure of an arrestin2-clathrin complex reveals a novel clathrin binding domain that modulates receptor trafficking. J Biol Chem 284(43):29860–29872

Kang Y, Zhou XE, Gao X, He Y, Liu W, Ishchenko A, Barty A, White TA, Yefanov O, Han GW, Xu Q, de Waal PW, Ke J, Tan MH, Zhang C, Moeller A, West GM, Pascal BD, Van Eps N, Caro LN, Vishnivetskiy SA, Lee RJ, Suino-Powell KM, Gu X, Pal K, Ma J, Zhi X, Boutet S, Williams GJ, Messerschmidt M, Gati C, Zatsepin NA, Wang D, James D, Basu S, Roy-Chowdhury S, Conrad CE, Coe J, Liu H, Lisova S, Kupitz C, Grotjohann I, Fromme R, Jiang Y, Tan M, Yang H, Li J, Wang M, Zheng Z, Li D, Howe N, Zhao Y, Standfuss J, Diederichs K, Dong Y, Potter CS, Carragher B, Caffrey M, Jiang H, Chapman HN, Spence JC, Fromme P, Weierstall U, Ernst OP, Katritch V, Gurevich VV, Griffin PR, Hubbell WL, Stevens RC, Cherezov V, Melcher K, Xu HE (2015) Crystal structure of rhodopsin bound to arrestin by femtosecond X-ray laser. Nature 523(7562):561–567

Kern RC, Kang DS, Benovic JL (2009) Arrestin2/clathrin interaction is regulated by key N- and C-terminal regions in arrestin2. Biochemistry 48(30):7190–7200

Kim YM, Benovic JL (2002) Differential roles of arrestin-2 interaction with clathrin and adaptor protein 2 in G protein-coupled receptor trafficking. J Biol Chem 277(34):30760–30768

Kim YJ, Hofmann KP, Ernst OP, Scheerer P, Choe HW, Sommer ME (2013) Crystal structure of pre-activated arrestin p44. Nature 497(7447):142–146

Kirchhausen T, Owen D, Harrison SC (2014) Molecular structure, function, and dynamics of clathrin-mediated membrane traffic. Cold Spring Harb Perspect Biol 6(5):a016725

Krupnick JG, Goodman OB, Keen JH, Benovic JL (1997) Arrestin/clathrin interaction. Localization of the clathrin binding domain of nonvisual arrestins to the carboxy terminus. J Biol Chem 272(23):15011–15016

Laporte SA, Oakley RH, Zhang J, Holt JA, Ferguson SS, Caron MG, Barak LS (1999) The beta2-adrenergic receptor/beta arrestin complex recruits the clathrin adaptor AP-2 during endocytosis. Proc Natl Acad Sci U S A 96(7):3712–3717

Laporte SA, Oakley RH, Holt JA, Barak LS, Caron MG (2000) The interaction of beta-arrestin with the AP-2 adaptor is required for the clustering of beta2-adrenergic receptor into clathrin-coated pits. J Biol Chem 275(30):23120–23126

Laporte SA, Miller WE, Kim KM, Caron MG (2002) Beta-arrestin/AP-2 interaction in G protein-coupled receptor internalization: identification of a beta-arrestin binging site in beta 2-adaptin. J Biol Chem 277(11):9247–9254

Lee MH, Appleton KM, Strungs EG, Kwon JY, Morinelli TA, Peterson YK, Laporte SA, Luttrell LM (2016) The conformational signature of β-arrestin2 predicts its trafficking and signalling functions. Nature 531(7596):665–668

Lemmon SK, Traub LM (2012) Getting in touch with the clathrin terminal domain. Traffic 13 (4):511–519

McMahon HT, Boucrot E (2011) Molecular mechanism and physiological functions of clathrin-mediated endocytosis. Nat Rev Mol Cell Biol 12(8):517–533

Milano SK, Pace HC, Kim YM, Brenner C, Benovic JL (2002) Scaffolding functions of arrestin-2 revealed by crystal structure and mutagenesis. Biochemistry 41(10):3321–3328

Moore CA, Milano SK, Benovic JL (2007) Regulation of receptor trafficking by GRKs and arrestins. Annu Rev Physiol 69:451–482

Nobles KN, Xiao K, Ahn S, Shukla AK, Lam CM, Rajagopal S, Strachan RT, Huang TY, Bressler EA, Hara MR, Shenoy SK, Gygi SP, Lefkowitz RJ (2011) Distinct phosphorylation sites on the β(2)-adrenergic receptor establish a barcode that encodes differential functions of β-arrestin. Sci Signal 4(185):ra51

Nuber S, Zabel U, Lorenz K, Nuber A, Milligan G, Tobin AB, Lohse MJ, Hoffmann C (2016) β-Arrestin biosensors reveal a rapid, receptor-dependent activation/deactivation cycle. Nature 531(7596):661–664

Oakley RH, Laporte SA, Holt JA, Caron MG, Barak LS (2000) Differential affinities of visual arrestin, beta arrestin1, and beta arrestin2 for G protein-coupled receptors delineate two major classes of receptors. J Biol Chem 275(22):17201–17210

Oakley RH, Laporte SA, Holt JA, Barak LS, Caron MG (2001) Molecular determinants underlying the formation of stable intracellular G protein-coupled receptor-beta-arrestin complexes after receptor endocytosis*. J Biol Chem 276(22):19452–19460

Owen DJ, Collins BM, Evans PR (2004) Adaptors for clathrin coats: structure and function. Annu Rev Cell Dev Biol 20:153–191

Palczewski K, Pulvermüller A, Buczylko J, Gutmann C, Hofmann KP (1991) Binding of inositol phosphates to arrestin. FEBS Lett 295(1–3):195–199

Praefcke GJ, Ford MG, Schmid EM, Olesen LE, Gallop JL, Peak-Chew SY, Vallis Y, Babu MM, Mills IG, McMahon HT (2004) Evolving nature of the AP2 alpha-appendage hub during clathrin-coated vesicle endocytosis. EMBO J 23(22):4371–4383

Santini F, Penn RB, Gagnon AW, Benovic JL, Keen JH (2000) Selective recruitment of arrestin-3 to clathrin coated pits upon stimulation of G protein-coupled receptors. J Cell Sci 113 (Pt 13):2463–2470

Santini F, Gaidarov I, Keen JH (2002) G protein-coupled receptor/arrestin3 modulation of the endocytic machinery. J Cell Biol 156(4):665–676

Schmid EM, McMahon HT (2007) Integrating molecular and network biology to decode endocytosis. Nature 448(7156):883–888

Schmid EM, Ford MG, Burtey A, Praefcke GJ, Peak-Chew SY, Mills IG, Benmerah A, McMahon HT (2006) Role of the AP2 beta-appendage hub in recruiting partners for clathrin-coated vesicle assembly. PLoS Biol 4(9):e262

Scott MG, Benmerah A, Muntaner O, Marullo S (2002) Recruitment of activated G protein-coupled receptors to pre-existing clathrin-coated pits in living cells. J Biol Chem 277(5):3552–3559

Shenoy SK, Lefkowitz RJ (2011) β-Arrestin-mediated receptor trafficking and signal transduction. Trends Pharmacol Sci 32(9):521–533

Shih W, Gallusser A, Kirchhausen T (1995) A clathrin-binding site in the hinge of the beta 2 chain of mammalian AP-2 complexes. J Biol Chem 270(52):31083–31090

Shukla AK, Manglik A, Kruse AC, Xiao K, Reis RI, Tseng WC, Staus DP, Hilger D, Uysal S, Huang LY, Paduch M, Tripathi-Shukla P, Koide A, Koide S, Weis WI, Kossiakoff AA, Kobilka BK, Lefkowitz RJ (2013) Structure of active β-arrestin-1 bound to a G-protein-coupled receptor phosphopeptide. Nature 497(7447):137–141

Sohlemann P, Hekma M, Puzicha M, Buchen C, Lohse MJ (1995) Binding of purified recombinant beta-arrestin to guanine-nucleotide-binding-protein-coupled receptors. Eur J Biochem 232(2):464–472

ter Haar E, Harrison SC, Kirchhausen T (2000) Peptide-in-groove interactions link target proteins to the beta-propeller of clathrin. Proc Natl Acad Sci 97(3):1096–1100

Traub LM (2009) Tickets to ride: selecting cargo for clathrin-regulated internalization. Nat Rev Mol Cell Biol 10(9):583–596

Traub LM, Bonifacino JS (2013) Cargo recognition in clathrin-mediated endocytosis. Cold Spring Harb Perspect Biol 5(11):a01670

Vishnivetskiy SA, Gimenez LE, Francis DJ, Hanson SM, Hubbell WL, Klug CS, Gurevich VV (2011) Few residues within an extensive binding interface drive receptor interaction and determine the specificity of arrestin proteins. J Biol Chem 286(27):24288–24299

Willox AK, Royle SJ (2012) Functional analysis of interaction sites on the N-terminal domain of clathrin heavy chain. Traffic 13(1):70–81

Yang F, Yu X, Liu C, Qu CX, Gong Z, Liu HD, Li FH, Wang HM, He DF, Yi F, Song C, Tian CL, Xiao KH, Wang JY, Sun JP (2015) Phospho-selective mechanisms of arrestin conformations and functions revealed by unnatural amino acid incorporation and (19)F-NMR. Nat Commun 6:8202

Zhang J, Barak LS, Anborgh PH, Laporte SA, Caron MG, Ferguson SS (1999) Cellular trafficking of G protein-coupled receptor/beta-arrestin endocytic complexes. J Biol Chem 274(16): 10999–11006

Zhou XE, Gao X, Barty A, Kang Y, He Y, Liu W, Ishchenko A, White TA, Yefanov O, Han GW, Xu Q, de Waal PW, Suino-Powell KM, Boutet S, Williams GJ, Wang M, Li D, Caffrey M, Chapman HN, Spence JC, Fromme P, Weierstall U, Stevens RC, Cherezov V, Melcher K, Xu HE (2016) X-ray laser diffraction for structure determination of the rhodopsin-arrestin complex. Sci Data 3:160021

Chapter 5
Arrestin-3: The Structural Basis of Lower Receptor Selectivity

Benjamin W. Spiller, Xuanzhi Zhan and Vsevolod V. Gurevich

Abstract Arrestin-3 (β-arrestin2) is the second member of the non-visual arrestin subfamily cloned. Together with arrestin-2 (β-arrestin1), it is responsible for arrestin function outside the visual system. Arrestins 2 and 3 are multi-functional proteins, originally identified as important for termination of signaling through G protein-coupled receptors (GPCRs), but now recognized as versatile regulators of signaling and trafficking of most GPCRs, as well as signaling switches controlling the balance between G protein-dependent and -independent GPCR signaling. Arrestin-3 is notably the least selective arrestin family member in that it binds a wide variety of GPCRs and that binding is less dependent on receptor activation and phosphorylation than for other arrestin subtypes. A recently determined arrestin-3 structure reveals that the basal conformation of arrestin-3 is similar to other arrestins, and that similar molecular interactions stabilize the basal state in all arrestins. A disruption of a structural element in the C-domain of arrestin-3 has been implicated as the cause of the reduced selectivity of arrestin-3. Here we compare the basal arrestin-3 structure with the recently determined structure of an activated receptor-bound arrestin-1.

Keywords Arrestin · Arrestin-3 · GPCR binding · Receptor specificity · Crystal structure · Receptor selectivity

---

B.W. Spiller (✉) · X. Zhan · V.V. Gurevich
Department of Pharmacology, Vanderbilt University, 2200 Pierce Avenue, Preston Research Building, Rm. 454, Nashville, TN 37232, USA
e-mail: benjamin.spiller@Vanderbilt.Edu

B.W. Spiller
Departments of Pharmacology and Microbiology and Immunology, Vanderbilt University, Nashville, TN 37232, USA

Present Address:
X. Zhan
Department of Chemistry, Tennessee Tech University, P.O. Box 5055, Cookeville, TN 38505, USA

V.V. Gurevich (ed.), *The Structural Basis of Arrestin Functions*,
DOI 10.1007/978-3-319-57553-7_5

Introduction

The four arrestin subtypes expressed in mammals can be broadly classified into visual and non-visual groups (Gurevich and Gurevich 2006a). The visual arrestins, arrestin-1 (rod arrestin) and arrestin-4 (cone arrestin) are expressed only in photoreceptors and bind rod and cone opsins (Strissel et al. 2006; Hanson et al. 2007; Nikonov et al. 2008). In contrast, arrestin-2 and arrestin-3 (also referred to as β-arrestin1 and β-arrestin2, respectively) are present in most cells, and interact with hundreds of G protein-coupled receptors (GPCRs) (Gurevich and Gurevich 2006b). In most cell types arrestin-2 is the major arrestin family member, being more abundant than arrestin-3 by ~10–20 fold (Gurevich et al. 2002, 2004). In addition to their roles in terminating GPCR signaling, arrestins 2 and 3 also serve as clathrin adaptors, thereby controlling GPCR endocytosis (Laporte et al. 1999; Oakley et al. 1999; Goodman et al. 1996), and as signaling hubs, where they scaffold and direct kinase networks, and as mediating ubiquitination of receptors and other signaling molecules [reviewed in Gurevich and Gurevich (2006b), DeWire et al. (2007), Jean-Charles et al. (2016)]. A key functional difference between arrestin-3 and other family members is its relative receptor promiscuity. Binding studies comparing different arrestins to different functional states of their cognate receptors (active phosphorylated, inactive phosphorylated, active unphosphorylated, and inactive unphosphorylated) revealed arrestin 3 to be significantly less specific than other arrestins. Arrestin-1 is the most specific and shows a >10 fold preference for active, phosphorylated, receptor over inactive, phosphorylated, receptor (Gurevich et al. 1993; Gurevich 1998). Both arrestin-2 and arrestin-3 are less selective, and show an ~2 fold preference for active, phosphorylated, receptor over inactive, phosphorylated, receptor, with arrestin-3 being the least selective (Gurevich et al. 1995). The requirement for receptor phosphorylation can be more effectively relieved through mutagenesis in arrestin-3 than in other arrestins, further supporting the conclusion that arrestin-3 is the most promiscuous arrestin (Gurevich and Gurevich 2006b).

In addition to being more promiscuous, arrestin-3 generally shows higher affinity for GPCRs, higher affinity for clathrin, as well as unique kinase scaffolding activity (Gurevich and Gurevich 2006b; Goodman et al. 1996; DeWire et al. 2007; Kohout et al. 2001; Oakley et al. 2000). Importantly, only arrestin-3, but not highly homologous arrestin-2, facilitates the activation of JNK family kinases (McDonald et al. 2000; Song et al. 2009a; Miller et al. 2001; Breitman et al. 2012; Zhan et al. 2011a, 2013, 2016; Kook et al. 2013). Despite these unique features, the two proteins are partially redundant because knockout a either non-visual arrestin is tolerated, but the double knockout is embryonic lethal (Kohout et al. 2001).

Arrestin 3 was the final isoform to have its structre reported, with arrestin 1, 2, and 4 published previously (Hirsch et al. 1999; Han et al. 2001; Kang et al. 2009; Milano et al. 2002, 2006; Sutton et al. 2005). Here we review the arrestin-3 structure, with the added benefit of comparison to the recent structure of the rhodopsin-arrestin-1 complex, which offers the first insight into an activated

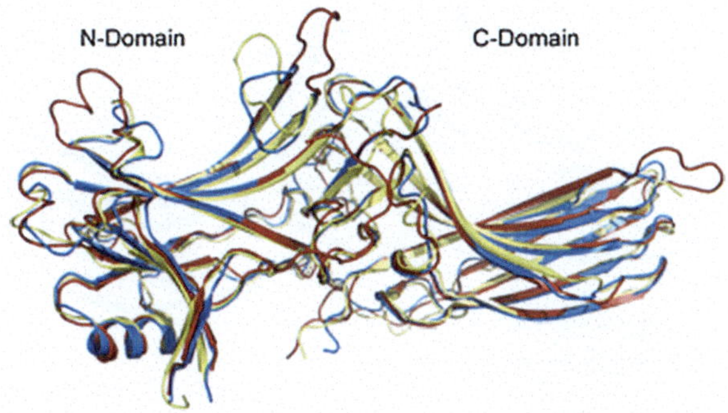

Fig. 5.1 Arrestin 3 is structurally similar to Arrestins 1 and 2. Arrestin 3 (pdb ID 3pd2) is shown in *blue*, Arrestin 1 (pdb ID 1cf1) is shown in *red*, and Arrestin 2 (pdb ID 1cf1) is shown in *yellow*. Arrestins are shown as a *ribbon diagram*, highlighting the conservation of all secondary structural elements

receptor-bound arrestin. The overall fold of arrestin-3 resembles that of the other family members, Fig. 5.1. The key features known to stabilize the basal state, the three-element interaction and the polar core, are present in the now expected form (Fig. 5.2), but the structure identified certain peculiarities characteristic for this subtype.

Arrestin-3 Structure

As expected, the arrestin-3 adopts a canonical arrestin fold (Figs. 5.1 and 5.2). It is composed of structurally homologous domains, the N-terminal and C-terminal domains, each of which is formed by a 7-stranded beta sandwich. The two domains are connected by a flexible linker and stabilized by inter-domain contacts. The extreme carboxy terminus traverses the C-domain to form critical interactions with the N-domain (three-element interaction, Fig. 5.2b). The last residue visible in the arrestin-3 structure is the arginine in the C-terminus that contributes a salt bridge to a critical functional component, the phosphate sensor (polar core, Fig. 5.2c). Additionally, three residues on this C-terminal peptide form part of the three-element interaction.

Intermolecular interactions hold arrestins in their basal conformations. Two sets of interactions known to be critical are the three-element interaction and the polar core (Fig. 5.2). Disruption of either of these interactions in known to reduce fidelity in the arrestin-receptor interaction, likely by decreasing the activation barrier for the transition to the activated state (Gurevich and Gurevich 2006b; Kovoor et al. 1999; Celver et al. 2002; Song et al. 2009b; Gurevich et al. 2011). The arrestin-3 structure confirmed that alterations to the polar core are not responsible for the increased promiscuity of arrestin-3. The polar core in arrestin-3 includes identical residues in

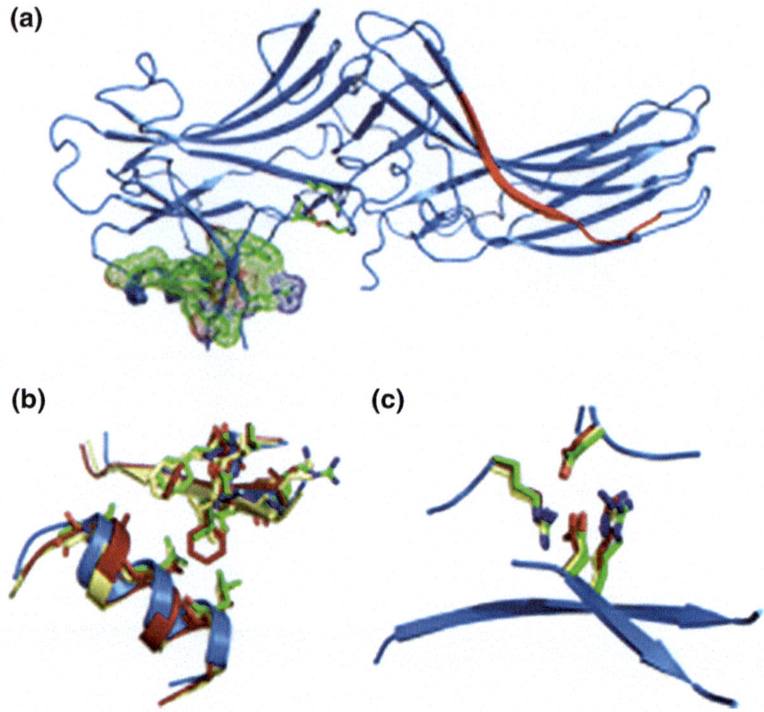

Fig. 5.2 Structurally important features are conserved in Arrestin 3. **a** Arrestin 3 with the conserved three-element interaction highlighted as a Van der Waals mesh and the polar core side chains rendered as stick figures. Additionally, the disrupted beta strand in the C-domain of Arrestin 3 is colored *red*. **b** and **c** Expanded overlays of Arrestins 1, 2, and 3, colored as in Fig. 5.1 and rendered from the coordinates listed in Fig. 5.1. Both views are rotated ~90° toward the reader, about a horizontal axis, relative to the view in **a**. **b** Conserved three-element interaction within three arrestins. **c** Near perfect conservation of the polar core among arrestins 1, 2, and 3. **b** Outermost strand of the arrestin 3 C-domain is also distorted relative to arrestin 2. The disruption is small, ~2 Å, but it disrupts hydrogen bonding to the adjacent strand and likely reduces the energy barrier to activation

nearly identical orientations to those observed in arrestin-1 and arrestin-2 (Fig. 5.2c). The polar core is essentially unchanged in arrestin-3, indicating that a different structural feature is responsible for the unique functional characteristics of arrestin-3.

The three-element interaction, although somewhat less well studied, is thought to serve as a molecular clamp that prevents release of the C-tail (Gurevich 1998; Vishnivetskiy et al. 2000). Release of the C-tail is a necessary step in arrestin activation (Palczewski et al. 1991; Kang et al. 2015; Shukla et al. 2013). In arrestin-3, the three-element interaction is present, but it is altered, as compared to arrestin-1 in three ways. The interaction involves three hydrophobic residues on each of three different secondary structural elements: β-strand I contributes Val8,

Ile9, and Phe10; α-helix I contributes Leu101, Leu105, and Leu109; and the C-tail contributes Phe386, Val387, and Phe388 (bovine arrestin-1 numbering). In arrestin-3, Val8 is replaced with an Arg, Ile9 is replaced with a Val, and Phe386 is replaced with an Ile (Fig. 5.2b). The possibility that these mutations may result in the C-tail being more weakly held in arrestin-3 than in arrestin-1 has not been tested experimentally. In any event, this change is likely not responsible for the entirety of the differences between the arrestins because arrestin-2 and arrestin-3 have identical three-element interactions (Fig. 5.2b), indicating another structural features must underlie the functional differences between the two non-visual arrestins.

An additional difference between arrestin-3 and arrestin-2 is a disruption in the packing of the C-domain. Residues 257–260, which form part of the outer most "strand" in the C-domain, are displaced relative to arrestin-2. The displacements are relatively small (1.3–2.0 Å), but are large enough to disrupt the β-sheet hydrogen bonding pattern, resulting in the loss of 2 hydrogen bonds (Zhan et al. 2011b). Despite the relatively small size of the distortion, it seems important for three reasons: it is substantially larger than the mean coordinate error of the structure; it occurs in a region known to be important for receptor binding; and the C-domain (including this strand) is essentially invariant in all arrestin structures (Hirsch et al. 1999; Han et al. 2001; Kang et al. 2009; Milano et al. 2002, 2006; Zhan et al. 2011b; Vishnivetskiy et al. 2004).

The importance of this distortion was tested by exchanging the strand with the homologous strand from arrestin-2. A bioluminescence resonance energy transfer-based arrestin recruitment experiment designed to determine the fraction of receptor-bound arrestin in the cell revealed that substitution of the distorted strand into arrestin-2 conferred arrestin-3 like features on arrestin-2 (Zhan et al. 2011b). β2AR binding of these chimeras and the wild type arrestins was compared in an arrestin recruitment assay (Zhan et al. 2011b). This assay revealed that distortion of the C-domain sheet promoted binding to inactive and active β2AR (Zhan et al. 2011b), a hallmark feature of arrestin-3 (Gurevich et al. 1995; Kovoor et al. 1999; Celver et al. 2002). Conversely, the introduction of the arrestin-2 residues into arrestin-3 reduced binding to unstimulated β2AR (Zhan et al. 2011b).

Comparisons to the Rhodopsin-Arrestin-1 Complex

The recently determined structure of rhodopsin bound to arrestin-1 (Kang et al. 2015) allows analysis of how differences in the three-element interaction and the disrupted β-strand in the C-domain might be important for the conformational changes associated with receptor binding. Figure 5.3 shows a comparison of receptor-bound arrestin-1 and the basal structure of arrestin-3. The most significant changes are the re-structuring of the finger loop (residues 64–77), the release of the C-tail from the N-domain, and the rotation of the C-domain relative to the N-domain. The finger loop intercalates into the rhodopsin, fitting into the cavity between the cytoplasmic ends of the transmemebrane helices that opens upon

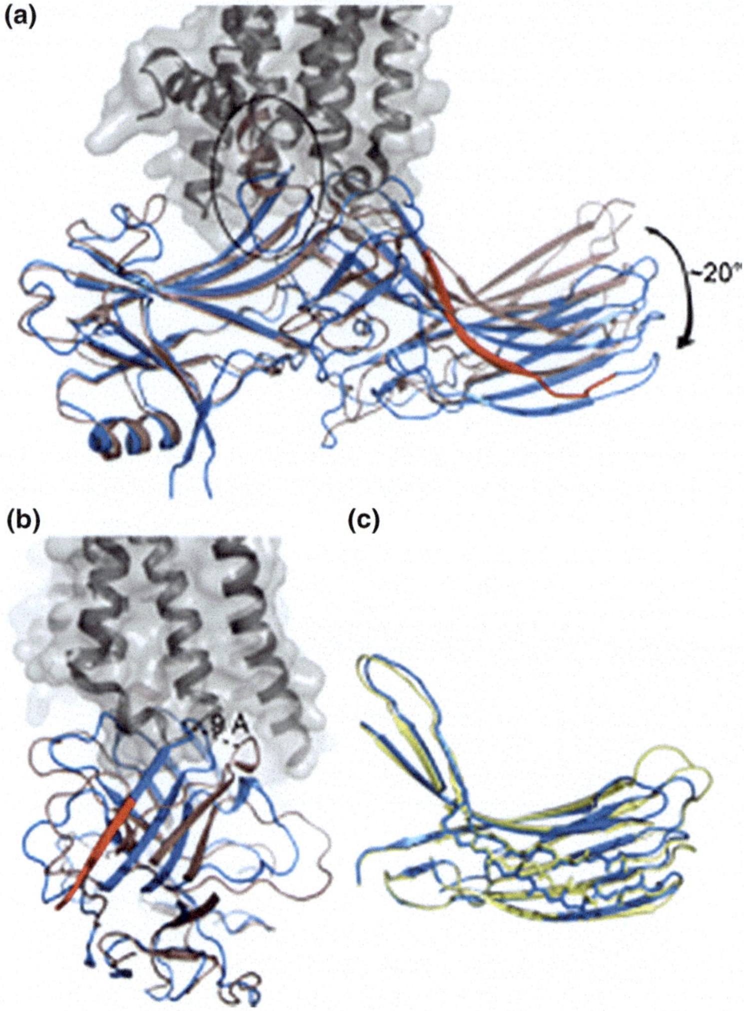

Fig. 5.3 Arrestin 3 is configured to transition to the receptor bound state. **a** Structural rearrangements that occur upon receptor binding. Arrestin 1 shown in salmon undergoes significant conformational changes upon rhodopsin (*grey*) binding (the arrestin-rhodopsin complex is rendered from (pdb 4zwj). The finger loop, highlighted with a *black ellipse*, undergoes a loop to helix transition and inserts into the helical bundle of the receptor. The C-domain rotates $\sim 20°$ relative to the N-domain. The arrestin 3 structure, overlayed based on an alignment to the N-domain, is shown in *blue*, with the outermost strand of the C-domain colored *red*. The homologous strand in arrestin 1 undergoes an ~ 9 Å displacement upon activation, as shown in **b**, in which the view is rotated 90° toward the reader about a vertical axis **c** an overlay of arrestin-1, yellow, with arrestin-3, blue, showing the connection between the finger loop and the loosely packed outer strand in arrestin-3. Loose packing of this strand may accomadate the conformational change associated with activation

GPCR activation (Farrens et al. 1996). This interaction, however, does not appear to require a substantial conformation change on the part of arrestin. The finger loop is the loop between the 4th and 5th strands of the N-domain and is structurally variable in arrestins. An interdomain conformational change is not likely required for it to adopt the active conformation.

The rotation of the C-domain by $\sim 20°$, in contrast, requires a complete reorganization of the interdomain interface and a coincident translation of the beta-sandwich. In arrestin-1, this reorganization results in a repacking of the outermost strand, such that it forms beta strand hydrogen bonds with the neighboring strand for only 7 residues, as compared to 11 residues in the basal state. If arrestin-3 undergoes a similar conformational change upon receptor binding, the loose packing of this strand in the basal state may better accommodate the conformational change associated with activation (Fig. 5.3c). This would explain apparently lower energy barrier, which results in high binding to inactive phosphorylated receptor and wide variety of GPCRs.

Acknowledgements This review is dedicated to the memory of Carsten Schubert. Carsten contributed extensively to the arrestin 1 and arrestin 2 structures. He was helpful and generous with his ideas and time. He passed away on his 51st birthday. May his memory be a comfort.

References

Breitman M, Kook S, Gimenez LE, Lizama BN, Palazzo MC, Gurevich EV, Gurevich VV (2012) Silent scaffolds: inhibition of c-Jun N-terminal kinase 3 activity in cell by dominant-negative arrestin-3 mutant. J Biol Chem 287:19653–19664

Celver J, Vishnivetskiy SA, Chavkin C, Gurevich VV (2002) Conservation of the phosphate-sensitive elements in the arrestin family of proteins. J Biol Chem 277:9043–9048

DeWire SM, Ahn S, Lefkowitz RJ, Shenoy SK (2007) β-arrestins and cell signaling. Annu Rev Physiol 69:483–510

Farrens DL, Altenbach C, Yang K, Hubbell WL, Khorana HG (1996) Requirement of rigid-body motion of transmembrane helices for light activation of rhodopsin. Science 274:768–770

Goodman OB, Krupnick JG, Santini F, Gurevich VV, Penn RB, Gagnon AW, Keen JH, Benovic JL (1996) Beta-arrestin acts as a clathrin adaptor in endocytosis of the beta2-adrenergic receptor. Nature 383:447–450

Gurevich VV (1998) The selectivity of visual arrestin for light-activated phosphorhodopsin is controlled by multiple nonredundant mechanisms. J Biol Chem 273:15501–15506

Gurevich EV, Gurevich VV (2006a) Arrestins: ubiquitous regulators of cellular signaling pathways. Genome Biol 7:236

Gurevich VV, Gurevich EV (2006b) The structural basis of arrestin-mediated regulation of G-protein-coupled receptors. Pharmacol Ther 110:465–502

Gurevich VV, Richardson RM, Kim CM, Hosey MM, Benovic JL (1993) Binding of wild type and chimeric arrestins to the m2 muscarinic cholinergic receptor. J Biol Chem 268: 16879–16882

Gurevich VV, Dion SB, Onorato JJ, Ptasienski J, Kim CM, Sterne-Marr R, Hosey MM, Benovic JL (1995) Arrestin interactions with G protein-coupled receptors. Direct binding studies of wild type and mutant arrestins with rhodopsin, beta 2-adrenergic, and m2 muscarinic cholinergic receptors. J Biol Chem 270:720–731

Gurevich EV, Benovic JL, Gurevich VV (2002) Arrestin2 and arrestin3 are differentially expressed in the rat brain during postnatal development. Neuroscience 109:421–436

Gurevich EV, Benovic JL, Gurevich VV (2004) Arrestin2 expression selectively increases during neural differentiation. J Neurochem 91:1404–1416

Gurevich VV, Hanson SM, Song X, Vishnivetskiy SA, Gurevich EV (2011) The functional cycle of visual arrestins in photoreceptor cells. Prog Retin Eye Res 30:405–430

Han M, Gurevich VV, Vishnivetskiy SA, Sigler PB, Schubert C (2001) Crystal structure of beta-arrestin at 1.9 A: possible mechanism of receptor binding and membrane translocation. Structure 9:869–880

Hanson SM, Gurevich EV, Vishnivetskiy SA, Ahmed MR, Song X, Gurevich VV (2007) Each rhodopsin molecule binds its own arrestin. Proc Natl Acad Sci U S A 104:3125–3128

Hirsch JA, Schubert C, Gurevich VV, Sigler PB (1999) The 2.8 A crystal structure of visual arrestin: a model for arrestin's regulation. Cell 97:257–269

Jean-Charles P-Y, Rajiv V, Shenoy SK (2016) Ubiquitin-related roles of β-arrestins in endocytic trafficking and signal transduction. J Cell Physiol 231:2071–2080

Kang DS, Kern RC, Puthenveedu MA, von Zastrow M, Williams JC, Benovic JL (2009) Structure of an arrestin2-clathrin complex reveals a novel clathrin binding domain that modulates receptor trafficking. J Biol Chem 284:29860–29872

Kang Y, Zhou XE, Gao X, He Y, Liu W, Ishchenko A, Barty A, White TA, Yefanov O, Han GW, Xu Q, de Waal PW, Ke J, Tan MHE, Zhang C, Moeller A, West GM, Pascal BD, Van Eps N, Caro LN, Vishnivetskiy SA, Lee RJ, Suino-Powell KM, Gu X, Pal K, Ma J, Zhi X, Boutet S, Williams GJ, Messerschmidt M, Gati C, Zatsepin NA, Wang D, James D, Basu S, Roy-Chowdhury S, Conrad CE, Coe J, Liu H, Lisova S, Kupitz C, Grotjohann I, Fromme R, Jiang Y, Tan M, Yang H, Li J, Wang M, Zheng Z, Li D, Howe N, Zhao Y, Standfuss J, Diederichs K, Dong Y, Potter CS, Carragher B, Caffrey M, Jiang H, Chapman HN, Spence JCH, Fromme P, Weierstall U, Ernst OP, Katritch V, Gurevich VV, Griffin PR, Hubbell WL, Stevens RC, Cherezov V, Melcher K, Xu HE (2015) Crystal structure of rhodopsin bound to arrestin by femtosecond X-ray laser. Nature 523:561–567

Kohout TA, Lin FS, Perry SJ, Conner DA, Lefkowitz RJ (2001) Beta-arrestin 1 and 2 differentially regulate heptahelical receptor signaling and trafficking. Proc Natl Acad Sci U S A 98: 1601–1606

Kook S, Zhan X, Kaoud TS, Dalby KN, Gurevich VV, Gurevich EV (2013) Arrestin-3 binds c-Jun N-terminal kinase 1 (JNK1) and JNK2 and facilitates the activation of these ubiquitous JNK isoforms in cells via scaffolding. J Biol Chem 288:37332–37342

Kovoor A, Celver J, Abdryashitov RI, Chavkin C, Gurevich VV (1999) Targeted construction of phosphorylation-independent beta-arrestin mutants with constitutive activity in cells. J Biol Chem 274:6831–6834

Laporte SA, Oakley RH, Zhang J, Holt JA, Ferguson SSG, Caron MG, Barak LS (1999) The β2-adrenergic receptor/βarrestin complex recruits the clathrin adaptor AP-2 during endocytosis. Proc Natl Acad Sci USA 96:3712–3717

McDonald PH, Chow CW, Miller WE, Laporte SA, Field ME, Lin FT, Davis RJ, Lefkowitz RJ (2000) Beta-arrestin 2: a receptor-regulated MAPK scaffold for the activation of JNK3. Science 290:1574–1577

Milano SK, Pace HC, Kim Y-M, Brenner C, Benovic JL (2002) Scaffolding functions of arrestin-2 revealed by crystal structure and mutagenesis. Biochemistry 41:3321–3328

Milano SK, Kim Y-M, Stefano FP, Benovic JL, Brenner C (2006) Nonvisual arrestin oligomerization and cellular localization are regulated by inositol hexakisphosphate binding. J Biol Chem 281:9812–9823

Miller WE, McDonald PH, Cai SF, Field ME, Davis RJ, Lefkowitz RJ (2001) Identification of a motif in the carboxyl terminus of beta -arrestin2 responsible for activation of JNK3. J Biol Chem 276:27770–27777

Nikonov SS, Brown BM, Davis JA, Zuniga FI, Bragin A, Pugh EN, Craft CM (2008) Mouse cones require an arrestin for normal inactivation of phototransduction. Neuron 59:462–474

Oakley RH, Laporte SA, Holt JA, Barak LS, Caron MG (1999) Association of beta-arrestin with G protein-coupled receptors during clathrin-mediated endocytosis dictates the profile of receptor resensitization. J Biol Chem 274:32248–32257

Oakley RH, Laporte SA, Holt JA, Caron MG, Barak LS (2000) Differential affinities of visual arrestin, beta arrestin1, and beta arrestin2 for G protein-coupled receptors delineate two major classes of receptors. J Biol Chem 275:17201–17210

Palczewski K, Pulvermüller A, Buczyłko J, Hofmann KP (1991) Phosphorylated rhodopsin and heparin induce similar conformational changes in arrestin. J Biol Chem 266:18649–18654

Shukla AK, Manglik A, Kruse AC, Xiao K, Reis RI, Tseng W-C, Staus DP, Hilger D, Uysal S, Huang L-Y, Paduch M, Tripathi-Shukla P, Koide A, Koide S, Weis WI, Kossiakoff AA, Kobilka BK, Lefkowitz RJ (2013) Structure of active β-arrestin-1 bound to a G-protein-coupled receptor phosphopeptide. Nature 497:137–141

Song X, Coffa S, Fu H, Gurevich V (2009a) How does arrestin assemble MAP kinases into a signaling complex? J Biol Chem

Song X, Vishnivetskiy SA, Gross OP, Emelianoff K, Mendez A, Chen J, Gurevich EV, Burns ME, Gurevich VV (2009b) Enhanced arrestin facilitates recovery and protects rods lacking rhodopsin phosphorylation. Curr Biol 19:700–705

Strissel KJ, Sokolov M, Trieu LH, Arshavsky VY (2006) Arrestin translocation is induced at a critical threshold of visual signaling and is superstoichiometric to bleached rhodopsin. J Neurosci 26:1146–1153

Sutton RB, Vishnivetskiy SA, Robert J, Hanson SM, Raman D, Knox BE, Kono M, Navarro J, Gurevich VV (2005) Crystal structure of cone arrestin at 2.3A: evolution of receptor specificity. J Mol Biol 354:1069–1080

Vishnivetskiy SA, Schubert C, Climaco GC, Gurevich YV, Velez MG, Gurevich VV (2000) An additional phosphate-binding element in arrestin molecule. Implications for the mechanism of arrestin activation. J Biol Chem 275:41049–41057

Vishnivetskiy SA, Hosey MM, Benovic JL, Gurevich VV (2004) Mapping the arrestin-receptor interface. Structural elements responsible for receptor specificity of arrestin proteins. J Biol Chem 279:1262–1268

Zhan X, Kaoud TS, Dalby KN, Gurevich VV (2011a) Nonvisual arrestins function as simple scaffolds assembling the MKK4-JNK3α2 signaling complex. Biochemistry 50:10520–10529

Zhan X, Gimenez LE, Gurevich VV, Spiller BW (2011b) Crystal structure of arrestin-3 reveals the basis of the difference in receptor binding between two non-visual subtypes. J Mol Biol 406:467–478

Zhan X, Kaoud TS, Kook S, Dalby KN, Gurevich VV (2013) JNK3 enzyme binding to arrestin-3 differentially affects the recruitment of upstream mitogen-activated protein (MAP) kinase kinases. J Biol Chem 288:28535–28547

Zhan X, Stoy H, Kaoud TS, Perry NA, Chen Q, Perez A, Els-Heindl S, Slagis JV, Iverson TM, Beck-Sickinger AG, Gurevich EV, Dalby KN, Gurevich VV (2016) Peptide mini-scaffold facilitates JNK3 activation in cells. Nat Publ Gr 6:21025

Chapter 6
Phosphate Sensor and Construction of Phosphorylation-Independent Arrestins

Reuven Wiener, Sergey A. Vishnivetskiy, Vsevolod V. Gurevich and Joel A. Hirsch

Abstract Arrestins preferentially bind active phosphorylated GPCRs. Receptor binding is associated with a global conformational change in arrestins. These findings lead to a model where arrestins have distinct sensors for the receptor-attached phosphates and active receptor conformation, and that simultaneous engagement of both sensors by corresponding parts of the receptor induces binding-associated conformational change. Receptor-attached phosphates perturb two intra-molecular interactions in arrestins that stabilize their basal conformation: the polar core between the two domains and the three-element interaction that anchors the arrestin C-tail. Indeed, mutations that disrupt those interactions yield "enhanced" mutants capable of binding active receptors regardless of their phosphorylation. Structural and functional characterization of these mutants lead us to propose an allosteric regulation model for arrestin. Further, it was proposed that these mutants can compensate for defects in GPCR phosphorylation, including those caused by mutations, thereby serving as tools for gene therapy of these gain-of-function GPCR mutations. This idea so far was tested only in the visual system, where partial compensation for lack of rhodopsin phosphorylation was reported. These proof-of-concept experiments suggested that this approach works, but more powerful phosphorylation-independent mutants are needed in photoreceptors using the fastest, most sensitive, and most demanding GPCR-driven signaling system.

Keywords Arrestin · GPCR · Phosphorylation · Congenital disorders · Gene therapy

R. Wiener · J.A. Hirsch (✉)
Department of Biochemistry and Molecular Biology, George S. Wise Faculty of Life Sciences, Tel Aviv University, 69978 Tel Aviv, Israel
e-mail: jhirsch@post.tau.ac.il

S.A. Vishnivetskiy · V.V. Gurevich (✉)
Department of Pharmacology, Vanderbilt University, 2200 Pierce Avenue, Preston Research Building, Rm. 454, Nashville, TN 37232, USA
e-mail: vsevolod.gurevich@vanderbilt.edu

V.V. Gurevich (ed.), *The Structural Basis of Arrestin Functions*,
DOI 10.1007/978-3-319-57553-7_6

Early Biochemistry and Structure-Function Characterization

Visual arrestin-1 (under the name of 48 kDa protein)[1] was first discovered as one of the retinal proteins that preferentially bind light-activated rhodopsin (Kuhn 1978). Later, Kuhn's group found that arrestin-1 binding is greatly enhanced by rhodopsin phosphorylation (Kuhn et al. 1984). These findings established arrestin specificity for active phosphorylated rhodopsin (P-Rh*), versus all other functional forms of rhodopsin: inactive unphosphorylated (Rh), active unphosphorylated (Rh*), and inactive phosphorylated (P-Rh). The finding that arrestin binding to P-Rh* has a steep Arrhenius activation energy ($\sim$39 kcal mol^{-1}) suggested that arrestin undergoes a global conformational change in the process (Schleicher et al. 1989). This implied that there must be a sophisticated molecular mechanism responsible for arrestin specificity, which allows binding-associated conformational rearrangements only when arrestin encounters P-Rh*.

With development of a femtomolar sensitive quantitative assay for the arrestin-rhodopsin interaction, it became clear that arrestin does specifically bind Rh* and inactive P-Rh, but its binding to P-Rh* is 10–20-fold higher than to either of these forms (Gurevich and Benovic 1993). This impressive binding differential suggested that the mechanism is more sophisticated than a simple cooperative two-site interaction. These findings led to a double-trigger model, which posited that arrestin has distinct sensors that detect (i) phosphorylation and (ii) the active state of the receptor, which mediate the binding to P-Rh and Rh*, respectively. The model explained the specificity for P-Rh* by the idea that arrestin acts as a coincidence detector; only when both sensors in arrestin are engaged by the receptor simultaneously (which only P-Rh* can do), all constraints are released, allowing arrestin to transition into a conformationally distinct high-affinity receptor-binding state (Gurevich and Gurevich 2004).

This double-trigger model stimulated a search for the two sensors. The expected difference between regular receptor-binding element and phosphate sensor was that the mutations in the former would decrease the binding to the receptor, whereas in the latter case, activating mutations that relieve the conformational constraints of arrestin would enable greater binding to non-preferred forms of rhodopsin. It was clearer what to look for in case of phosphate sensor: phosphates are negatively charged at physiological pH, so the sensor as a phosphate-binding element must have positive charges. At that stage, there were no structures, but all arrestins cloned by that time had one linear stretch rich in positive charges (Fig. 6.1). Naturally, lysines and arginines in that stretch were targeted first (Gurevich and Benovic 1995). Quite a few charge neutralization mutations in that region reduced

[1]Here we use the systematic names of arrestin proteins: arrestin-1 (historic names S-antigen, 48 kDa protein, visual or rod arrestin), arrestin-2 (β-arrestin or β-arrestin1), arrestin-3 (β-arrestin2 or hTHY-ARRX), and arrestin-4 (cone or X-arrestin; for unclear reasons its gene is called "arrestin 3" in the HUGO database).

B. taurus arr1 (161)	ED**K**IP**KK**SSV**R**LLI**RK**
B. taurus arr2 (155)	EE**K**I**HKR**NSV**R**LVI**RK**
B. taurus arr3 (156)	EE**K**S**HKR**NSV**R**LVI**RK**
H. sapiens arr4 (152)	EETVS**KR**DYV**R**LVV**RK**
D. melanogaster arr2 (152)	DD**R**Q**HKR**SMVSLVI**KK**
C. erythrocephala arr2 (152)	DD**R**Q**HKR**SMVSLVI**KK**
L. migratoria arr2 (161)	EE**K**G**HKR**SAVTLAI**KK**
L. polyphemus arr (160)	DE**K**P**HKR**NSVSMAI**RK**
D. melanogaster arr1 (152)	CD**R**S**HRR**STINLGI**RK**

Fig. 6.1 Conserved motif rich in positive charges in arrestins. Aligned sequences (the number of the first residues is shown in parentheses) of cow (*Bos taurus*) arrestins 1, 2, and 3, human (*Homo sapiens*) arrestin-4, Drosophila melanogaster sensory arrestins 1 and 2, blowfly (*Calliphora erythrocephala*) arrestin-2, locust (*Locusta mogratoria*) arrestin-2, and horseshoe crab (*Limulus polyphemus*) arrestin. Positively charged residues are shown in *bold*. Note that this 16-residue element contains 5–6 positive charges, most of which are highly conserved in evolution. Interestingly, a homologous element capable of phosphate binding was found in an unrelated protein ataxin-7 (Mushegian et al. 2000)

arrestin-1 binding to both P-Rh and P-Rh*, identifying those charges as phosphate-binding residues. One mutant Arg175Gln, behaved in a strikingly different manner: it demonstrated much higher binding to Rh*, which is consistent with Arg175 being the phosphate sensor (Gurevich and Benovic 1995). However, the binding of Arg175Gln mutant to inactive P-Rh and even P-Rh* was also enhanced. Thus, it was necessary to demonstrate that Arg175 actually binds phosphates, as phosphate sensor was expected to, rather than doing something different. Luckily, by that time a somewhat naively designed mini-arrestin, [residues 1–191 now known to be the arrestin N-domain (Granzin et al. 1998; Hirsch et al. 1999)] was described, which still binds best to P-Rh*, but where that binding appears to be the sum of its binding to inactive P-Rh and Rh* (Gurevich and Benovic 1992, 1993). In the context of that short form, the Arg175Gln mutation simply reduced the binding to both P-Rh and P-Rh*, but did not affect Rh* binding, demonstrating that Arg175 does interact with phosphates. Subsequent replacement of Arg175 with all other 19 residues showed that charge elimination or reversal "pre-activated" arrestin-1, enhancing its binding to Rh*, whereas conservative Arg175Lys substitution that preserved the charge did not (Gurevich and Benovic 1997). These data suggested that in arrestin-1 Arg175 has an intra-molecular negatively charged partner, and that the salt bridge between Arg175 and this putative partner acts as a phosphate sensor: receptor-attached phosphates neutralize the charge of Arg175, thereby breaking that bridge, which "tells" the rest of the molecule that the phosphates are in place (Gurevich and Gurevich 2004; Gurevich and Benovic 1997).

Naturally, this proposal called for a search for that putative negatively charged partner. Arrestin-1 is a 404 residue protein with 57 negatively charged Asp and Glu side chains. Thus, the partner of Arg175 could have been identified either by "carpet bombing"—mutating each one of these negative charges individually, or by solving the structure that would reveal the neighbors of Arg175. Two crystal structures of arrestin-1 were solved within the next two years (Granzin et al. 1998; Hirsch et al. 1999). Both identified the same three "suspects": Asp30, Asp296, and Asp303 in the vicinity of Arg175. Interestingly, these residues along with Arg382, form the "polar core" right on the interface of the two arrestin domains (Hirsch et al. 1999) (Fig. 6.2). The functional significance of this arrangement was underscored not only by the presence of Arg175 there, but also by the fact that these charges were buried, whereas charged residues in soluble proteins are usually located at the protein-solvent surface, due to their hydrophilicity. Exhaustive mutagenesis of these residues established that the salt bridge between Arg175 and Asp296 acts as the key phosphate sensor in arrestin-1. Charge reversal of either residue greatly increased arrestin-1 binding to Rh*, whereas simultaneous reversal of both charges, restoring the salt bridge in the opposite configuration, yielded arrestin-1 with essentially wild type (WT) selectivity for P-Rh* (Hirsch et al. 1999; Vishnivetskiy et al. 1999).

These results unambiguously identified the phosphate sensor and established the mechanism of its function. In addition, the data solved a mystery in the field: how can just two non-visual arrestins present in vertebrates "serve" hundreds of different GPCRs with minimal sequence conservation of their cytoplasmic faces? The Arg-Asp salt bridge mechanism of phosphate recognition is sequence context-independent, so that arrestins can respond to the presence of receptor-attached phosphates regardless of the nature of surrounding residues (Vishnivetskiy et al. 1999). Thus, it was not surprising that virtually identical polar core was found in other subtypes: arrestin-2 (Han et al. 2001; Milano et al. 2002), arrestin-3 (Zhan et al. 2011), and arrestin-4 (Sutton et al. 2005).

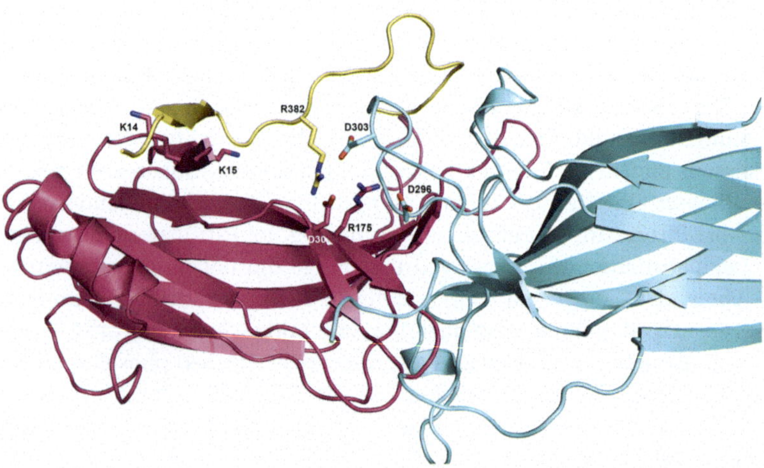

Fig. 6.2 Structure of basal WT arrestin. The N-domain is depicted in magenta while the C-domain is *cyan*. The C-tail is colored *yellow*. Residues mentioned in the text are shown, in particular those residues that are part of the polar core. Based on PDB 1CF1 (Hirsch et al. 1999)

Numerous lines of evidence suggest that for high-affinity arrestin binding, the receptor requires more than one phosphate (Gurevich et al. 1995; Mendez et al. 2000; Vishnivetskiy et al. 2007; Azevedo et al. 2015), whereas the postulated mechanism of the phosphate sensor function suggests that it should respond to a single phosphate. Subsequent mutagenesis revealed that there is more to the phosphate sensor than the polar core. In addition to several phosphate-binding residues identified earlier (Gurevich and Benovic 1995), which the crystal structures showed to be located in the β-strand X (Hirsch et al. 1999), two lysines in the β-strand I (Lys 14 and Lys15 in bovine arrestin-1) (Fig. 6.2) were identified as phosphate-binding residues (Vishnivetskiy et al. 2000). The elimination of these charges by mutagenesis turned out to be the most potent binding-reducing mutations described (Vishnivetskiy et al. 2000). Interestingly, in the context of phosphorylation-independent mutants this substitution did not affect arrestin binding to P-Rh* much (Vishnivetskiy et al. 2000).

Based on these data and localization of these two lysines on the surface of arrestin-1 (Fig. 6.2), a more sophisticated model of phosphate sensing was proposed. Exposed Lys14–15 on the β-strand I were proposed to "greet" receptor-attached phosphates first and guide them to the polar core (Vishnivetskiy et al. 2000). Thus, their function is indispensable in WT arrestin-1, but unnecessary in mutants where the polar core is already destabilized by mutations. The distance between these lysines, other phosphate-binding residues (Gurevich and Benovic 1995), and the polar core (Hirsch et al. 1999) shows that a single phosphate cannot reach all of these elements at the same time. This physical distance (16 Å) suggests why multiple [three in the case of rhodopsin (Mendez et al. 2000; Vishnivetskiy et al. 2007)] receptor-attached phosphates are needed for high-affinity arrestin binding and termination of G protein-mediated signaling.

The identification of the two lysines in the β-strand I as phosphate-binding residues also explained how receptor binding releases the arrestin C-tail from its basal position. The release of the arrestin C-tail by P-Rh* and poly-anions, such as heparin, was discovered long before crystal structures (Palczewski et al. 1991). Truncation (Gurevich and Benovic 1992, 1993), alternative splicing (Smith et al. 1994), or perturbation by a triple alanine (called 3A) substitution of the hydrophobic residues that anchor it to the N-domain (Gurevich 1998) facilitates the binding to any active form of rhodopsin, phosphorylated and unphosphorylated. For the two lysines in neighboring positions in the β-strand I to bind phosphates, this β-strand has to be deformed. This would likely move hydrophobic residues that mediate its interaction with the C-tail out of their basal position, facilitating the release of the arrestin C-tail (Vishnivetskiy et al. 2000). Thus, in addition to destabilizing the polar core, receptor-attached phosphates also break the three-element interaction that holds the C-tail in place (Hirsch et al. 1999). This explains why the deletion or detachment of the C-tail, similarly to the polar core mutations, makes arrestin phosphorylation-independent.

Towards an Allosteric Model for Arrestin

These collective findings suggest a way to frame arrestin regulation and the functional consequences in a classical allosteric scheme. Two extreme states can be postulated, namely a basal, largely inactive (*T* for tense) form and the active form (*R* for relaxed), that are in equilibrium:

$$T \leftrightarrow R.$$

Allosteric modulators then are the phosphorylated C-terminal tail and the activated receptor. The action of both modulators shifts an increased ensemble percentage of arrestin into the *R* state. The activation energy barrier between states is large, in keeping with the well-documented Arrhenius activation energy, i.e. a temperature-dependence in arrestin binding to receptor, noted earlier. The *R* state will be competent for specific, high affinity binding to activated receptor (P-Rh*). Thus, the *R* state would not be expected to have exactly the same conformation as seen in a receptor:arrestin complex (Kang et al. 2015). The various mutants discussed above then have increased population shifting between states.

Evidence for this model was obtained when we structurally characterized purified constitutively active arrestin mutants. Circular dichroism (CD) spectra of WT arrestin and three constitutively active mutants, i.e. Arg175Glu, 3A and ΔV8, arrestin proteolyzed by *Staphylococcus aureus* V8 protease, equivalent to p44, were taken at 4 °C (Fig. 6.3a). All four spectra show a negative band at 216 nm corresponding to β-sheet structures (Johnson 1990), consistent with the β-sheet structures that were presented in the crystal structures (Granzin et al. 1998; Hirsch et al. 1999). Notably, the WT arrestin CD spectrum was distinguished by a shoulder at 190 nm, detected also by Palczewski et al. (1992), which does not appear in the constitutively active mutants. This spectral signature, indicating a structural difference may confer the selectivity of WT to the phosphorylated light active rhodopsin (P-Rh*) that is compromised in the constitutively active mutants. We also measured the CD spectra of WT and mutants at different temperatures. Increasing temperature increments leads to an amplification of the negative band at 216 nm, as well as the positive band around 190 nm for both WT and mutants. Notably, the shoulder at 190 nm that characterized WT at 4 °C disappears as temperature increases, and the WT spectrum comes to resemble the constitutively activated mutants. Thus, WT arrestin at 35 °C has an almost identical spectrum as Arg175Glu at 4 °C (Fig. 6.3b). This shoulder reappears when the sample is cooled back to 4 °C (data not shown), indicating that it is a reversible structural feature, in agreement with our postulated equilibrium model.

In order to correlate these structural findings with functional data, we investigated the effect of temperature on arrestin binding. WT and two constitutively active mutants, namely Arg175Glu and 3A were tested for binding to light-activated and phosphorylated rhodopsin (P-Rh*) and to non-phosphorylated

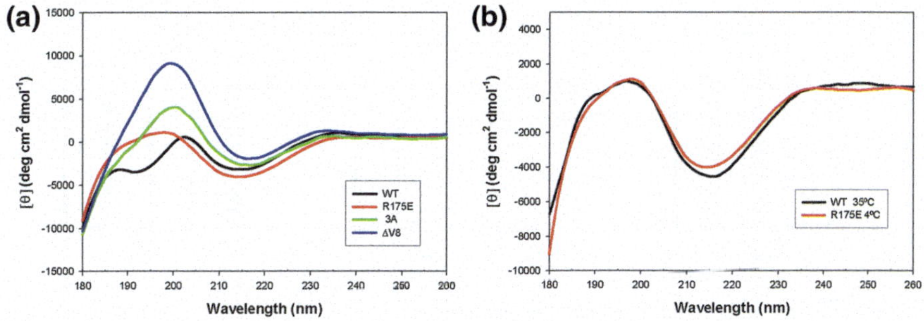

Fig. 6.3 CD spectra of WT arrestin and constitutively active mutants. All CD spectra were collected over the range of 260–180 nm at a scan rate of 1 nm s^{-1}. For all measurements, a cell with 0.1 mm path length was used. Each spectrum is an average of 5 scans. Accurate concentration of protein was obtained by monitoring the absorption of the 190 nm band using the corresponding extinction coefficient (Δ_{190} = 10560 L cm^{-1} mol^{-1} $amide^{-1}$) (Palczewski et al. 1992). **a** All spectra show bands at 216 nm consistent with β-sheet structure. As opposed to the constitutively active mutants, the WT has a shoulder at 190 nm. All the spectra shown in this panel were recorded at 4 °C. **b** A comparison of CD spectra between WT at 35 °C and the constitutively active mutant, R175E, at 4 °C. At these temperatures the WT and the R175E mutant have almost identical CD spectra

light activated rhodopsin (Rh*) at five different temperatures: 0, 20, 37, 40 and 45 °C (Fig. 6.4a, b). WT shows a steep increase in binding to P-Rh* as temperature increased until 37 °C. Binding at 37 °C to P-Rh* was 25-fold higher than at 0 °C. However, temperature effects for Arg175Glu mutant binding are moderate while for 3A there is almost no effect in comparison to WT. Moreover, the quantity of WT that bound to either Rh*-P or to Rh* at 37 °C was very similar to the amount of 3A mutant that bound to Rh*-P or to Rh* at 0 °C, respectively. Finally, the selectivity of WT for P-Rh* at physiological temperatures is similar to the constitutively active mutants at 4 °C (Fig. 6.4c).

These measurements correlate well with the CD data that display similar temperature effects (Fig. 6.3). Notably, WT arrestin has a shoulder around 190 nm in contrast to the constitutively active mutants, which may suggest a specific structural conformation, important for the WT arrestin inactive state. Indeed, functional data show minimal binding of arrestin to Rh*-P at 0 °C in comparison to the constitutively active mutants (Fig. 6.4a), indicating that the CD spectrum of WT at 4 °C corresponds to predominantly the *T* form of arrestin.

Thermal energy input to WT arrestin diminishes the differences between the CD spectrum of WT arrestin and the CD spectra of its constitutively active mutants, such that at 35 °C WT strongly resembles a mutant spectrum taken at 4 °C. The shoulder that characterizes the WT arrestin CD spectrum at 4 °C vanishes with temperature increases, suggesting some shift of the T form population towards the *R* form. Functional data of arrestin strongly support this conclusion. Accordingly, WT arrestin at 37 °C behaves like constitutively active mutant at 0 °C, i.e. WT arrestin binds P-Rh* like constitutively active mutant at 0 °C. Functional

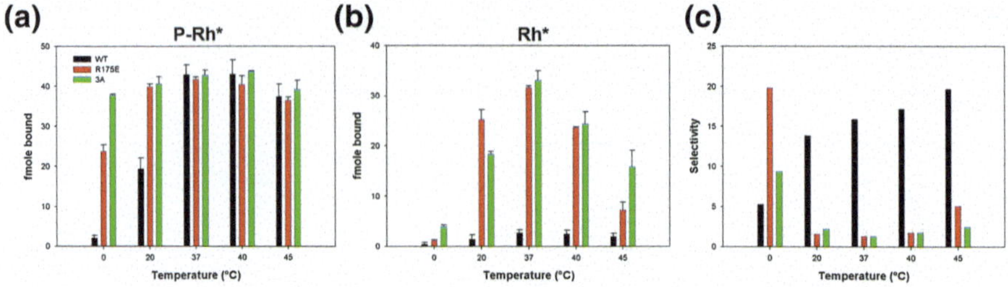

Fig. 6.4 Functional effects of temperature on arrestin binding to different states of rhodopsin. **a** Arrestin binding to P-Rh*. **b** Arrestin binding to Rh*. Binding experiments were performed as described (Vishnivetskiy et al. 2000). **c** Temperature effects on selectivity of WT arrestin and constitutively active mutants. Selectivity was defined as the ratio between binding to P-Rh* to Rh*

experiments with WT and constitutively active mutants at different temperatures demonstrate that the selectivity of WT arrestin for Rh*-P versus Rh* rises with temperature in contrast to the constitutively active mutants. Thus, at 37 °C WT arrestin is three fold more specific to Rh*-P than at 0 °C. In terms of selectivity, the mutants have maximal selectivity at 0 °C versus 37 °C for WT.

Warming the constitutively active mutants does not significantly increase the binding to Rh*-P but rather to Rh*. Therefore, we conclude that for the constitutively active mutants, the phosphorylated C-terminal segment of rhodopsin is sufficient to enable binding to rhodopsin regardless of the temperature. Nevertheless, binding to Rh* remains temperature-dependent, since the constitutively active mutants still need to undergo conformational changes, driven by thermal energy in the absence of the phosphorylated C-terminal segment of rhodopsin in order to bind Rh*. In contrast to the constitutively active mutants, WT arrestin requires both temperature and the presence of the phosphorylated C-terminal segment of rhodopsin in order to bind rhodopsin (low binding to Rh*-P at 0 °C and to Rh* at 37 °C).

As postulated above, an equilibrium pertains between the *T* and *R* states, which is appreciably shifted in favor of the *T* state. This form, we suggest, has a relatively rigid conformation, stabilized both by the intact polar core and the C tail interactions (Vishnivetskiy et al. 1999). Upon interaction of arrestin with an allosteric effector, the phosphorylated C-terminal segment of rhodopsin, the latter pushes the equilibrium in favor of the active *R* state. The phosphorylated C-terminal segment of rhodopsin shifts the equilibrium by destabilizing arrestin's inactive conformation (Vishnivetskiy et al. 2000). In addition, thermal energy also pushes the equilibrium towards the active state and indeed, physiological temperature is required in order to obtain maximal binding of arrestin to P-Rh* in vitro. At 0 °C, the natural allosteric effector cannot provide sufficient push to move the equilibrium in favor of the active state (very low binding to P-Rh* at 0 °C). Subsequently, the active conformation is stabilized by binding to the receptor, and as a result further pushes the equilibrium towards the active state. In contrast with WT arrestin, the mutations in constitutively active forms destabilize arrestin's inactive state, most probably by disrupting the polar core or the C tail interactions, and as a result push the equilibrium in favor of the active state. Therefore, they do not require the phosphorylated C-terminal segment of rhodopsin, and can bind Rh*.

This framework appears to be in line with the more recent structural studies that have provided important insights into the conformation of constitutive arrestin mutants like p44 and R175E. Two crystal structures of p44 have been reported. One report, by Granzin and coworkers, described a structure that was remarkably similar to the basal, inactive WT structures (Granzin et al. 2012). The second report, by Sommers and coworkers, revealed p44 with a significant change in conformation, most notably, a rigid body C-domain shift and specific changes in the polar core (Kim et al. 2013). In addition, a crystal structure of R175E was reported last year, revealing conformations similar to the conformationally novel p44, but with a smaller inter-domain twist, suggesting that this is the structure of the activation intermediate (Granzin et al. 2015) (see Chap. 12 for details). All of these findings can be reconciled and organized by our thermodynamic allosteric model. The constitutive mutants sample both *T* and *R* states, residing more often than WT in the *R* state. Hence, upon crystallization, it is possible to capture in a given crystal form either the *T* or *R* state. Importantly, the p44 structure described in the *R* state had been incubated and crystallized with opsin, indicating that receptor assists in activating or shifting the p44 into that precise conformation. It appears that R175E is more shifted or destabilized than p44, again in line with our observations above.

Applications

All the experiments described above were performed on visual arrestin-1. However, the structural features identified by arrestin-1 mutations are conserved in the arrestin family at the level of sequence (Gurevich and Gurevich 2006) and 3D structures (Granzin et al. 1998; Hirsch et al. 1999; Han et al. 2001; Milano et al. 2002; Zhan et al. 2011; Sutton et al. 2005). Thus, it is hardly surprising that homologous mutations in other arrestin subtypes yield a similar biochemical phenotype: tight binding to unphosphorylated receptors and suppression of the G protein signaling without receptor phosphorylation (Gurevich et al. 1997; Kovoor et al. 1999; Celver et al. 2002). It was generally concluded that the mechanism of activation is conserved in the arrestin family (Gurevich and Gurevich 2004, 2006), raising two questions: are there generally applicable methods of generating "enhanced" arrestins and might they have translational use?

Humans have ~800 different GPCR subtypes. Naturally, there are quite a few genetic errors in these receptors, many of which lead to congenital disorders (Schoneberg et al. 2004; Stoy and Gurevich 2015). Mutations in genes fall into two broad categories: loss-of-function and gain-of-function. In case of loss-of-function mutations, a strategy for gene therapy is clear: restoration of the WT protein sequence should solve the problem. Moreover, loss-of-function mutations are usually recessive, as enough functional protein is often generated from the normal second allele. Cases of haplo-insufficiency are rare, especially among signaling proteins, such as GPCRs.

In contrast, gain-of-function mutations are dominant, as mutant protein generates excessive signaling even in the presence of a perfectly normal product from the second allele. Until recently, it has not been obvious how to approach these mutations therapeutically. Today there may be options, at least conceptually. At the simplest level, faulty genes can now be edited to correct the errors [reviewed in Stoy and Gurevich (2015)]. However, such a strategy would make sense in stem cells generating rapidly dividing cells (e.g., skin, lung epithelium, etc.), but would hardly work in non-dividing cells, such as neurons, where the gene in every cell has to be edited. Alternatively, ribozymes specifically attacking mutant mRNA can be engineered. However, in many cases such as frame-shift and point mutations, normal and faulty mRNA differ by just one base (Fuchs et al. 1995). It is quite challenging to design ribozymes that combine this level of precision with high enough efficiency to significantly reduce excessive signaling. Therefore, in the case of mutated GPCRs, a compensational approach has been proposed: counteract excessive receptor activity with expression of an enhanced arrestin that has a greater than normal ability to suppress signaling (Song et al. 2009). Similar to the ribozyme strategy, this approach requires expression of the desired construct in the majority of affected cells.

Functionally, gain-of-function mutations in GPCRs fall into two categories: those that cause increased constitutive activity (signaling in the absence of activating agonists) and those that reduce the ability of GRK/arrestin system to quench G protein activation, such as defects in receptor phosphorylation (Mendez et al. 2000; Chen et al. 1995, 1999). Conceivably, in the latter case, expression of a phosphorylation-independent arrestin mutant could compensate and return signaling closer to normal.

The idea of compensational gene therapy has been experimentally tested to date only in the visual system. Excessive rhodopsin signaling in rod photoreceptors due to the absence of cognate GRK1 (Chen et al. 1999), arrestin-1 (Xu et al. 1997), the absence (Chen et al. 1995) or insufficient number of phosphorylation sites (Mendez et al. 2000) invariably leads to prolonged photo-responses. This excessive signaling results in night blindness (Fuchs et al. 1995; Yamamoto et al. 1997) and eventually leads to retinal degeneration. Transgenic mouse lines expressing an arrestin-1 3A mutant with higher than normal ability to bind Rh* were created (Nair et al. 2005). Transgenic mice were then bred in a GRK1 (rhodopsin kinase) knockout background and directly compared with WT arrestin-1 mice. The success of these proof-of-concept experiments was partial. On the one hand, enhanced arrestin-1 facilitated photoresponse recovery and improved photoreceptor survival (Song et al. 2009). On the other hand, the time of half recovery of "compensated" photoreceptors was still many times longer than that in WT mice (Song et al. 2009). Moreover, lower ($\sim$50% of WT) expression of the enhanced mutant worked better than higher ($\sim$240% of WT) expression (Song et al. 2009), and high levels of enhanced mutant turned out to be detrimental for photoreceptor health and survival

(Song et al. 2013). Independent experiments also showed that even relatively low affinity of arrestin-1 for clathrin adaptor AP2 can be detrimental for the rod health, whereas elimination of the AP2 binding site makes the mutant harmless and allows it to compensate for constitutive activity of rhodopsin mutant (Moaven et al. 2013). Hence, these pioneering experiments suggested that to be therapeutically usable the enhanced versions of arrestin-1 should be further improved in two ways. The mutant should have a (i) higher affinity for Rh* and (ii) lack an AP2 binding site. New and improved phosphorylation-independent versions of arrestin-1 need to be tested experimentally to determine how far the compensation can be pushed in the visual system (Vishnivetskiy et al. 2013).

However, rod photoresponse is the fastest and most sensitive GPCR-driven signaling system. Rods respond to single photons (Baylor et al. 1979) and turn off within 250 ms (Mendez et al. 2000), i.e., many times faster than any other GPCR-elicited response. Thus, an approach that yields partial compensation in rods is likely to ensure full compensation in other cells. Non-visual arrestins can be pre-activated by mutations homologous to those that pre-activate arrestin-1 (Kovoor et al. 1999; Celver et al. 2002). The key challenge is that while arrestin-1 is highly selective for rhodopsin, non-visual arrestins are quite promiscuous, binding many GPCRs with comparable affinity (Gimenez et al. 2012; Barak et al. 1997). Most cells express numerous GPCR subtypes, only one of which is likely to be an overactive GPCR mutant. Thus, a promiscuous enhanced non-visual arrestin meant to suppress excessive signaling by the mutant will likely affect the signaling by the other GPCRs in the same cell, causing unwanted side effects. Having generated the rhodopsin-specific arrestin-1, evolution demonstrated that high receptor specificity of arrestin can be achieved. Approaches to the problem of receptor selectivity of non-visual arrestins is discussed in more detail in Chap. 9.

To summarize, we now know where receptor-attached phosphates bind in arrestins and how they promote arrestin “activation”, i.e., a global conformational rearrangement necessary for the transition into high-affinity receptor-binding state. Several mutations targeting intra-molecular interactions that hold arrestins in their basal conformation were shown to facilitate the activation, yielding mutants that can bind any active receptor, phosphorylated or not. These mutants appear to have the ability to compensate for excessive signaling of GPCR with gain-of-function mutations, but their full therapeutic potential needs to be further explored.

References

Azevedo AW, Doan T, Moaven H, Sokal I, Baameur F, Vishnivetskiy SA, Homan KT, Tesmer JJ, Gurevich VV, Chen J, Rieke F (2015) C-terminal threonines and serines play distinct roles in the desensitization of rhodopsin, a G protein-coupled receptor. Elife 4:05981

Barak LS, Ferguson SS, Zhang J, Caron MG (1997) A beta-arrestin/green fluorescent protein biosensor for detecting G protein-coupled receptor activation. J Biol Chem 272:27497–27500

Baylor DA, Lamb TD, Yau KW (1979) Responses of retinal rods to single photons. J Physiol 288:613–634

Celver J, Vishnivetskiy SA, Chavkin C, Gurevich VV (2002) Conservation of the phosphate-sensitive elements in the arrestin family of proteins. J Biol Chem 277:9043–9048
Chen J, Makino CL, Peachey NS, Baylor DA, Simon MI (1995) Mechanisms of rhodopsin inactivation in vivo as revealed by a COOH-terminal truncation mutant. Science 267:374–377
Chen CK, Burns ME, Spencer M, Niemi GA, Chen J, Hurley JB, Baylor DA, Simon MI (1999) Abnormal photoresponses and light-induced apoptosis in rods lacking rhodopsin kinase. Proc Natl Acad Sci U S A 96:3718–3722
Fuchs S, Nakazawa M, Maw M, Tamai M, Oguchi Y, Gal A (1995) A homozygous 1-base pair deletion in the arrestin gene is a frequent cause of Oguchi disease in Japanese. Nat Genet 10:360–362
Gimenez LE, Vishnivetskiy SA, Baameur F, Gurevich VV (2012) Manipulation of very few receptor discriminator residues greatly enhances receptor specificity of non-visual arrestins. J Biol Chem 287:29495–29505
Granzin J, Wilden U, Choe HW, Labahn J, Krafft B, Buldt G (1998) X-ray crystal structure of arrestin from bovine rod outer segments. Nature 391:918–921
Granzin J, Cousin A, Weirauch M, Schlesinger R, Buldt G, Batra-Safferling R (2012) Crystal structure of p44, a constitutively active splice variant of visual arrestin. J Mol Biol 416: 611–618
Granzin J, Stadler A, Cousin A, Schlesinger R, Batra-Safferling R (2015) Structural evidence for the role of polar core residue Arg175 in arrestin activation. Sci Rep 5:15808
Gurevich VV (1998) The selectivity of visual arrestin for light-activated phosphorhodopsin is controlled by multiple nonredundant mechanisms. J Biol Chem 273:15501–15506
Gurevich VV, Benovic JL (1992) Cell-free expression of visual arrestin. Truncation mutagenesis identifies multiple domains involved in rhodopsin interaction. J Biol Chem 267:21919–21923
Gurevich VV, Benovic JL (1993) Visual arrestin interaction with rhodopsin: sequential multisite binding ensures strict selectivity towards light-activated phosphorylated rhodopsin. J Biol Chem 268:11628–11638
Gurevich VV, Benovic JL (1995) Visual arrestin binding to rhodopsin: diverse functional roles of positively charged residues within the phosphorylation-recognition region of arrestin. J Biol Chem 270:6010–6016
Gurevich VV, Benovic JL (1997) Mechanism of phosphorylation-recognition by visual arrestin and the transition of arrestin into a high affinity binding state. Mol Pharmacol 51:161–169
Gurevich VV, Gurevich EV (2004) The molecular acrobatics of arrestin activation. Trends Pharmacol Sci 25:105–111
Gurevich VV, Gurevich EV (2006) The structural basis of arrestin-mediated regulation of G protein-coupled receptors. Pharm Ther 110:465–502
Gurevich VV, Dion SB, Onorato JJ, Ptasienski J, Kim CM, Sterne-Marr R, Hosey MM, Benovic JL (1995) Arrestin interaction with G protein-coupled receptors. Direct binding studies of wild type and mutant arrestins with rhodopsin, b2-adrenergic, and m2 muscarinic cholinergic receptors. J Biol Chem 270:720–731
Gurevich VV, Pals-Rylaarsdam R, Benovic JL, Hosey MM, Onorato JJ (1997) Agonist-receptor-arrestin, an alternative ternary complex with high agonist affinity. J Biol Chem 272:28849–28852
Han M, Gurevich VV, Vishnivetskiy SA, Sigler PB, Schubert C (2001) Crystal structure of beta-arrestin at 1.9 A: possible mechanism of receptor binding and membrane translocation. Structure 9:869–880
Hirsch JA, Schubert C, Gurevich VV, Sigler PB (1999) The 2.8 A crystal structure of visual arrestin: a model for arrestin's regulation. Cell 97:257–269
Johnson WC Jr (1990) Protein secondary structure and circular dichroism: a practical guide. Proteins 7:205–214
Kang Y, Zhou XE, Gao X, He Y, Liu W, Ishchenko A, Barty A, White TA, Yefanov O, Han GW, Xu Q, de Waal PW, Ke J, Tan MHE, Zhang C, Moeller A, West GM, Van Eps N, Caro LN, Vishnivetskiy SA, Lee RJ, Suino-Powell KM, Gu X, Pal K, Ma J, Zhi X, Boutet S, Williams GJ, Messerschmidt M, Gati C, Zatsepin NA, Wang D, James D, Basu S,

Roy-Chowdhuty S, Conrad S, Coe J, Liu H, Lisova S, Kupitz C, Grotjohann I, Fromme R, Jiang Y, Tan M, Yang H, Li J, Wang M, Zheng Z, Li D, Zhao Y, Standfuss J, Diederichs K, Dong Y, Potter CS, Carragher B, Caffrey M, Jiang H, Chapman HN, Spence JCH, Fromme P, Weierstall U, Ernst OP, Katritch V, Gurevich VV, Griffin PR, Hubbell WL, Stevens RC, Cherezov V, Melcher K, Xu HE (2015) Crystal structure of rhodopsin bound to arrestin determined by femtosecond X-ray laser. Nature 523:561–567

Kim YJ, Hofmann KP, Ernst OP, Scheerer P, Choe HW, Sommer ME (2013) Crystal structure of pre-activated arrestin p44. Nature 497:142–146

Kovoor A, Celver J, Abdryashitov RI, Chavkin C, Gurevich VV (1999) Targeted construction of phosphorylation-independent b-arrestin mutants with constitutive activity in cells. J Biol Chem 274:6831–6834

Kuhn H (1978) Light-regulated binding of rhodopsin kinase and other proteins to cattle photoreceptor membranes. Biochemistry 17:4389–4395

Kuhn H, Hall SW, Wilden U (1984) Light-induced binding of 48-kDa protein to photoreceptor membranes is highly enhanced by phosphorylation of rhodopsin. FEBS Lett 176:473–478

Mendez A, Burns ME, Roca A, Lem J, Wu LW, Simon MI, Baylor DA, Chen J (2000) Rapid and reproducible deactivation of rhodopsin requires multiple phosphorylation sites. Neuron 28:153–164

Milano SK, Pace HC, Kim YM, Brenner C, Benovic JL (2002) Scaffolding functions of arrestin-2 revealed by crystal structure and mutagenesis. Biochemistry 41:3321–3328

Moaven H, Koike Y, Jao CC, Gurevich VV, Langen R, Chen J (2013) Visual arrestin interaction with clathrin adaptor AP-2 regulates photoreceptor survival in the vertebrate retina. Proc Natl Acad Sci U S A 110:9463–9468

Mushegian AR, Vishnivetskiy SA, Gurevich VV (2000) Conserved phosphoprotein interaction motif is functionally interchangeable between ataxin-7 and arrestins. Biochemistry 39: 6809–6813

Nair KS, Hanson SM, Mendez A, Gurevich EV, Kennedy MJ, Shestopalov VI, Vishnivetskiy SA, Chen J, Hurley JB, Gurevich VV, Slepak VZ (2005) Light-dependent redistribution of arrestin in vertebrate rods is an energy-independent process governed by protein-protein interactions. Neuron 46

Palczewski K, Pulvermuller A, Buczylko J, Hofmann KP (1991) Phosphorylated rhodopsin and heparin induce similar conformational changes in arrestin. J Biol Chem 266:18649–18654

Palczewski K, Riazance-Lawrence JH, Johnson WC Jr (1992) Structural properties of arrestin studied by chemical modification and circular dichroism. Biochemistry 31:3902–3906

Schleicher A, Kuhn H, Hofmann KP (1989) Kinetics, binding constant, and activation energy of the 48-kDa protein-rhodopsin complex by extra-metarhodopsin II. Biochemistry 28: 1770–1775

Schoneberg T, Schulz A, Biebermann H, Hermsdorf T, Rompler H, Sangkuhl K (2004) Mutant G-protein-coupled receptors as a cause of human diseases. Pharmacol Ther 104:173–206

Smith WC, Milam AH, Dugger D, Arendt A, Hargrave PA, Palczewski K (1994) A splice variant of arrestin. Molecular cloning and localization in bovine retina. J Biol Chem 269:15407–15410

Song X, Vishnivetskiy SA, Gross OP, Emelianoff K, Mendez A, Chen J, Gurevich EV, Burns ME, Gurevich VV (2009) Enhanced arrestin facilitates recovery and protects rod photoreceptors deficient in rhodopsin phosphorylation. Curr Biol 19:700–705

Song X, Seo J, Baameur F, Vishnivetskiy SA, Chen Q, Kook S, Kim M, Brooks EK, Altenbach C, Hong Y, Hanson SM, Palazzo MC, Chen J, Hubbell WL, Gurevich EV, Gurevich VV (2013) Rapid degeneration of rod photoreceptors expressing self-association-deficient arrestin-1 mutant. Cell Signal 25:2613–2624

Stoy H, Gurevich VV (2015) How genetic errors in GPCRs affect their function: possible therapeutic strategies. Genes Dis 2:108–132

Sutton RB, Vishnivetskiy SA, Robert J, Hanson SM, Raman D, Knox BE, Kono M, Navarro J, Gurevich VV (2005) Crystal structure of cone arrestin at 2.3 Å: evolution of receptor specificity. J Mol Biol 354:1069–1080

Vishnivetskiy SA, Paz CL, Schubert C, Hirsch JA, Sigler PB, Gurevich VV (1999) How does arrestin respond to the phosphorylated state of rhodopsin? J Biol Chem 274:11451–11454

Vishnivetskiy SA, Schubert C, Climaco GC, Gurevich YV, Velez M-G, Gurevich VV (2000) An additional phosphate-binding element in arrestin molecule: implications for the mechanism of arrestin activation. J Biol Chem 275:41049–41057

Vishnivetskiy SA, Raman D, Wei J, Kennedy MJ, Hurley JB, Gurevich VV (2007) Regulation of arrestin binding by rhodopsin phosphorylation level. J Biol Chem 282:32075–32083

Vishnivetskiy SA, Chen Q, Palazzo MC, Brooks EK, Altenbach C, Iverson TM, Hubbell WL, Gurevich VV (2013) Engineering visual arrestin-1 with special functional characteristics. J Biol Chem 288:11741–11750

Xu J, Dodd RL, Makino CL, Simon MI, Baylor DA, Chen J (1997) Prolonged photoresponses in transgenic mouse rods lacking arrestin. Nature 389:505–509

Yamamoto S, Sippel KC, Berson EL, Dryja TP (1997) Defects in the rhodopsin kinase gene in the Oguchi form of stationary night blindness. Nat Genet 15:175–178

Zhan X, Gimenez LE, Gurevich VV, Spiller BW (2011) Crystal structure of arrestin-3 reveals the basis of the difference in receptor binding between two non-visual arrestins. J Mol Biol 406:467–478

Chapter 7
Comprehensive Analysis of the Role of Arrestin Residues in Receptor Binding

Raphael S. Haider, Aurélien Rizk, Gebhard F.X. Schertler and Martin K. Ostermaier

Abstract The molecular mechanism of arrestin–G protein-coupled receptor (GPCR) complex formation has been analyzed using inter alia spectroscopic, mutagenesis and structure determination methods. The latest crystal structure elucidating one conformation of the rhodopsin-bound arrestin-1 complex confirmed our structural model of an overall complex conformation and experimentally based assumptions about single residues located at the direct binding interface. Comprehensive analyses of single amino acid contributions in arrestin-1 and other methodologies expand our knowledge of structure–function relationships for receptor recognition and arrestin–GPCR complex formation. With a lot of progress made in the last few years, we review most recent literature covering the roles of essential arrestin-1 residues for pre-activation, receptor recognition and receptor binding. Mechanistic understanding of these processes at the atomic level is key for the identification of receptor conformations triggering given signaling pathways and the creation of novel tools for biased ligand drug discovery.

Keywords Arrestin · Rhodopsin · Protein–protein interaction · Mutagenesis · GPCR

R.S. Haider · G.F.X. Schertler
Laboratory of Biomolecular Research, Paul Scherrer Institute, 5232 Villigen, Switzerland

A. Rizk · M.K. Ostermaier (✉)
InterAx Biotech AG, PARK InnovAARE, 5234 Villigen, Switzerland
e-mail: ostermaier@interaxbiotech.com

G.F.X. Schertler
Department of Biology, ETH Zurich, Wolfgang-Pauli-Str. 27, 8093 Zurich, Switzerland

V.V. Gurevich (ed.), *The Structural Basis of Arrestin Functions*,
DOI 10.1007/978-3-319-57553-7_7

Introduction: The Role of Arrestin in GPCR Trafficking and Signaling

Arrestins form a family of homologous proteins regulating G protein-coupled receptor (GPCR) signaling. Arrestin-1 and arrestin-4 are expressed in the visual system while arrestin-2 (or ß-arrestin1) and arrestin-3 (or ß-arrestin2) are expressed ubiquitously in the body (Lohse and Hoffmann 2014). Arrestins were first discovered as desensitizing proteins of G protein signaling, which is induced by ligand binding to GPCRs (Kuhn and Wilden 1987; Benovic et al. 1987). After ligand-induced activation of GPCRs, GPCR kinases (GRKs) phosphorylate specific residues of GPCRs, which leads to the recruitment of arrestins to active phosphorylated receptors. G protein activation is partially inhibited by receptor phosphorylation itself (Benovic et al. 1987). Until recently it was additionally assumed that G protein signaling was completely inhibited after arrestin recruitment to the receptor (Lefkowitz 1998). The notion that sustained G protein signaling can occur for some GPCRs of a certain class after internalization (Ferrandon et al. 2009; Feinstein et al. 2013; Calebiro et al. 2009; Irannejad et al. 2013) stimulated a study that showed for the first time evidence that some GPCRs can form super-complexes composed of G protein and ß-arrestin1 explaining sustained G protein signaling (Thomsen et al. 2016).

However, inhibition of G protein activation for the majority of GPCRs is not the only role of arrestins in GPCR signaling. ß-arrestin1 and ß-arrestin2 also recruit the adaptor protein 2 (AP-2), as well as clathrin, in order to facilitate receptor internalization (Tian et al. 2014). Visual arrestins (arrestin-1 and arrestin-4) only interact marginally with AP-2 and hence do not trigger receptor internalization (Laporte et al. 2000). Internalized receptors are then either dephosphorylated and recycled to the membrane or degraded in lysosomes (Tian et al. 2014). The rate of re-sensitization through recycling or down regulation by degradation determines the amount of receptors available on the membrane and thus affects the G protein signaling response to subsequent stimuli.

For some receptors, such as the $ß_2$-adrenoreceptor, the interaction with ß-arrestin is transient. ß-arrestins and receptors co-localize in clathrin-coated pits near the cell surface and dissociate before the receptors travel to endosomes. Receptors following this pattern are termed class A receptors and typically undergo rapid recycling after internalization. In contrast, class B receptors, such as the angiotensin receptor AT_{1A}, show stronger binding to ß-arrestins with co-localization still detectable in endosomes (Oakley et al. 1999, 2000). Receptors of this class typically recycle much slower to the cell surface (DeWire et al. 2007; Pierce and Lefkowitz 2001) and some of them have been reported to promote sustained G protein signaling after receptor internalization (Ferrandon et al. 2009; Feinstein et al. 2013; Calebiro et al. 2009; Thomsen et al. 2016). In addition to their role in desensitization and trafficking, arrestins interact with many signaling proteins. ß-arrestins interact with sarcoma family (Src) tyrosine kinases and act as scaffolds for the mitogen-activated protein kinase (MAPK) cascades, including the extracellular signal-regulated kinase (ERK) cascade and the c-jun N-terminal kinase 3 (JNK3) cascade (Luttrell 2013).

The finding that GPCRs trigger G protein-dependent signaling and ß-arrestin-dependent trafficking and signaling opened up new strategies for drug discovery. Searching for ligands preferentially activating G protein or arrestin pathways, so-called biased ligands, might indeed lead to new drugs with reduced side effects, tolerance and toxicity (Luttrell 2013). There is also great potential in looking for ligands preferentially activating given arrestin-dependent signaling pathways or preferentially triggering recycling or degradation of a particular receptor. Indeed, we hypothesize that there might be an even higher complexity of biased signaling for certain class B receptors that internalize via arrestin and subsequently show recycling, degradation and/or simultaneous sustained G protein signaling.

Here we discuss a general framework for arrestin–GPCR complex formation, using the interactions between arrestin-1 and the GPCR rhodopsin as a model, which is so far the most studied. The photoreceptor rhodopsin is comprised of the apo-receptor opsin and its Schiff base-bound natural ligand, retinal. Retinal serves as an inverse agonist in its 11-*cis* conformation and triggers rhodopsin activation subsequently to its light dependent isomerization to all-*trans* retinal. The all-trans-retinal–bound activated conformation of rhodopsin induces G protein signaling (Arshavsky et al. 1994) and is susceptible to intracellular phosphorylation (Mendez et al. 2000; Vishnivetskiy et al. 2013a). Only one minute after its activation, the Schiff base-linked all-*trans* retinal dissociates from the receptor, which will further reside in its phosphorylated opsin conformation until dephosphorylation takes place and the binding of another retinal molecule occurs (Hofmann et al. 1992).

Although rhodopsin, as the dim-light photoreceptor, is characterized by a quite unique activation mechanism, it also shares a multitude of characteristics with other GPCRs. Not only because of these shared attributes, but also because it is the only GPCR that can be isolated and crystallized directly from natural sources (Unger et al. 1997), rhodopsin became a model system for structural analyses of GPCRs early on. Until 2007 it was the only GPCR with a known high resolution crystal structure and the most abundantly used receptor to explain processes, which are shared by the entire GPCR family (Deupi 2014). The first crystal structures of disease-causing GPCR mutants have been obtained from rhodopsin mutants G90D (Singhal et al. 2013) and T94I (Singhal et al. 2016) explaining the structure-function relationship for congenital stationary night blindness. Also the first crystal complex structure of arrestin with a GPCR was elucidated using rhodopsin and arrestin-1 (Kang et al. 2015).

An in-depth understanding of the molecular mechanism by which arrestin is activated and binds a cognate GPCR will enable the identification of receptor conformations leading to specific pathways, as well as to efficient and more precise discovery of biased ligands. Here we present the current knowledge on arrestin recognition of active receptors and the roles of distinct arrestin residues in receptor binding.

Molecular Mechanism of Arrestin-1 Engagement with Rhodopsin Revealed by Scanning Mutagenesis

Scanning Mutagenesis as a Tool for GPCR Pharmacology Elucidation

Site-directed scanning mutagenesis is known for circa 35 years, but has been recognized more recently as enabling technology to study GPCR pharmacology [reviewed in: Heydenreich et al. (2015)], such as GPCR signaling via G proteins (Flock et al. 2015; Sun et al. 2015), arrestin activation (Ostermaier et al. 2014a, b; Peterhans et al. 2016), or GPCR drug discovery (Congreve et al. 2011, 2014; Robertson et al. 2011; Rich et al. 2011; Chen et al. 2012; Congreve and Marshall 2010). The most frequently seen specialization is alanine scanning, which means the replacement of each amino acid of the native protein individually with alanine (Sun et al. 2013). Depending on the objective of the study the resulting GPCR mutant or signaling protein mutant is tested for its thermostability (Heydenreich et al. 2015; Sun et al. 2015) or altered ability to interact with its corresponding GPCR (Sun et al. 2015; Ostermaier et al. 2014a; Peterhans et al. 2016). Alanine scanning of $G\alpha_{i1}$ and arrestin-1 has led to comprehensive functional maps on top of corresponding 3D protein structures bringing structure and function at single amino acid resolution in direct relationship (Sun et al. 2015; Ostermaier et al. 2014a; Peterhans et al. 2016). In combination with crystal structures of rhodopsin, individual G protein subunits, trimeric G proteins, arrestin-1 or complex structures thereof, key amino acid residue triggers for protein activation and protein–protein interaction have been identified. The high sensitivity of employed assays allows exploration of differences in protein stability or GPCR–protein interaction induced by the presence or absence of GTP or the natural opsin agonist all-trans retinal (Sun et al. 2015; Peterhans et al. 2016). Scanning mutagenesis of arrestin-1 has enabled to 'read' ligand-dependent GPCR conformations (Peterhans et al. 2016) similar to FRET-, BRET- or NMR-based methods utilizing arrestin-2 or arrestin-3 as 'biosensors' (Nuber et al. 2016; Lee et al. 2016; Yang et al. 2015). Although rational design of arrestin-1 mutants over the course of more than 20 years has finally lead to the structure elucidation of the rhodopsin–arrestin-1 complex (Kang et al. 2015), we would like to point out compelling advantages of scanning mutagenesis versus rational mutant design to reveal molecular mechanisms of arrestin-1 action, such as the unbiased comprehensiveness in the analysis of single residue positions, performance in speed to grasp the larger 'picture' of molecular action, and the high detail level regarding quantitative comparison of single residue contributions. Scanning mutagenesis with subsequent combination of further selected single point mutations has been preferred historically (Warne et al. 2008; Lebon and Tate 2011; White et al. 2012; Hollenstein et al. 2013) over evolutionary approaches (Egloff et al. 2014) to obtain crystal structures of GPCRs. We believe that scanning mutagenesis remains an essential and complementary method to elucidate the pharmacological properties of GPCRs and the first step of arrestin-dependent signaling that is activation and binding of arrestin to its cognate GPCR.

Phosphorylation Defines the Nature of GPCR–Arrestin Interactions

Arrestin recruitment to the corresponding GPCR is a process comprised of multiple distinct events. Every step influences and, in fact, defines the nature of the ultimately formed complex. The mechanism of receptor activation does not only facilitate the dissociation of the trimeric G protein but may also result in the formation of several different conformational states of the receptor. These distinct conformations affect the affinity towards every intracellular binding partner, such as receptor kinases (GRKs), scaffolding proteins and arrestins (Moore et al. 2007; Mary et al. 2012). Moreover, there is strong evidence that different activated conformations of the GPCR lead to targeted phosphorylation by certain kinases only (Nobles et al. 2011). Thus, a receptor, activated via different processes or ligands, may show different phosphorylation patterns depending on its conformation as well as the affinity towards a distinct set of receptor kinases and particular kinetics of phosphorylation. These so-called "phosphorylation barcodes" as well as the particular receptor conformations have been shown to play a definite role in arrestin recruitment and affect further arrestin functions (Yang et al. 2015).

Inactive Arrestin-1 and Its Key Regulation Sites

The inactive conformation of the arrestin-1 protein is maintained by several stabilizing, intramolecular interactions, which lead to a structural self-inhibition of the protein (Hirsch et al. 1999; Han et al. 2001). First and foremost, the anchoring of the C-terminus at the N-domain of arrestin, held in place by the so-called "three-element" interaction (Gurevich and Gurevich 2013). The "three-element" interaction is formed by hydrophobic residues of β-strand I (H10, V11 and F13) and α-helix I (L107 and L111), both located in the N-domain of arrestin, in order to form a strong interaction with phenylalanine residues located in the distal C-terminal region (F375, F377 and F380) (Ostermaier et al. 2014a; Peterhans et al. 2016). This entrapment of the rather flexible arrestin-1 C-tail (residues 372–404) is also one of the most important interactions to fall apart during activation, thus enabling the commencement of further structural changes (Gurevich and Gurevich 2013). Mutational studies showed that upon substitution of these residues by alanines the protein exhibits a higher affinity towards unphosphorylated rhodopsin species (like Rh*) (Peterhans et al. 2016) and higher complex stability in general, if bound to the GPCR (Ostermaier et al. 2014a). Moreover, the phosphorylation-independent rhodopsin-binding of the "triple-A" arrestin-1 mutant (F375A, V376A, F377A)

(Gurevich 1998) can be interpreted as a direct consequence of the disturbance of the "three-element" interaction.

The most important interaction stabilizing the orientation of the N- and C-domains towards each other is called the polar core (Hirsch et al. 1999), which is highly conserved in the arrestin family. The term polar core describes a cluster of solvent-excluded, polar amino acids forming an extended hydrogen bond network, embedded between the N- and C-domains of the protein. Due to these interactions, the two domains are held in place as a trademark of the inactive arrestin conformation. The key interactions are formed by two aspartate residues, located in the gate loop (residues 296–305), a functional part of the lariat loop [residues 274–304 (Hirsch et al. 1999)], at position 296 and 303 (Han et al. 2001). In the inactive form those two residues pose as the main counterions for R175, located in β-strand X at the interface between the N- and C-domain (Granzin et al. 2015; Gurevich and Benovic 1995). In a similar orientation, the residue D30 facilitates further interactions between R175 and the crucial entrapment of the C-terminal R382 residue from inside the N-domain, to further fasten the C-tail onto the inactive arrestin conformation (Ostermaier et al. 2014b). The inactive as well as the pre-activated conformation of arrestin-1 are depicted in Fig. 7.1.

Structural studies of arrestin-1 mutants, lacking polar amino acids in this cluster, and activated β-arrestin1 show that, upon breakdown of the polar core interactions, a substantial rotation of the N- and C-domain relative to each other takes place. With this interlobe rotation of 20°–21° (Kim et al. 2013; Shukla et al. 2013), alongside further rearrangements of receptor interacting loops located in the central crest of the arrestin molecule, arrestin becomes capable of high affinity GPCR binding (Kang et al. 2015). The driving force responsible for these dramatic conformational changes is the breakdown of the polar core, initiated by releasing the C-tail from the described intramolecular interactions, namely the "three-element" interaction as well as the polar core itself. For this process to take place, an association of the inactive arrestin and a phosphorylated GPCR C-tail is needed, in order to establish high-affinity interactions that are able to force open the C-tail restraint inside arrestin (Vishnivetskiy et al. 2000). The manner in which this so-called "C-tail exchange" takes place is highly dependent on the number and location of phosphorylated residues in the C-terminal domain of the GPCR (Yang et al. 2015; Moore et al. 2007), defining a conformational need for distinct phosphorylation patterns.

Pre-activation of Arrestin-1 by C-Tail Exchange Mechanism

Phosphorylation of intracellular elemets of activated GPCRs occurs after dissociation of the primary signaling adaptor protein—the trimeric G protein. The discovery of the GPCR kinase 1 (GRK1), formerly known as rhodopsin kinase, was supplying the first evidence that phosphorylation of the GPCR may affect further processing, as well as signaling (Weller et al. 1975). This initial discovery is now

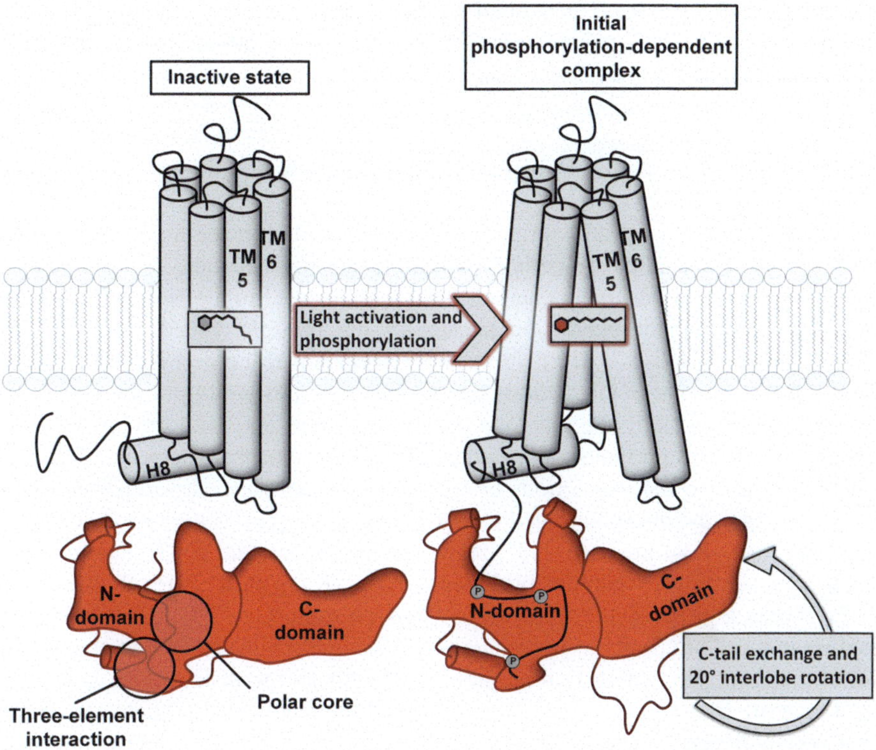

Fig. 7.1 Comparison of the inactive and pre-activated arrestin-1 conformation. In the inactive conformation, the two tight intramolecular interactions, namely the polar core and the "three-element" interaction, stabilize arrestin-1. Upon light activation and phosphorylation of rhodopsin, arrestin-1 engages the phosphorylated C-tail of the receptor. By disruption of the polar core as well as the "three-element" interaction, the C-tail of arrestin-1 is dislocated and sterically replaced by the phosphorylated GPCR C-tail. Arrestin-1 changes its conformation while performing an interlobe rotation of 20°. Arrestin-1 is now pre-activated and capable of high affinity binding

extended by the realization that not only seven different GRKs may alter the phosphorylation state of a given GPCR, but also different other cytoplasmic kinases, such as protein kinase A (PKA). Interestingly, each kinase generates a distinct phosphorylation pattern at the intracellular surface of the GPCR, leading to major differences between phosphorylation patterns generated by either GRKs or PKA (Tran et al. 2004) but also between patterns created by different members of the GRK family (Nobles et al. 2011). GPCR phosphorylation is essential for structural rearrangement of the arrestin protein, in order to reach a conformation suitable for high-affinity receptor binding. Thus, differences in receptor phosphorylation have been hypothesized to affect the affinity of the GPCR–arrestin complex and might even lead to differences in arrestin-mediated receptor trafficking and processing. Latest studies revealed strong evidence for the existence of so-called "phosphorylation barcodes" generated by a variety of kinases and the ability of arrestins to decipher this phosphorylation inscribed information, in order to facilitate different signaling outcomes (Yang et al. 2015).

Establishment of an Initial Phosphorylation-Dependent Binding

The unique ability of the arrestin protein to detect and embrace a phosphorylated GPCR C-tail is mediated by certain phosphate-sensing residues. A multitude of solvent-exposed and positively charged residues is located in the N-domain, which are likely predominant interaction sites with the phosphorylated GPCR C-tail residues. In the inactive conformation, these seem not to be accessible for intermolecular charge-charge interactions due to the steric positioning of the gate loop and the C-tail. Especially the gate loop positioning in the inactive form hides R175, a residue which was previously considered crucial for phosphate detection (Kovoor et al. 1999), due to the interactions of the polar core network. Nevertheless, when arrestin engages the phosphorylated GPCR C-tail, this region is the first to respond. Structural analysis of pre-activated arrestin species (Kim et al. 2013) and arrestin–GPCR (or C-terminal phosphopeptide) complexes (Kang et al. 2015; Shukla et al. 2013) reveal a shift of the gate loop. With phosphorylated GPCR C-terminal residues in close proximity to it, residue K300 recognizes this change of charges as part of the loop. This interaction might then be sufficient to dislocate the gate loop (Shukla et al. 2013). Due to this rearrangement, the charge-charge network of the polar core collapses, while the gate loop is undergoing a dramatic reorientation of up to 180°. The alanine mutant K300A is characterized as one of the arrestin-1 mutants with the lowest affinity towards phosphorylated and light-activated rhodopsin (Ostermaier et al. 2014a), underlining the importance of this single residue for receptor recognition. The new positioning of the gate loop is stabilized by further charged interactions, established between D303 and R288 (located in the lariat loop) as well as R291 and D138 [located in the middle loop (residues 133–142)]. The gate loop is now clipped between the lariat and middle loop and reveals the polar core. The latest structures of receptor- or phosphopeptide-bound arrestins reveal that there is in fact no interaction to be detected between R175 and phosphate moieties of the GPCR C-tail (Kang et al. 2015; Shukla et al. 2013), it becomes clear that R175 serves as an important anchor of the polar core, stabilizing the inactive arrestin-1 conformation, but not as a phosphate-interacting residue per se.

The disruption of the polar core, facilitated by the rotation of the gate loop, triggers several other structural changes, which lead to pre-activation of the arrestin protein. Due to a withdrawal of negatively charged residues (D303 and D296) from the polar core region as the gate loop rotates towards the middle loop, the major C-tail anchoring residue R382 is also freed from the polar core. Furthermore, it was shown that R29, a residue located in the N-domain between the polar core and "three-element" sites, interacts with other residues of the GPCR C-tail. These interactions lead to a collapse of further hydrogen bonds to C-terminal residues, namely F380, E379 and F378 (Granzin et al. 2015). Characterized not only as a highly important residue for receptor recognition via scanning mutagenesis, but in fact as the residue with the most negative impact on receptor binding when substituted with alanine (Ostermaier et al. 2014a; Peterhans et al. 2016), R29 is one of the most critical residues controlling the arrestin C-tail release as a phosphate sensor.

As the arrestin C-tail gains flexibility with every collapsed interaction, more and more positively charged residues inside the arrestin N-domain become solvent-exposed. In its phosphorylated and light-activated state, rhodopsin usually exhibits three phosphorylated residues, namely S334, S338 and S343 (Ohguro et al. 1996). Now the computational model of arrestin with the rhodopsin C-tail, relying on the latest crystal structure, is showing the phosphate interacting residues of arrestin in the final complex. According to this model, the phosphorylated S334 residue of the receptor interacts with K150 and K166, two lysine residues located at the tip of the N-domain. Beside the gate loop residue K300, K15, and R171 are assumed to interact with the phosphorylated S338 residue, whereas K14, Y25 and K110 are in reach to interact with the phosphorylated S343 (all residue positions have been adjusted to the corresponding positions of bovine arrestin-1 and rhodopsin) (Kang et al. 2015). Similar conclusions, regarding the layout of possible phosphate interacting residue clusters, but for a different species of both rhodopsin and arrestin-1, were drawn in a previously published review article (Ostermaier et al. 2014b). With the positioning of these residues, a positively charged basin is formed inside the N-domain, in which the phosphorylated GPCR C-tail can neatly be embraced (as depicted in Fig. 7.2). These interactions lead to the disruption of the "three-element" interaction, ultimately replacing the arrestin C-tail with the phosphorylated GPCR C-tail. The substitution of any residues involved in the hydrophobic "three-element" interaction with alanine strongly enhances the stability of the GPCR–arrestin complex. With H10A, V11A, F13A, L107A and L111A exhibiting very high values for receptor complex stability, it is interesting that the alanine mutants F375A and F377A as well as F380A enhance receptor binding to an even larger extent (Ostermaier et al. 2014a). This again underlines the importance of the "three-element" interaction site in anchoring the arrestin C-tail and stabilizing the inactive conformation of the protein. With the collapse of both intramolecular interaction sites, the polar core and the "three-element" site, facilitating the dislocation of the arrestin C-tail, the protein reorganizes and enters a pre-activated conformation, now capable of high-affinity complex formation with the receptor.

We propose the described process of initial complex formation between arrestin-1 and rhodopsin via the C-tail exchange mechanism as a framework for initial phosphorylation-dependent interactions of other arrestins with their GPCR (Ostermaier et al. 2014b). One more recent example supporting this hypothesis is the structural architecture of a chimeric ß$_2$-adrenoreceptor/vasopressin type 2 receptor in complex with a "hanging" ß-arrestin1, which has been derived by single-particle electron microscopy (Shukla et al. 2014). We would like to point out the relevance of slight differences in arrestin–GPCR interactions for subsequent cellular processes even if we assume at least one fundamentally conserved process

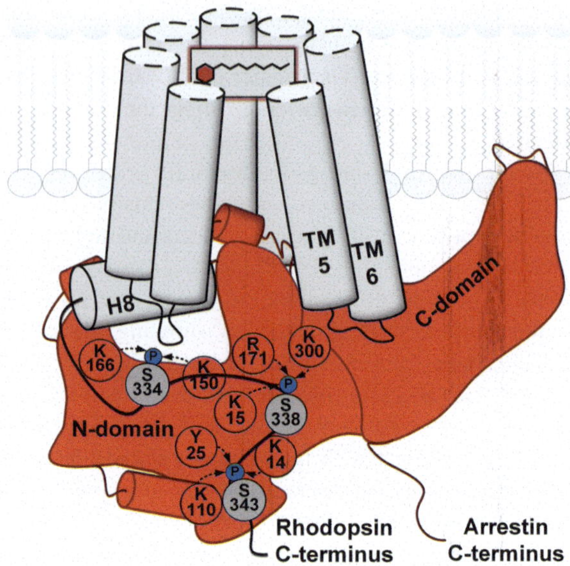

Fig. 7.2 Arrestin-1 in its rhodopsin bound conformation. The cytoplasmic interface of rhodopsin as well as arrestin-1 in its receptor-bound conformation are depicted. The phosphorylated residues of the rhodopsin C-tail as well as the interacting arrestin-1 residues are *highlighted*. Residue positions have been adjusted for bovine arrestin-1 and rhodopsin

for complex formation across the four arrestins with phosphorylated GPCRs that consists of "fishing" arrestin via C-tail exchange, initial engagement with receptor and subsequent formation of a low- or high-affinity complex.

Activation and Phosphorylation States of the GPCR Influence the Conformation of the Final Complex

Phylogenetic analyses have revealed that the human genome harbors more than 800 genes encoding GPCRs (Fredriksson et al. 2003). As every single receptor is characterized by its own mode of activation and intracellular receptor phosphorylation, arrestins have to be very versatile to be able to interact with even a majority of GPCRs. Especially the phosphorylation of GPCR C-tails may differ widely, as it is not only highly dependent on the kinases which are co-expressed by the receptor-bearing cell type but also due to sequence variations of the receptor-specific C-terminal residues. Interestingly, phosphate-interacting residues in arrestin-1 seem to vary if arrestin-1 is bound to light-activated and phosphorylated rhodopsin or if bound to phosphorylated opsin (Peterhans et al. 2016) (Fig. 7.3). We hypothesize that the phosphorylated opsin state of the receptor forms after light activation of the retinal-bound rhodopsin, as the subsequent hydrolysis of the Schiff-base linkage enables the retinal molecule to leave the binding cavity inside the GPCR (Hofmann et al. 1992). As phosphate-sensing residues, which interact with the phosphorylated C-tail of light-activated rhodopsin, were already discussed above, it was shown by scanning mutagenesis that the residues K300, K14 and K15 are important for arrestin-1 binding to either receptor state. The complex stability of arrestin-1 and phosphorylated opsin, on the other hand, is heavily decreased if the positive charges on residues R18, K55, R56, R81, K150, K166 or R171 are removed. Residues, which are shown to be important for the binding to phosphorylated opsin, mark a distinctly different pattern of positive charges in the arrestin-1 protein, if compared to light-activated and phosphorylated rhodopsin binding. Whereas both clusters of phosphate-interacting residues are located in the N-domain of arrestin-1, it seems that the conformational change of the receptor, which goes in hand with the dissociation of retinal, may lead to a translocation of the GPCR C-tail, which still stays attached to arrestin-1. As the phosphorylated opsin C-tail rather interacts with residues along the cup-shaped region on top of the arrestin-1 N-domain, instead of associating with residues across it (including residues, which are located near the N-terminus, like K110), these findings indicate that the activation state of the receptor may influence the conformation, orientation and the positioning of interacting residues in the GPCR–arrestin complex. A comparison of the activated and phosphorylated rhodopsin and putative phosphorylated opsin-bound states is shown in Fig. 7.3.

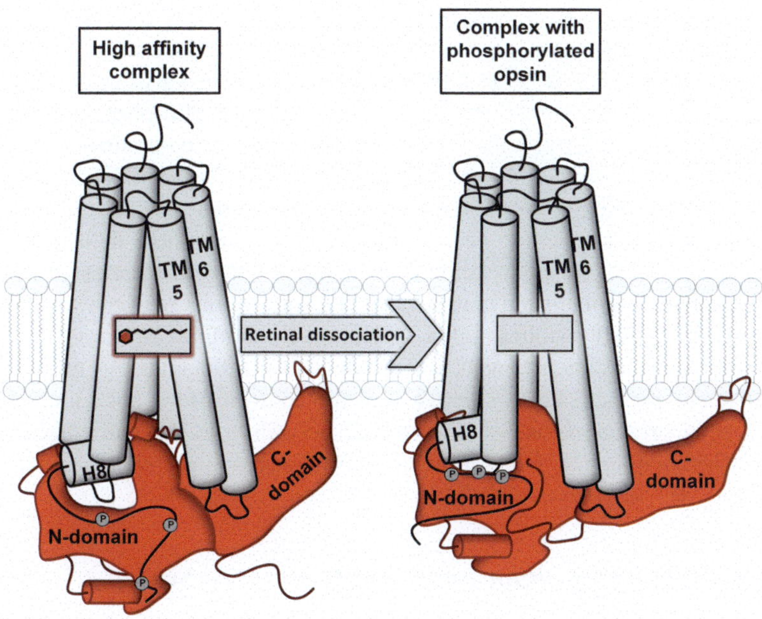

Fig. 7.3 Comparison of arrestin-1 bound to either activated and phosphorylated rhodopsin or to phosphorylated opsin. As scanning mutagenesis revealed different phosphate-interacting arrestin-1 residues for the binding of activated and phosphorylated rhodopsin as well as phosphorylated opsin, the putative conformational states of receptor binding were depicted. As the affinity of arrestin-1 for phosphorylated opsin is generally weaker if compared to phosphorylated and activated rhodopsin, the illustration shows one of many possible layouts, rather than a definite conformation

Similar conclusions were also drawn based on site-directed fluorescence studies of arrestin-1 binding to different receptor states (Sommer et al. 2012). One substantial difference between arrestin-1 binding to light-activated and phosphorylated rhodopsin and the phosphorylated apo-receptor might be the rotation of the gate loop, which only seems necessary for arrestin-1 association with ligand-bound receptor states. Without this intramolecular reorientation, it is likely that the arrestin-1 C-tail stays attached to the polar core and "three-element" interaction in a basal state-like fashion, while being bound to phosphorylated opsin. This distinct conformation would then hinder phosphate-interacting residues across the arrestin-1 N-domain to interact with the phosphorylated opsin C-tail and provoke a different binding pattern (Peterhans et al. 2016). The concealment of these main phosphate-interacting arrestin-1 residues may also explain earlier findings, indicating that arrestin-1 binding to the apo-form of the receptor is more dependent on higher phosphorylation levels of the C-terminus as compared to arrestin-1 binding to light-activated rhodopsin (Vishnivetskiy et al. 2007).

With tremendous efforts being made to unravel the mechanism of biased arrestin signaling and the various functions and pathways in which arrestin may serve as a mediator, NMR studies now revealed the existence of a variety of distinct arrestin conformations induced by different GPCR phosphorylation patterns (Yang et al. 2015). As already discussed, phosphorylation patterns of GPCR C-tails gain their diversity through the particular activation state, phosphorylation by different intracellular kinases, as well as GPCR-specific sequence variances. With every phosphorylation pattern attracting a distinct set of arrestin phosphorylation-sensing residues (Nobles et al. 2011), the GPCR-arrestin complex materializes in a unique conformation, characterized also by the affinity towards intracellular binding partners (Yang et al. 2015). The fashion in which activated GPCR trafficking and non-canonical GPCR signaling occurs, may even be much more defined by slight differences in GPCR phosphorylation then assumed before (Nuber et al. 2016).

The culminating conclusion can only be that interactions between arrestins and phosphorylated GPCR residues highly depend on a multitude of parameters defined by the receptor and its environment. It now becomes clear that there is no such thing as a universal pre-activated or active arrestin conformation, but rather distinct intramolecular arrangements, which organize according to the particular receptor sequence, activation state and C-terminal phosphorylation pattern.

Active GPCR Recognition on a Single Residue Scale

As the arrestin phosphate sensors engage the phosphorylated C-tail of the GPCR, a second set of putative GPCR-interacting residues establishes the formation of a high-affinity complex. With the crystal structure of the rhodopsin-arrestin-1 complex at our disposal (Kang et al. 2015), it is now possible to precisely describe the interface of the complex on a single residue scale.

GPCR Activation-Sensing Residues of Arrestin-1

Pre-activation of the arrestin molecule induces several significant structural re-arrangements affecting not only the C-tail of arrestin but also the positioning of intramolecular loops and a substantial inter-domain rotation. These changes of the tertiary structure of arrestin are thought to expose distinct activation-sensing residues and enable the protein to engage the activated GPCR. Beside the movement of the gate loop as one of the hallmark processes in arrestin pre-activation, especially the displacement of the middle loop as well as the finger loop (residues 68–78) are of major importance to the recognition of receptor activation (Kang et al. 2015). Already shown in the crystal structure of the truncated and pre-active arrestin mutant p44 (Granzin et al. 2012), the middle loop undergoes a shift upon arrestin pre-activation, thus moving away from the receptor binding interface and towards the N-domain (Vishnivetskiy et al. 2013b). This shift might be a direct result of interactions between the dislocated gate loop and the middle loop itself.

In the inactive conformation, the middle loop is held in place by a salt bridge between K141 and D71, the latter located in the finger loop. This interaction is disrupted for high-affinity receptor binding, which explains the enhanced receptor binding capabilities of the K141A mutant (Ostermaier et al. 2014a; Peterhans et al. 2016). While mutations in the middle loop affect the binding of arrestin to activated and phosphorylated rhodopsin, more drastic changes have been observed for the affinity towards non-preferred receptor species (Vishnivetskiy et al. 2013b). Thus, alanine substitutions in the middle loop region of arrestin generally enhance the binding to phosphorylated opsin (Peterhans et al. 2016), indicating that a functional middle loop determines the specificity of arrestin towards the agonist-bound and activated receptor conformation. Whereas rhodopsin in its activated and phosphorylated conformation is able to interact with the middle loop in order to facilitate its rotation away from the binding interface, other receptor conformations might fail to provoke a necessary rotation of the gate loop which would enable such a structural re-arrangement. These findings mark the middle loop, in its entirety, as an activation-sensing component of arrestin. While binding of activated and phosphorylated rhodopsin, the intracellular loop 2 (ICL 2) of rhodopsin can neatly enter a cleft between the middle loop and the lariat loop, which is generated due to the interlobe rotation of arrestin and the relocation of the two loops (Kang et al. 2015).

Pre-activation of the arrestin molecule also influences the secondary and tertiary structure of the finger loop. The β-strands bordering the loop extend upon re-arrangement of the arrestin molecule to enable a stretching of its conformation towards the receptor (Granzin et al. 2012). The arrestin-1 finger loop is able to engage rhodopsin at a cavity, which opens up due to the conformational change upon receptor activation. In this sense, the shape of the finger loop acts as an indirect activation sensor, as it is only establishing high-affinity binding upon activation-dependent opening of the cytoplasmic receptor crevice. When bound to the receptor, the finger loop adapts a short α-helical conformation (Feuerstein et al. 2009; Kim et al. 2012) in order to interact with the C-terminus of transmembrane (TM) helix 7 as well as the N-terminus of helix 8 of the receptor (Kang et al. 2015).

The residues which are involved in these interactions as well as the structural integrity of the finger loop (Q70, V75, M76, G77 and L78) are crucial for receptor binding and heavily decrease the affinity towards different receptor species upon alanine substitution (Peterhans et al. 2016). The structural fit of the finger loop inside the positively charged cytoplasmic cavity of rhodopsin is additionally supported by a charge complementation effect, as the finger loop features three negatively charged residues, namely E71, D72 and D74 (Kang et al. 2015).

The β-strand following the finger loop (residues 79–86), on the other hand, establishes an additional binding interface between arrestin and the receptor, as it interacts with TM5, TM6 and the connective ICL3. X-ray crystallography validates the positioning of residues F86 and D83 in close proximity to the receptor, thus they are thought to enable these interactions (Kang et al. 2015). Furthermore, the cytoplasmic side of TM5 seems to be enclosed in the central loop area of arrestin, as it is also interacting with residues located in the lariat loop of arrestin [especially Y251 (Ostermaier et al. 2014a)]. These interactions are, again, dependent on the characteristic outward movements of TM5 and TM6, which happen upon rhodopsin activation. In the case of β-arrestin1, hydrogen-deuterium exchange mass spectrometry and chemical cross-linking studies are consistent with a very similar engagement of the corresponding finger, middle and lariat loop for the binding of the $ß_2$-adrenoreceptor chimera (Shukla et al. 2014).

Whereas arrestin pre-activation is highly dependent upon the establishment and location of distinct phosphorylation patterns at the GPCR C-tail, activation-sensing is rather governed by steric barriers, which have to be overcome for the formation of a high-affinity complex. Both, the pre-activation of arrestin as well as the activation of the GPCR, are perquisites, which prime the two proteins by arranging the essential binding interface.

Membrane Interactions of Arrestin-1 and Binding Stoichiometry

The C-edge is another region of arrestin-1, which is thought to equip the protein with characteristics crucial for high-affinity receptor binding (Ostermaier et al. 2014a). Although it is not located in close proximity to any receptor residues of the putative binding interface between arrestin-1 and rhodopsin, alanine substitutions in the solvent-exposed loops of the C-edge strongly influence binding affinities towards various receptor species (Peterhans et al. 2016). A cluster of uncharged or hydrophobic residues, which are comprised within the 344-loop (L338, L339, L342 and S345) as well as the 200-loop (F197, M198 and S199), were shown to influence arrestin-1 binding to the activated and phosphorylated receptor. Due to the recent elucidation of the complex crystal structure and the reported interlobe rotation of arrestin-1, as well as the asymmetric nature of the rhodopsin–arrestin-1 interaction, it is now possible to associate the location of these residues with the hydrophobic interior of the phospholipid membrane. Arrestin-1 in its receptor-bound conformation might thus be able to penetrate the cell membrane and enable an additional anchoring of the complex in the phospholipid bilayer, adjacent to the receptor

(Kang et al. 2015). Interestingly, for phosphorylated opsin binding of arrestin-1, alanine mutations of the 344-loop increase the affinity towards the receptor. Similar findings show, that the residues V268, N271, S272 and L274, located within the 270-loop of the C-edge, strongly influence the specificity of arrestin-1 towards different receptor species. These residues have been shown to increase the affinity of arrestin for the activated and phosphorylated receptor upon alanine substitution while decreasing its affinity towards phosphorylated opsin (Peterhans et al. 2016). As the re-positioning of the C-edge is dependent on the rotation of the N- and C-domain against each other, these findings again imply the existence of highly different arrestin-1 binding conformations for particular receptor activation states.

Still, the exact function of the C-edge and other structures at the outline of the arrestin-1 molecule are not completely understood. There are several hypotheses, which explain the influence of these regions on receptor binding with the existence of further binding interfaces or varying binding stoichiometry. Some studies suggest that the C-edge of arrestin in its receptor-bound conformation might interact with an adjacent receptor implying a process in which arrestin is able to bind to rhodopsin in a two-to-one stoichiometry (Sommer et al. 2012). Furthermore, the 160-loop of arrestin-1 might be seen as an additional binding interface for rhodopsin, as it was shown that it interacts with the central cytoplasmic cavity of rhodopsin (Sinha et al. 2014).

Pharmacological Relevance of Phosphobarcodes for Arrestin-Dependent Signaling

How arrestin-1 interacts with rhodopsin is by far most in-depth studied compared to any other arrestin–GPCR interactions. On top of that studies on the interaction of arrestin-1 with disease-causing rhodopsin mutants have revealed that G90D rhodopsin, which causes congenital stationary night blindness, is phosphorylated to a slightly higher degree than wild type rhodopsin, but demonstrated reduced ability to bind arrestin-1, especially in its phosphorylated opsin state (Vishnivetskiy et al. 2013a; Singhal et al. 2013). Thus the arrestin-1–rhodopsin interaction serves as a framework and guidance towards elucidation of the pharmacologically impactful, diverse and ubiquitous ß-arrestin–GPCR complexes. Research on arrestin-1 had a considerable lead in time over ß-arrestins dating back to the mid 1970s with purification and characterization of the protein (Wacker et al. 1977; Dorey and Faure 1977). In the early 1990s Vsevolod Gurevich has started mutagenesis studies on arrestin-1 and continued in his group until key elements in the activation mechanism of arrestin-1 have been revealed [reviewed in (Gurevich et al. 2011)]. At the same time one of the two ß-arrestins has been discovered as G protein-inhibiting factor (Lohse et al. 1990). ß-arrestins were first considered to only play a role in receptor desensitization and internalization (Ferguson et al. 1996). It was then discovered that they can also activate G protein-independent signaling pathways (Luttrell and Lefkowitz 2002; Charest and Bouvier 2003) and that some ligands activate preferentially one or the other pathway, which resulted in the concept of functional selectivity or biased signaling (Azzi et al. 2003; Shenoy and Lefkowitz 2005). Now, recent evidence suggests that there is much more diversity than a simple dichotomy between G protein-dependent and ß-arrestin-dependent signaling. An analysis of SII and TRV027, two angiotensin AT_1 receptor ligands biased towards ß-arrestin signaling, showed that they lead to distinct patterns of kinase phosphorylation and gene expression (Santos et al. 2015). While it is possible that this diversity is the result of yet unknown effectors, it was also shown that SII and TRV027 lead to distinct conformational changes of ß-arrestins and that this could account for the observed dissimilarity in signaling patterns. In addition, a study using fluorescein arsenical hairpin bioluminescence resonance energy transfer reporters to monitor conformational changes in ß-arrestin2 showed that information about ligand-induced receptor conformation is encoded in changes in ß-arrestin2 conformation (Lee et al. 2016). This provides further insights into how receptors can use ß-arrestins for different purposes and highlights the role of ß-arrestins in reading receptor states and in consecutively encoding this information as distinct ß-arrestin conformations. Moreover, it was shown in recent years that ß-arrestin1 and ß-arrestin2 have non-overlapping and sometimes antagonistic functions and that they can be differentially recruited depending on receptor, ligand and receptor phosphorylation (Srivastava et al. 2015). Together, these findings show the diversity and pharmacological potential of arrestin-dependent signaling and the need of further work in understanding arrestin-receptor interaction that could lead to novel tools for the discovery of more functional selective ligands.

Acknowledgements We thank Dr. Xavier Deupi for fruitful discussions.

References

Arshavsky VY, Dumke CL, Zhu Y, Artemyev NO, Skiba NP, Hamm HE, Bownds MD (1994) Regulation of transducin GTPase activity in bovine rod outer segments. J Biol Chem 269:19882–19887

Azzi M, Charest PG, Angers S, Rousseau G, Kohout T, Bouvier M, Pineyro G (2003) Beta-arrestin-mediated activation of MAPK by inverse agonists reveals distinct active conformations for G protein-coupled receptors. Proc Natl Acad Sci U S A 100:11406–11411

Benovic JL, Kuhn H, Weyand I, Codina J, Caron MG, Lefkowitz RJ (1987) Functional desensitization of the isolated beta-adrenergic receptor by the beta-adrenergic receptor kinase: potential role of an analog of the retinal protein arrestin (48-kDa protein). Proc Natl Acad Sci USA 84:8879–8882

Calebiro D, Nikolaev VO, Gagliani MC, de Filippis T, Dees C, Tacchetti C, Persani L, Lohse MJ (2009) Persistent cAMP-signals triggered by internalized G-protein-coupled receptors. PLoS Biol 7:e1000172

Charest PG, Bouvier M (2003) Palmitoylation of the V2 vasopressin receptor carboxyl tail enhances beta-arrestin recruitment leading to efficient receptor endocytosis and ERK1/2 activation. J Biol Chem 278:41541–41551

Chen L, Jin L, Zhou N (2012) An update of novel screening methods for GPCR in drug discovery. Expert Opin Drug Discov 7:791–806

Congreve M, Marshall F (2010) The impact of GPCR structures on pharmacology and structure-based drug design. Br J Pharmacol 159:986–996

Congreve M, Langmead CJ, Mason JS, Marshall FH (2011) Progress in structure based drug design for G protein-coupled receptors. J Med Chem 54:4283–4311

Congreve M, Dias JM, Marshall FH (2014) Structure-based drug design for G protein-coupled receptors. Prog Med Chem 53:1–63

Deupi X (2014) Relevance of rhodopsin studies for GPCR activation. Biochim Biophys Acta 1837:674–682

DeWire SM, Ahn S, Lefkowitz RJ, Shenoy SK (2007) Beta-arrestins and cell signaling. Annu Rev Physiol 69:483–510

Dorey C, Faure JP (1977) Isolation and characterization of a retinal antigen inducing experimental autoimmune uveo-retinitis. Ann Immunol (Paris) 128:229–232

Egloff P, Hillenbrand M, Klenk C, Batyuk A, Heine P, Balada S, Schlinkmann KM, Scott DJ, Schutz M, Pluckthun A (2014) Structure of signaling-competent neurotensin receptor 1 obtained by directed evolution in Escherichia coli. Proc Natl Acad Sci U S A 111:E655–E662

Feinstein TN, Yui N, Webber MJ, Wehbi VL, Stevenson HP, King JD Jr, Hallows KR, Brown D, Bouley R, Vilardaga JP (2013) Noncanonical control of vasopressin receptor type 2 signaling by retromer and arrestin. J Biol Chem 288:27849–27860

Ferguson SS, Downey WE III, Colapietro AM, Barak LS, Menard L, Caron MG (1996) Role of beta-arrestin in mediating agonist-promoted G protein-coupled receptor internalization. Science 271:363–366

Ferrandon S, Feinstein TN, Castro M, Wang B, Bouley R, Potts JT, Gardella TJ, Vilardaga JP (2009) Sustained cyclic AMP production by parathyroid hormone receptor endocytosis. Nat Chem Biol 5:734–742

Feuerstein SE, Pulvermuller A, Hartmann R, Granzin J, Stoldt M, Henklein P, Ernst OP, Heck M, Willbold D, Koenig BW (2009) Helix formation in arrestin accompanies recognition of photoactivated rhodopsin. Biochemistry 48:10733–10742

Flock T, Ravarani CN, Sun D, Venkatakrishnan AJ, Kayikci M, Tate CG, Veprintsev DB, Babu MM (2015) Universal allosteric mechanism for Galpha activation by GPCRs. Nature 524:173–179

Fredriksson R, Lagerstrom MC, Lundin LG, Schioth HB (2003) The G-protein-coupled receptors in the human genome form five main families. Phylogenetic analysis, paralogon groups, and fingerprints. Mol Pharmacol 63:1256–1272

Granzin J, Cousin A, Weirauch M, Schlesinger R, Buldt G, Batra-Safferling R (2012) Crystal structure of p44, a constitutively active splice variant of visual arrestin. J Mol Biol 416:611–618

Granzin J, Stadler A, Cousin A, Schlesinger R, Batra-Safferling R (2015) Structural evidence for the role of polar core residue Arg175 in arrestin activation. Sci Rep 5:15808

Gurevich VV (1998) The selectivity of visual arrestin for light-activated phosphorhodopsin is controlled by multiple nonredundant mechanisms. J Biol Chem 273:15501–15506

Gurevich VV, Benovic JL (1995) Visual arrestin binding to rhodopsin. Diverse functional roles of positively charged residues within the phosphorylation-recognition region of arrestin. J Biol Chem 270:6010–6016

Gurevich VV, Gurevich EV (2013) Structural determinants of arrestin functions. Prog Mol Biol Transl Sci 118:57–92

Gurevich VV, Hanson SM, Song X, Vishnivetskiy SA, Gurevich EV (2011) The functional cycle of visual arrestins in photoreceptor cells. Prog Retinal Eye Res 30:405–430

Han M, Gurevich VV, Vishnivetskiy SA, Sigler PB, Schubert C (2001) Crystal structure of beta-arrestin at 1.9 Å: possible mechanism of receptor binding and membrane translocation. Structure 9:869–880

Heydenreich FM, Vuckovic Z, Matkovic M, Veprintsev DB (2015) Stabilization of G protein-coupled receptors by point mutations. Front Pharmacol 6:82

Hirsch JA, Schubert C, Gurevich VV, Sigler PB (1999) The 2.8 Å crystal structure of visual arrestin: a model for arrestin's regulation. Cell 97:257–269

Hofmann KP, Pulvermuller A, Buczylko J, Van Hooser P, Palczewski K (1992) The role of arrestin and retinoids in the regeneration pathway of rhodopsin. J Biol Chem 267:15701–15706

Hollenstein K, Kean J, Bortolato A, Cheng RK, Dore AS, Jazayeri A, Cooke RM, Weir M, Marshall FH (2013) Structure of class B GPCR corticotropin-releasing factor receptor 1. Nature 499:438–443

Irannejad R, Tomshine JC, Tomshine JR, Chevalier M, Mahoney JP, Steyaert J, Rasmussen SG, Sunahara RK, El-Samad H, Huang B, von Zastrow M (2013) Conformational biosensors reveal GPCR signalling from endosomes. Nature 495:534–538

Kang Y, Zhou XE, Gao X, He Y, Liu W, Ishchenko A, Barty A, White TA, Yefanov O, Han GW, Xu Q, de Waal PW, Ke J, Tan MH, Zhang C, Moeller A, West GM, Pascal BD, Van Eps N, Caro LN, Vishnivetskiy SA, Lee RJ, Suino-Powell KM, Gu X, Pal K, Ma J, Zhi X, Boutet S, Williams GJ, Messerschmidt M, Gati C, Zatsepin NA, Wang D, James D, Basu S, Roy-Chowdhury S, Conrad CE, Coe J, Liu H, Lisova S, Kupitz C, Grotjohann I, Fromme R, Jiang Y, Tan M, Yang H, Li J, Wang M, Zheng Z, Li D, Howe N, Zhao Y, Standfuss J, Diederichs K, Dong Y, Potter CS, Carragher B, Caffrey M, Jiang H, Chapman HN, Spence JC, Fromme P, Weierstall U, Ernst OP, Katritch V, Gurevich VV, Griffin PR, Hubbell WL, Stevens RC, Cherezov V, Melcher K, Xu HE (2015) Crystal structure of rhodopsin bound to arrestin by femtosecond X-ray laser. Nature 523:561–567

Kim M, Vishnivetskiy SA, Van Eps N, Alexander NS, Cleghorn WM, Zhan X, Hanson SM, Morizumi T, Ernst OP, Meiler J, Gurevich VV, Hubbell WL (2012) Conformation of receptor-bound visual arrestin. Proc Natl Acad Sci U S A 109:18407–18412

Kim YJ, Hofmann KP, Ernst OP, Scheerer P, Choe HW, Sommer ME (2013) Crystal structure of pre-activated arrestin p44. Nature 497:142–146

Kovoor A, Celver J, Abdryashitov RI, Chavkin C, Gurevich VV (1999) Targeted construction of phosphorylation-independent beta-arrestin mutants with constitutive activity in cells. J Biol Chem 274:6831–6834

Kuhn H, Wilden U (1987) Deactivation of photoactivated rhodopsin by rhodopsin-kinase and arrestin. J Recept Res 7:283–298

Laporte SA, Oakley RH, Holt JA, Barak LS, Caron MG (2000) The interaction of beta-arrestin with the AP-2 adaptor is required for the clustering of beta 2-adrenergic receptor into clathrin-coated pits. J Biol Chem 275:23120–23126

Lebon G, Tate CG (2011) Structure of the adenosine-bound conformation of the human adenosine A(2A) receptor. Med Sci (Paris) 27:926–928

Lee MH, Appleton KM, Strungs EG, Kwon JY, Morinelli TA, Peterson YK, Laporte SA, Luttrell LM (2016) The conformational signature of beta-arrestin2 predicts its trafficking and signalling functions. Nature 531:665–668

Lefkowitz RJ (1998) G protein-coupled receptors. III. New roles for receptor kinases and beta-arrestins in receptor signaling and desensitization. J Biol Chem 273:18677–18680

Lohse MJ, Hoffmann C (2014) Arrestin interactions with G protein-coupled receptors. Handb Exp Pharmacol 219:15–56

Lohse MJ, Benovic JL, Codina J, Caron MG, Lefkowitz RJ (1990) Beta-arrestin: a protein that regulates beta-adrenergic receptor function. Science 248:1547–1550

Luttrell LM (2013) Arrestin pathways as drug targets. Prog Mol Biol Transl Sci 118:469–497

Luttrell LM, Lefkowitz RJ (2002) The role of beta-arrestins in the termination and transduction of G-protein-coupled receptor signals. J Cell Sci 115:455–465

Mary S, Damian M, Louet M, Floquet N, Fehrentz JA, Marie J, Martinez J, Baneres JL (2012) Ligands and signaling proteins govern the conformational landscape explored by a G protein-coupled receptor. Proc Natl Acad Sci U S A 109:8304–8309

Mendez A, Burns ME, Roca A, Lem J, Wu LW, Simon MI, Baylor DA, Chen J (2000) Rapid and reproducible deactivation of rhodopsin requires multiple phosphorylation sites. Neuron 28:153–164

Moore CA, Milano SK, Benovic JL (2007) Regulation of receptor trafficking by GRKs and arrestins. Annu Rev Physiol 69:451–482

Nobles KN, Xiao K, Ahn S, Shukla AK, Lam CM, Rajagopal S, Strachan RT, Huang TY, Bressler EA, Hara MR, Shenoy SK, Gygi SP, Lefkowitz RJ (2011) Distinct phosphorylation sites on the beta(2)-adrenergic receptor establish a barcode that encodes differential functions of beta-arrestin. Sci Signal 4:ra51

Nuber S, Zabel U, Lorenz K, Nuber A, Milligan G, Tobin AB, Lohse MJ, Hoffmann C (2016) Beta-arrestin biosensors reveal a rapid, receptor-dependent activation/deactivation cycle. Nature 531:661–664

Oakley RH, Laporte SA, Holt JA, Barak LS, Caron MG (1999) Association of beta-arrestin with G protein-coupled receptors during clathrin-mediated endocytosis dictates the profile of receptor resensitization. J Biol Chem 274:32248–32257

Oakley RH, Laporte SA, Holt JA, Caron MG, Barak LS (2000) Differential affinities of visual arrestin, beta arrestin1, and beta arrestin2 for G protein-coupled receptors delineate two major classes of receptors. J Biol Chem 275:17201–17210

Ohguro H, Rudnicka-Nawrot M, Buczylko J, Zhao X, Taylor JA, Walsh KA, Palczewski K (1996) Structural and enzymatic aspects of rhodopsin phosphorylation. J Biol Chem 271:5215–5224

Ostermaier MK, Peterhans C, Jaussi R, Deupi X, Standfuss J (2014a) Functional map of arrestin-1 at single amino acid resolution. Proc Natl Acad Sci U S A 111:1825–1830

Ostermaier MK, Schertler GF, Standfuss J (2014b) Molecular mechanism of phosphorylation-dependent arrestin activation. Curr Opin Struct Biol 29:143–151

Peterhans C, Lally CC, Ostermaier MK, Sommer ME, Standfuss J (2016) Functional map of arrestin binding to phosphorylated opsin, with and without agonist. Sci Rep 6:28686

Pierce KL, Lefkowitz RJ (2001) Classical and new roles of beta-arrestins in the regulation of G-protein-coupled receptors. Nat Rev Neurosci 2:727–733

Rich RL, Errey J, Marshall F, Myszka DG (2011) Biacore analysis with stabilized G-protein-coupled receptors. Anal Biochem 409:267–272

Robertson N, Jazayeri A, Errey J, Baig A, Hurrell E, Zhukov A, Langmead CJ, Weir M, Marshall FH (2011) The properties of thermostabilised G protein-coupled receptors (StaRs) and their use in drug discovery. Neuropharmacology 60:36–44

Santos GA, Duarte DA, Parreiras ESLT, Teixeira FR, Silva-Rocha R, Oliveira EB, Bouvier M, Costa-Neto CM (2015) Comparative analyses of downstream signal transduction targets modulated after activation of the AT1 receptor by two beta-arrestin-biased agonists. Front Pharmacol 6:131

Shenoy SK, Lefkowitz RJ (2005) Seven-transmembrane receptor signaling through beta-arrestin. Sci STKE 2005:cm10

Shukla AK, Manglik A, Kruse AC, Xiao K, Reis RI, Tseng WC, Staus DP, Hilger D, Uysal S, Huang LY, Paduch M, Tripathi-Shukla P, Koide A, Koide S, Weis WI, Kossiakoff AA, Kobilka BK, Lefkowitz RJ (2013) Structure of active beta-arrestin-1 bound to a G-protein-coupled receptor phosphopeptide. Nature 497:137–141

Shukla AK, Westfield GH, Xiao K, Reis RI, Huang LY, Tripathi-Shukla P, Qian J, Li S, Blanc A, Oleskie AN, Dosey AM, Su M, Liang CR, Gu LL, Shan JM, Chen X, Hanna R, Choi M, Yao XJ, Klink BU, Kahsai AW, Sidhu SS, Koide S, Penczek PA, Kossiakoff AA, Woods VL Jr, Kobilka BK, Skiniotis G, Lefkowitz RJ (2014) Visualization of arrestin recruitment by a G-protein-coupled receptor. Nature 512:218–222

Singhal A, Ostermaier MK, Vishnivetskiy SA, Panneels V, Homan KT, Tesmer JJ, Veprintsev D, Deupi X, Gurevich VV, Schertler GF, Standfuss J (2013) Insights into congenital stationary night blindness based on the structure of G90D rhodopsin. EMBO Rep 14:520–526

Singhal A, Guo Y, Matkovic M, Schertler G, Deupi X, Yan EC, Standfuss J (2016) Structural role of the T94I rhodopsin mutation in congenital stationary night blindness. EMBO Rep 17:1431–1440

Sinha A, Jones Brunette AM, Fay JF, Schafer CT, Farrens DL (2014) Rhodopsin TM6 can interact with two separate and distinct sites on arrestin: evidence for structural plasticity and multiple docking modes in arrestin-rhodopsin binding. Biochemistry 53:3294–3307

Sommer ME, Hofmann KP, Heck M (2012) Distinct loops in arrestin differentially regulate ligand binding within the GPCR opsin. Nat Commun 3:995
Srivastava A, Gupta B, Gupta C, Shukla AK (2015) Emerging functional divergence of beta-arrestin isoforms in GPCR function. Trends Endocrinol Metab 26:628–642
Sun D, Ostermaier MK, Heydenreich FM, Mayer D, Jaussi R, Standfuss J, Veprintsev DB (2013) AAscan, PCRdesign and MutantChecker: a suite of programs for primer design and sequence analysis for high-throughput scanning mutagenesis. PLoS ONE 8:e78878
Sun D, Flock T, Deupi X, Maeda S, Matkovic M, Mendieta S, Mayer D, Dawson RJ, Schertler GF, Babu MM, Veprintsev DB (2015) Probing Galphai1 protein activation at single-amino acid resolution. Nat Struct Mol Biol 22:686–694
Thomsen AR, Plouffe B, Cahill TJ III, Shukla AK, Tarrasch JT, Dosey AM, Kahsai AW, Strachan RT, Pani B, Mahoney JP, Huang L, Breton B, Heydenreich FM, Sunahara RK, Skiniotis G, Bouvier M, Lefkowitz RJ (2016) GPCR-G protein-beta-arrestin super-complex mediates sustained G protein signaling. Cell 166:907–919
Tian X, Kang DS, Benovic JL (2014) Beta-arrestins and G protein-coupled receptor trafficking. Handb Exp Pharmacol 219:173–186
Tran TM, Friedman J, Qunaibi E, Baameur F, Moore RH, Clark RB (2004) Characterization of agonist stimulation of cAMP-dependent protein kinase and G protein-coupled receptor kinase phosphorylation of the beta2-adrenergic receptor using phosphoserine-specific antibodies. Mol Pharmacol 65:196–206
Unger VM, Hargrave PA, Baldwin JM, Schertler GF (1997) Arrangement of rhodopsin transmembrane alpha-helices. Nature 389:203–206
Vishnivetskiy SA, Schubert C, Climaco GC, Gurevich YV, Velez MG, Gurevich VV (2000) An additional phosphate-binding element in arrestin molecule. Implications for the mechanism of arrestin activation. J Biol Chem 275:41049–41057
Vishnivetskiy SA, Raman D, Wei J, Kennedy MJ, Hurley JB, Gurevich VV (2007) Regulation of arrestin binding by rhodopsin phosphorylation level. J Biol Chem 282:32075–32083
Vishnivetskiy SA, Ostermaier MK, Singhal A, Panneels V, Homan KT, Glukhova A, Sligar SG, Tesmer JJ, Schertler GF, Standfuss J, Gurevich VV (2013a) Constitutively active rhodopsin mutants causing night blindness are effectively phosphorylated by GRKs but differ in arrestin-1 binding. Cell Signal 25:2155–2162
Vishnivetskiy SA, Baameur F, Findley KR, Gurevich VV (2013b) Critical role of the central 139-loop in stability and binding selectivity of arrestin-1. J Biol Chem 288:11741–11750
Wacker WB, Donoso LA, Kalsow CM, Yankeelov JA Jr, Organisciak DT (1977) Experimental allergic uveitis. Isolation, characterization, and localization of a soluble uveitopathogenic antigen from bovine retina. J Immunol 119:1949–1958
Warne T, Serrano-Vega MJ, Baker JG, Moukhametzianov R, Edwards PC, Henderson R, Leslie AG, Tate CG, Schertler GF (2008) Structure of a beta1-adrenergic G-protein-coupled receptor. Nature 454:486–491
Weller M, Virmaux N, Mandel P (1975) Light-stimulated phosphorylation of rhodopsin in the retina: the presence of a protein kinase that is specific for photobleached rhodopsin. Proc Natl Acad Sci U S A 72:381–385
White JF, Noinaj N, Shibata Y, Love J, Kloss B, Xu F, Gvozdenovic-Jeremic J, Shah P, Shiloach J, Tate CG, Grisshammer R (2012) Structure of the agonist-bound neurotensin receptor. Nature 490:508–513
Yang F, Yu X, Liu C, Qu CX, Gong Z, Liu HD, Li FH, Wang HM, He DF, Yi F, Song C, Tian CL, Xiao KH, Wang JY, Sun JP (2015) Phospho-selective mechanisms of arrestin conformations and functions revealed by unnatural amino acid incorporation and (19)F-NMR. Nat Commun 6:8202

Chapter 8
How Arrestin Recognizes and Binds Active GPCRs

Martha E. Sommer

Abstract Arrestins are structurally flexible and functionally versatile proteins that regulate the activity of hundreds of different G protein-coupled receptors (GPCRs). A hallmark of arrestin function is that these proteins are only activated for binding the active receptor upon interaction with receptor-attached phosphate groups. Recent years have yielded crystal structures of pre-activated arrestins and arrestin in complex with an active receptor, which provide insight into the arrestin activation mechanism. At the same time, functional studies indicate that arrestin employs different binding modes along the path to tight receptor binding, and the structure of the arrestin-receptor complex is modulated by the activation and phosphorylation state of the active receptor. In this chapter we discuss our current understanding of the receptor-binding mechanism of arrestin, from the initial interaction with the phosphorylated receptor to the structural transformation required for tight binding to the active receptor.

Keywords Arrestin · Rhodopsin · GPCR · Fluorescence spectroscopy · Mutagenesis

About 800 different G protein-coupled receptors (**GPCR**s) are expressed in the human body and play a central role in multiple sensory and physiological systems. These receptors bind a wide variety of ligands and share a common structure of seven transmembrane (TM) helices. The receptor exists in a conformational equilibrium comprised of multiple states of varying degrees of activity. In the basal inactive state, the helical bundle is bound together by several hydrogen bond networks and electrostatic interactions. The binding of agonist favours an active receptor conformation, in which the cytoplasmic face of the receptor is open for interaction with cytosolic proteins. The binding partners of the receptor are namely G protein, GPCR kinase (GRK), and arrestin.

M.E. Sommer (✉)
Institute of Medical Physics and Biophysics (CC2),
Charité Medical University, Berlin, Germany
e-mail: martha.sommer@charite.de

V.V. Gurevich (ed.), *The Structural Basis of Arrestin Functions*,
DOI 10.1007/978-3-319-57553-7_8

G proteins consist of three subunits, termed α, β and γ, and different subunit combinations give rise to about 20 distinct G proteins. Binding of the heterotrimeric G protein to the active receptor induces nucleotide exchange within the G protein and dissociation of the subunits, which then interact with other components of the cell signalling machinery. In essence, the receptor transduces the signal of agonist binding across the cell membrane, and the G protein amplifies this signal within the cytoplasm.

Seven different GRKs exist, each of which phosphorylates a different subset of GPCRs. The kinase activity of GRKs is stimulated by interaction with activated GPCRs. GRKs phosphorylate several serine and threonine residues within the cytoplasmic C-terminal tail of the receptor. Some receptors are also phosphorylated on their cytoplasmic loops. Only three receptor-attached phosphates are required to stimulate arrestin binding to the active GPCR (Vishnivetskiy et al. 2007), yet most GPCRs contain 7 or more phosphorylation sites. The extra sites confer another level of regulation, as different phosphorylation patterns (also called "barcodes") control the multiple functions of arrestins [reviewed in Reiter et al. (2012)]. Different phosphorylation barcodes arise from different C-terminal primary sequence, tissue-specific or ligand-selected GRK activity, or sequential rounds of receptor activation.

The arrestin family is composed of only four members. Arrestin-1 and arrestin-4, respectively called rod and cone arrestin, are expressed primarily in the retina and interact with light-sensitive GPCRs called opsins. Arrestin-2 and arrestin-3, respectively called β-arrestin 1 and β-arrestin 2, are expressed ubiquitously and interact with hundreds of different GPCRs. This attribute means the β-arrestins must be versatile and relatively promiscuous binding partners. A primary role of arrestin binding is to block G protein coupling and thereby stop GPCR signalling. The β-arrestins are additionally able to induce receptor internalization by recruiting clathrin and other elements of the cellular internalization machinery, as well as mediate their own signalling cascades by scaffolding signalling kinases. In recent years evidence has emerged that the pattern of receptor phosphorylation affects both the conformation of bound arrestin (Shenoy et al. 2009; Nobles et al. 2011) as well as the stability of the arrestin-GPCR complex (Zindel et al. 2014; Oakley et al. 2001; Tohgo et al. 2003). These factors presumably control arrestin-mediated receptor internalization and trafficking (Oakley et al. 2001; Zindel et al. 2014) and arrestin-mediated MAP kinase activity (Tohgo et al. 2003; Ren et al. 2005; Nobles et al. 2011) [reviewed in Tobin et al. (2008)].

Arrestins are composed of two cup-like domains composed of β-sheets, called the N- and C-domain (Fig. 8.1). The tips of the domains are capped with flexible loops, which compose the **central crest** (finger loop, middle loop and C-loop) and the **C-edge** (344-loop and 197-loop). The interdomain interface is stabilized by hydrophobic interactions and hydrogen-bond networks. These include the polar core (Han et al. 2001), the YKS(N)D(A) motif (Kim et al. 2013), the middle loop (*a.k.a.* 139-loop) (Vishnivetskiy et al. 2013), and the long Loop 17-18 (Kim et al. 2013; Han et al. 2001), which winds between the two domains and makes multiple contacts with both the N- and C-domains. Loop 17-18 also includes the gate loop, which is a functional part of the polar core (Fig. 8.1). A third functionally critical domain of arrestin is the auto-inhibitory C-terminal tail (**C-tail**). The *distal C-tail* contains a large number of acidic residues and is highly dynamic, as it is not visualized in any known structure of arrestin. In contrast, the *proximal C-tail* is stably anchored to the body of the N-domain by the hydrophobic 3-element interaction (Vishnivetskiy et al. 2000) and a salt bridge within the polar core. Together, the interdomain interface and the C-tail restrict flexibility within arrestin, namely of the central crest loops and between the N- and C-domains, and thereby prevent interaction with the active receptor.

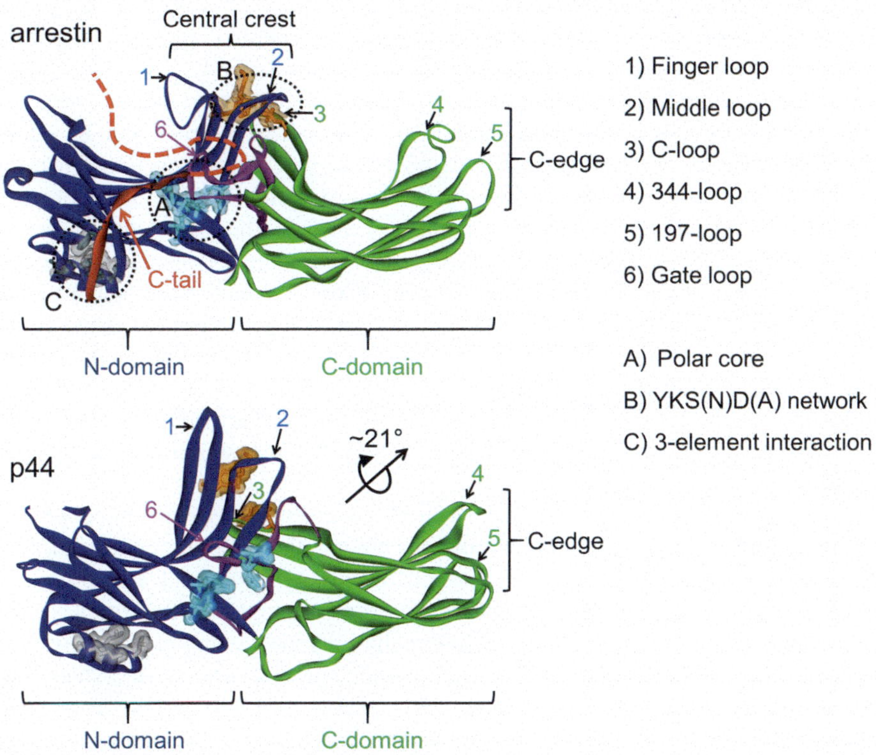

Fig. 8.1 Crystal structures of basal arrestin (PDB code 1CF1, molecule D) and pre-activated arrestin p44 (PDB code 4J2Q, molecule B) (Kim et al. 2013; Hirsch et al. 1999). The N-domain is coloured *blue* and the C-domain is coloured *green*. The C-tail of arrestin is *red*, and Loop 17-18 is *magenta*. Note that only the proximal C-tail is resolved in the arrestin crystal structure, and the unresolved dynamic distal portion is illustrated with a *red dashed line*. The loop linking the C-domain to the C-tail is also not resolved in the arrestin crystal structure. Notable loops that are discussed in the text are labelled with numbers, which are defined in the legend to the *right*. Important intramolecular interactions that stabilize the basal state are illustrated by sidechains with Van der Waal surfaces and labelled with capital letters (defined in legend). Note how these networks are broken in the p44 structure, which results in a ~21° rotation of the C-domain compared to its orientation in the basal arrestin

It has long been believed that arrestin activation entails a displacement of the C-tail by the phosphorylated receptor C-terminus (**Rpp**), based on changes in protease susceptibility (Palczewski et al. 1991a, c), distances measured between spin-labels (Hanson et al. 2006; Vishnivetskiy et al. 2010), and changes in the NMR spectrum of arrestin (Zhuang et al. 2010, 2013). This theory is consistent with the phosphorylation-independent receptor binding behaviour of **p44**, a naturally occurring splice variant of arrestin-1 that lacks the entire C-tail (Palczewski et al. 1994; Pulvermüller et al. 1997). Supposedly C-tail displacement by the Rpp breaks the polar core and loosens the intramolecular interactions that hold arrestin in an inactive state. Two crystal structures of pre-active arrestin, p44 (Kim et al. 2013) and arrestin-2 bound to a peptide analogue of the Rpp (Shukla et al. 2013), indicate the conformational changes that occur in arrestin upon full C-tail displacement (Fig. 8.1). These changes include a twisting displacement of the gate loop that breaks the polar core, release of the central crest loops from their restricted basal conformations, and a 21° rotation of the two domains relative to one another.

In the recently published crystal structure of arrestin-1 in complex with an active GPCR (Kang et al. 2015), it is clear how these structural changes facilitate arrestin binding of the active receptor (discussed in more detail below). The complex consists of a constitutively active mutant of human rhodopsin fused on its N-terminus to T4 lysozyme and fused on its C-terminus to a constitutively active mutant of mouse arrestin-1 via a 15 amino-acid linker. Although the receptor in this structure lacks ligand (opsin), it is obviously in an active state (Ops*). Likewise, the arrestin in this structure shows all the hallmarks of activation seen in the pre-activated arrestin structures (Kim et al. 2013; Shukla et al. 2013), even though the receptor in the complex is not phosphorylated. Despite the artificial nature of the Ops*/arrestin-1 fusion complex structure, it provides a first glimpse of arrestin bound tightly to an active receptor.

This chapter focuses on the molecular mechanisms by which arrestin couples to an active, phosphorylated GPCR. We discuss the initial interaction of arrestin with the phosphorylated receptor, which we call the pre-complex, and the conformational changes in arrestin involved in formation of the high-affinity complex. We have gained insight into these functionally distinct complexes using the long-studied GPCR rhodopsin and its binding partner arrestin-1, which are expressed in the rod cells of the retina and mediate dim-light vision. Dark-state inactive rhodopsin consists of the aporeceptor opsin and a covalently attached inverse agonist, 11-*cis*-retinal. Light absorption converts the ligand to the agonist all-*trans*-retinal, which results in the active receptor form Metarhodopsin II (**Meta II**). Meta II is phosphorylated by GRK1 on its cytoplasmic C-terminal tail. As an experimental model of the pre-complex, we study the interaction of arrestin-1 with phosphorylated inactive forms of the receptor, specifically dark-state phosphorylated rhodopsin (**Rho-P**) and the phosphorylated aporeceptor opsin (**Ops-P**). Using these receptor forms, we can experimentally probe arrestin-receptor interactions that occur before full arrestin activation and formation of the high-affinity complex. Transition to the high-affinity complex is induced by irradiation of the sample with visible light, which converts Rho-P to Meta II-P. Methods we have used to study arrestin-receptor interactions include site-directed fluorescence spectroscopy, Fourier transform infrared spectroscopy (FTIR), and alanine scan mutagenesis. The functional insights we gain from these studies are interpreted within the context of the crystal structures of different arrestins in basal (Hirsch et al. 1999; Granzin et al. 1998; Han et al. 2001; Granzin et al. 2012), pre-activated (Granzin et al. 2015; Kim et al. 2013; Shukla et al. 2013), and fully activated states (Kang et al. 2015). This approach has yielded valuable insights that allow us to propose a mechanistic model of arrestin activation and receptor coupling.

The Pre-complex

The initial interaction of the phosphorylated receptor and arrestin recruits arrestin to the active receptor and activates arrestin for binding the helical bundle of the receptor. Currently very little is known regarding the structure and organization of the pre-complex. Experimental evidence suggests the primary sites of interaction are the phosphorylated receptor C-terminus (Gurevich and Benovic 1993, 1995; Vishnivetskiy et al. 2000; Hanson and Gurevich 2006; Peterhans et al. 2016) and the membrane (Lally et al. 2017). The pre-complex was first proposed by Alexander Pulvermüller, Klaus Peter Hofmann and colleagues (IMPB Charité, Berlin) based on kinetic binding experiments comparing arrestin-1 and p44 (Schröder et al. 2002). This proposal was based on the conclusion that p44 can bypass the initial binding step, which serves normally to displace the C-tail of arrestin. Subsequently two independent studies have presented time-resolved fluorescence data supporting a multi-step binding mechanism, in which arrestin is recruited to the receptor before undergoing a significant conformational change (Kirchberg et al. 2011; Nuber et al. 2016). We interpret these two stages to represent, first formation of the pre-complex, and second, transition to the high-affinity complex.

It has been long known that arrestins intrinsically bind poly-anions, including heparin and inositol phosphate (Palczewski et al. 1991b), as well as receptor-attached phosphate groups [but not simple anions like phosphate (Wilson and Copeland 1997)]. The concave surfaces of all four arrestin subtypes are studded with basic residues that could serve as binding sites for poly-anions. Crystallographic analysis indicated that two molecules of IP6 bind arrestin-2, one each within the concave surfaces of the N- and C-domains (Milano et al. 2006). IP6 binding facilitates homo- and hetero-dimerization of arrestin-2 and arrestin-3. IP6 also binds arrestin-1, although with much lower affinity [K_D in the range of 0.2–160 μM (Palczewski et al. 1991b; Wilson and Copeland 1997; Zhuang et al. 2010)]. NMR analysis indicated that IP6 binds within the N-domain of arrestin-1 (Zhuang et al. 2010). This NMR analysis did not detect any binding of IP6 within the C-domain, although site-directed fluorescence experiments suggest a second low-affinity IP6 binding site, most likely in the C-domain (Sommer, unpublished work).

Mutagenesis has long been applied to arrestin to identify potential receptor-interacting sites, and to differentiate which sites are sensitive to the phosphorylation and activation states of the receptor [reviewed in (Gurevich et al. 2011)]. Alanine scan mutagenesis is an unbiased and comprehensive method that has been recently applied to arrestin-1 by the group of Joerg Standfuss (PSI, Switzerland) to identify sites of interaction and functionally important areas that undergo structural changes during arrestin activation and receptor binding (Ostermaier et al. 2014a; Peterhans et al. 2016). For the most part, the alanine-scan studies confirmed the findings from previous mutagenesis studies (Ostermaier et al. 2014a, b). In collaboration with the Standfuss group, we found evidence that the phosphorylated receptor C-terminus binds arrestin differently in the pre-complex and the high-affinity complex (Fig. 8.2). For the interaction of arrestin-1 with Ops-P, which mirrors the pre-complex, mutation of ten different basic residues within the cup of the N-domain significantly decreased arrestin-1 binding to Ops-P.

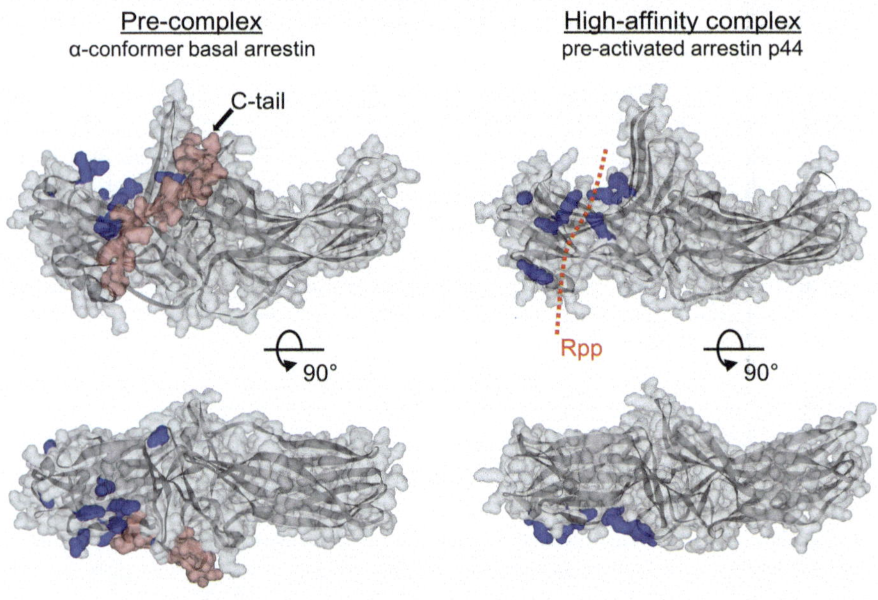

Fig. 8.2 Different "Rpp-binding footprints" for the pre-complex and the high-affinity complex. Alanine scan mutagenesis identified different sets of positively charged residues used in binding the phosphorylated receptor C-terminus (Rpp) in binding inactive Ops-P (pre-complex, *left*) and light-activated Meta II-P (high-affinity complex, *right*) (Peterhans et al. 2016). The implicated residues are plotted in *blue* on the structures of basal arrestin (α-conformer, PDB code 1CF1) for the pre-complex and p44 (PDB code 4J2Q). Note that for basal arrestin, the distal C-tail (*rose*) does not block access to the implicated phosphate-binding residues, which is consistent with our hypothesis that only the proximal C-tail is displaced in the pre-complex. The putative Rpp-binding crevice that was identified in the crystal structure of activated arrestin-2 bound to a peptide analogue of the Rpp (Shukla et al. 2013) is indicated by a *dashed red line*. Lateral views of arrestin and p44 are shown in the *top panels*, and top-down views (showing receptor binding interface) are presented in the *bottom panels*

The large number of implicated phospho-sensing residues is congruent with the fact that the affinity of arrestin-1 for Ops-P is directly proportional to the degree of receptor phosphorylation (i.e. the more phosphates per receptor, the higher the affinity) (Vishnivetskiy et al. 2007). This binding mode stands in contrast to that used for Meta II-P, in which fewer residues along the lateral side of the N-domain are implicated (Fig. 8.2). The Rpp binding mode utilized by Meta II-P is discussed in more detail in the section below.

If Rpp binds only within the cup of the N-domain, then it is possible that only the distal C-tail of arrestin is displaced in the pre-complex. This assumption is supported by site-directed fluorescence experiments using a mutant designed to monitor the relative position of the gate loop and, by extension, the state of the polar core (Kim et al. 2013; Peterhans et al. 2016). These experiments suggest that the polar core is not broken and thus the proximal C-tail is likely not displaced when arrestin is bound to Ops-P in a pre-complex. Only upon activation of the receptor and transition to the high-affinity complex is the polar core broken. Furthermore, site-directed fluorescence experiments employing mutants that report on other conformational changes associated with activation (e.g. release of central crest loops and interdomain rotation) suggest that arrestin is bound to the receptor in a conformational state resembling the basal state in the pre-complex (Lally and Sommer, unpublished work). This supposition is consistent with differences seen in the functional maps for pre-complex and high-affinity complex derived from alanine-scan mutagenesis (Peterhans et al. 2016). Collectively the fluorescence and mutagenesis data argue against the common assumption that Rpp binding in the pre-complex displaces the entire C-tail and induces activating conformational changes in arrestin (Shukla et al. 2013; Kang et al. 2015; Schröder et al. 2002).

The other site of interaction in the pre-complex is the anchoring of the C-edge of arrestin in the membrane. This interaction has recently been characterized by our group using site-directed fluorescence spectroscopy (Lally et al. 2017). Briefly, the proximity of bimane fluorophores placed at specific sites on loops within the C-edge to the membrane was probed using spin-labelled fatty acids, which were incorporated into native rod outer segment membranes containing phosphorylated rhodopsin. Spin-labels quench bimane fluorescence when in close proximity, and fatty acids with spin label at different positions on the acyl chain allowed the differentiation of deep and shallow membrane anchoring. Our results clearly show that the 344-loop embeds deep within the hydrophobic layer of the membrane when arrestin is bound to dark-state Rho-P in a pre-complex. Furthermore, membrane anchoring is dependent on the presence of phosphorylated receptor, suggesting that Rpp binding either activates the C-edge of arrestin for membrane anchoring or is required to recruit arrestin to the membrane. The relative levels of quenching for neighbouring sites suggest an extended 344-loop conformation (Lally et al. 2017), similar as in the "α-conformer" of the basal arrestin-1 structure (Hirsch et al. 1999).

Membrane anchoring in the pre-complex involves the embedding of leucine residues on the 344-loop within the hydrophobic interior of the membrane. Intriguingly, rhodopsin flips negatively charged acidic phospholipids to the cytoplasmic side upon activation (Hessel et al. 2000, 2001), and acidic phospholipids have been shown by us and others to be necessary for arrestin-1 binding of light-activated phosphorylated rhodopsin (Sommer et al. 2006; Bayburt et al. 2011). Furthermore, we have failed to detect any pre-complex formation of arrestin-1 with dark-state Rho-P in the absence of acidic phospholipids (Sommer, unpublished work). These data indicate that acidic phospholipids are required for arrestin to interact with the phosphorylated receptor. Notably, several basic residues line the C-domain (e.g. K235, K236, K238, K267, K330, and K332 in bovine arrestin-1), which we hypothesize help attract the C-edge of arrestin for membrane anchoring.

To summarize the experimental evidence, in the pre-complex arrestin is bound by the Rpp within the cup of the N-domain and by the membrane at the C-edge (Fig. 8.3). Given the relatively deep insertion of the membrane anchor, the peripheral engagement of the Rpp, and the fact that the central crest loops do not engage the receptor, arrestin is likely loosely associated with the receptor in the pre-complex. A great degree of rotational freedom and a highly dynamic complex is expected. Interestingly, a "hanging" interaction mode of arrestin-2 with a chimeric phosphorylated GPCR was observed by negative-stain electron microscopy (Shukla et al. 2014). The authors of this study hypothesized this complex represents tethering of arrestin solely by the Rpp. Due to the artificial nature of the Fab-stabilized

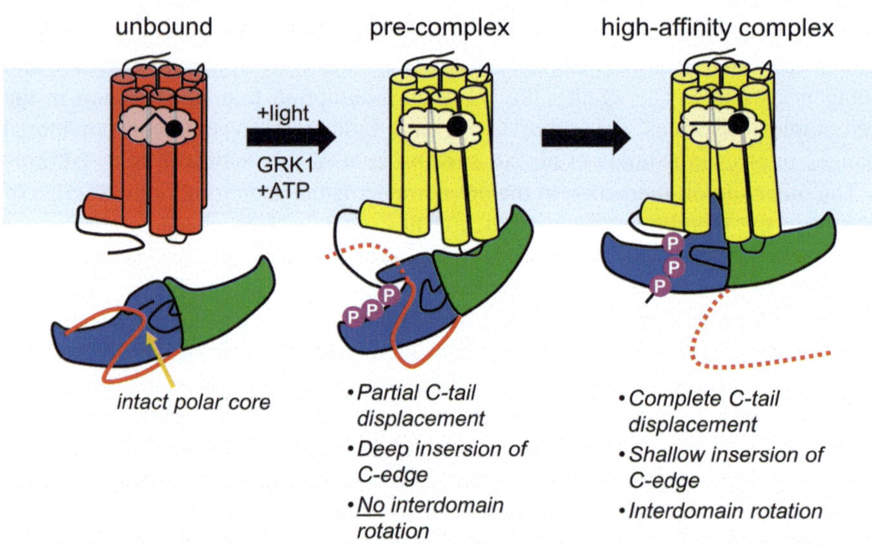

Fig. 8.3 Model of the pre-complex and high-affinity complex. Binding of arrestin to rhodopsin is illustrated. Basal arrestin (*blue* N-domain and *green* C-domain) has an intact polar core and does not interact with dark-state rhodopsin (*red*). Upon receptor light-activation (Meta II, *yellow*) and phosphorylation by GRK, arrestin interacts in a pre-complex that displaces the distal C-tail and embeds the C-edge membrane deep in the membrane. However, no significant activating conformational changes (e.g. interdomain rotation) take place. Upon transition to the high-affinity complex, the Rpp moves from the cup of the N-domain to the putative Rpp-binding trench located along the side of the N-domain. This movement displaces the entire C-tail and results in significant conformational changes linked to interdomain rotation

complexes and the absence of membrane (receptors were solubilized in a neutral detergent), we postulate the visualized complexes represent the pre-complex without the benefit of membrane anchoring. Hence future investigations of the overall organization and structure of the pre-complex should include membrane or a membrane mimic in order to engage the C-edge of arrestin.

Notably, pre-complex formation as the initial interaction of arrestin and the receptor is fairly nonspecific and likely similar for all GPCRs. The interaction with the Rpp and the membrane is primarily electrostatic and hydrophobic, respectively. The nonspecific nature of these interactions allow arrestins to interact with a wide variety of GPCRs, since active receptors are normally phosphorylated in their active state, and all receptors reside in a membrane. This attribute makes sense for the β-arrestins (arrestin-2 and arrestin-3), which couple to hundreds of different GPCRs.

However, the functional relevance of the pre-complex is not limited to being an obligate step before forming the high-affinity complex. Arrestin also forms complexes with phosphorylated inactive receptors, which occur when the receptor returns to an inactive state after ligand dissociation and before receptor dephosphorylation occurs (Lee et al. 2010; Sommer et al. 2014; Zhuang et al. 2013). Inactive receptors in the vicinity of active receptors can also be phosphorylated by activated GRKs, a phenomenon called high-gain phosphorylation (Shi et al. 2005; Binder et al. 1990, 1996). The interaction of arrestin with inactive phosphorylated receptors could play a significant physiological role. For example, in the visual system binding of arrestin-1 to Ops-P could quench the residual G protein-activating ability of opsin (Sommer et al. 2014; Zhuang et al. 2013). In addition, our group discovered that arrestin-1 facilitates the uptake of all-*trans*-retinal in Ops-P (Sommer et al. 2012), which could limit the build-up of potentially toxic levels of retinal in rod cells in bright continuous light (Sommer et al. 2014). The fact that most GPCRs have seven or more phosphorylation sites suggests the interaction of arrestins with highly phosphorylated aporeceptors might be prevalent within the larger GPCR family and could play a role in regulating ligand binding affinity (Gurevich et al. 1997).

Transition to the High-Affinity Complex

Pre-complex formation serves to recruit arrestin to the receptor, where arrestin is anchored at the N-domain by the Rpp and at the C-edge by the membrane (Fig. 8.3). Although arrestin is not fully activated in the pre-complex, the displacement of the distal C-tail increases flexibility and mobility of the central crest loops. Importantly, pre-complex formation brings the receptor-binding elements of arrestin close to the helical core of the receptor. Hence the two proteins can interact with one another as they sample their individual conformational spaces. Transition from the pre-complex to the high-affinity complex involves a substantial intramolecular conversion in arrestin, which mutually stabilizes the active form of the receptor. Agonist binding favours an active conformation of the receptor, and engagement by arrestin further stabilizes this active conformation. These conclusions are supported by spectroscopic studies of arrestin-1 binding to light-activated phosphorylated rhodopsin, which are briefly summarized below.

Over twenty five years ago, Klaus Peter Hofmann and co-workers developed an absorbance-based assay called "extra Meta II" to observe the formation of the high-affinity arrestin-1/Meta II-P complex (Schleicher et al. 1989). Light-activation of rhodopsin results in an equilibrium of two species, the active state Meta II and its inactive precursor Meta I. At low temperature and high pH (e.g. 2 °C and pH 8), the Meta I photoproduct is heavily favoured. When the receptor is phosphorylated and arrestin-1 interacts, Meta II is stabilized at the expense of Meta I. In other words, arrestin-1 binding shifts the Meta I ↔ Meta II equilibrium in favour of Meta II. Importantly, the extra Meta II assay can be used to quantitatively describe the formation of the high-affinity arrestin-receptor complex. Arrestin-1 activation and formation of the high-affinity complex entails a high activation energy (140 kJ/mol). For p44, the energy required for Meta II-P binding is halved (70 kJ/mol) (Pulvermüller et al. 1997), suggesting that about half the energy of activation is involved in releasing the intramolecular restrictions of the C-tail. At sufficiently high concentrations, Meta II-P binding by arrestin-1 and p44 occur at the same rate, which suggests that both arrestin-1 and p44 undergo the same intramolecular conversion, and that this step is rate-limiting for formation of the high-affinity complex (Schröder et al. 2002). Our group has recently characterized this interaction using FTIR spectroscopy (Beyriere et al. 2015). The observed spectral signatures confirmed the stabilization of the active Meta II species by arrestin-1 and additionally indicated a loss of beta-sheet, most likely arising from changes in the secondary structure of arrestin-1. Importantly, Meta II stabilization and beta-sheet loss in arrestin occurred at the same rate, indicating that both proteins mutually stabilize one another in their active states during formation of the high-affinity complex. Moreover, p44 and arrestin showed the same rates of Meta II binding and spectral signatures of beta-sheet loss, indicating that they undergo the same structural transition during high-affinity complex formation (Beyriere et al. 2015).

The comparison of the structure of basal arrestin to that in the Ops*/arrestin-1 fusion complex indicates the intramolecular conversions that arrestin undergoes during tight receptor binding. Below we summarize the major conformational changes, and how these changes facilitate coupling to the active phosphorylated receptor:

Displacement of the proximal C-tail—In the pre-complex, the Rpp is bound within the cup of the N-domain, and transition to the high-affinity complex is accompanied by a movement of the Rpp to a positively charged binding trench on the lateral side of the N-domain (Fig. 8.3). This trench is normally obscured by the proximal C-tail, and the Rpp presumably displaces the C-tail. A peptide analogue of the Rpp has been observed by X-ray protein crystallography to bind within this trench (Shukla et al. 2013). This crystal structure shows that peptide-attached phosphate groups interact with many of the residues identified as phospho-sensors for Meta II-P binding (e.g. K14, K15, R29, K110, K300) (Fig. 8.2) (Peterhans et al. 2016). Displacement of the C-tail and binding of the Rpp induces a forward twisting movement of the gate loop, which breaks the polar core and brings a phosphate-binding lysine residue on the gate loop (K300 in arrestin-1, K294 in arrestin-2) into the Rpp-binding trench.

Interdomain rotation—The binding of the Rpp within the putative binding trench is associated with a movement of the gate loop. This conformational change is transmitted along the entire length of Loop 17-18, which winds between the N- and C-domains and forms multiple contacts that stabilize the basal-state interdomain interface (Fig. 8.1). Loop 17-18 is pulled down and away from the central crest region, thereby causing a lateral displacement of the middle loop away from the body of arrestin (Kim et al. 2012; Kang et al. 2015) and a movement of the C-loop (Y247-Y254 in arrestin-1) down and away from the central crest. These movements break the YKS(N)D(A) network, which stabilizes basal arrestin by linking the C-loop (Y247-Y254 in arrestin-1) within the C-domain to the middle and finger loops within the N-domain. Along with the breaking of the polar core, these conformational rearrangements dramatically increase flexibility between the two domains of arrestin. Hence a 21° rotation of the C-domain is observed in the crystal structures of pre-activated and receptor-bound arrestin (Kim et al. 2013; Shukla et al. 2013; Kang et al. 2015). Interdomain rotation has two major effects on the arrestin structure: a crevice opens within the central crest, and the C-edge adopts a different orientation and conformation. These changes facilitate arrestin binding to the receptor and the adjacent membrane.

Opening of receptor-binding crevice—The lateral displacement of the middle loop and the downward movement of the C-loop opens a cleft that accommodates intercellular loop 2 (CL2) of the receptor. In the Ops*/arrestin-1 fusion complex structure, CL2 forms a short helix that makes hydrophobic and van der Waals contacts with the middle loop and C-loops of arrestin. In addition to this interaction, the downward displacement Loop 17-18 opens space for the TM5/6 bundle. In the crystal structure of the complex, a few hydrogen bond contacts are present between TM5/6 and arrestin. In essence, coupling to the helical bundle of the receptor is a direct result of interdomain rotation that allows a deformation of the flexible loops of arrestin. Furthermore, the sparseness of specific hydrogen bonds or salt bridges within the arrestin-receptor interface hints at how the β-arrestins are able to bind so many different GPCRs. Notably, the interdomain hydrogen bond networks of arrestin-2 and arrestin-3 are much weaker than in arrestin-1 (Kim et al. 2013), meaning these arrestins are already "half-way on" (Gurevich et al. 2011) and explains their much lower dependence on receptor phosphorylation for activation and receptor binding.

Engagement of the finger loop—In the pre-complex, the displacement of the distal C-tail by the Rpp likely increases mobility of the finger loop (Hanson et al. 2008). The flexible finger loop is positioned close to the open cytoplasmic face of the receptor, where it can explore the binding cavity. It is possible that binding of the finger loop and stabilization of the active state of the receptor follows a similar stepwise and mutual reduction of conformational space as recently described for G protein (Elgeti et al. 2013). In the Ops*/arrestin-1 fusion complex structure, the finger loop is observed to adopt a helical conformation within the cytoplasmic crevice of the active receptor (Kang et al. 2015). A similar binding mode was observed in a crystal structure of Ops* bound to a peptide analogue of the finger loop (Szczepek et al. 2014). In this structure, specific hydrogen bonding is observed between the finger loop and critical functional motifs on the receptor (e.g. NPxxY $(x)_{5,6}$F and E(D)RY), which are highly conserved among GPCRs. Consistently, the finger loop sequence is highly conserved among all four arrestins. Engagement of the finger loop is crucial for stabilizing the active form of the receptor (Sommer et al. 2012) and hence formation of the high affinity complex.

Membrane anchoring of the C-edge—Rotation of the C-domain results in a different alignment of the C-edge that changes how this functional domain interacts with the membrane as compared to the pre-complex. In the pre-complex, the 344-loop adopts an extended conformation and is deeply inserted into the hydrophobic layer of the membrane. In the high-affinity complex, this loop adopts a folded conformation as seen in the p44 structure and interacts more shallowly with the membrane. In addition, the rotation of the C-domain allows the 197-loop to engage the membrane in the high-affinity complex. These differences in C-edge membrane engagement in the pre-complex and high-affinity complex were clearly observed in site-directed fluorescence experiments (Lally et al. 2017) and are consistent with differences in functional maps for Meta II-P binding and Ops-P binding derived from alanine scan mutagenesis (Peterhans et al. 2016).

In summary, formation of the arrestin-receptor complex occurs by a multistep mechanism (Fig. 8.3). Initial electrostatic attraction of arrestin to the Rpp and the negatively charged membrane surface recruits arrestin to the membrane, where the N-domain binds the Rpp and the C-edge anchors within the membrane. This interaction brings the central crest of arrestin within close proximity of the receptor helical bundle, so that the two proteins can interact and mutually affect one another. Transition to the high-affinity complex involves multiple conformational rearrangements in arrestin, which allow arrestin to both specifically stabilize the active form of the receptor as well as flexibly accommodate the cytoplasmic face of the receptor. This flexibility is likely critical for making the β-arrestins versatile binding partners for so many different GPCRs.

References

Bayburt TH, Vishnivetskiy SA, McLean MA, Morizumi T, Huang CC, Tesmer JJ, Ernst OP, Sligar SG, Gurevich VV (2011) Monomeric rhodopsin is sufficient for normal rhodopsin kinase (GRK1) phosphorylation and arrestin-1 binding. J Biol Chem 286(2):1420–1428

Beyriere F, Sommer ME, Szczepek M, Bartl FJ, Hofmann KP, Heck M, Ritter E (2015) Formation and decay of the arrestin-rhodopsin complex in native disc membranes. J Biol Chem 290 (20):12919–12928

Binder BM, Biernbaum MS, Bownds MD (1990) Light activation of one rhodopsin molecule causes the phosphorylation of hundreds of others. A reaction observed in electropermeabilized frog rod outer segments exposed to dim illumination. J Biol Chem 265(25):15333–15340

Binder BM, O'Connor TM, Bownds MD, Arshavsky VY (1996) Phosphorylation of non-bleached rhodopsin in intact retinas and living frogs. J Biol Chem 271(33):19826–19830

Elgeti M, Rose AS, Bartl FJ, Hildebrand PW, Hofmann KP, Heck M (2013) Precision vs flexibility in GPCR signaling. J Am Chem Soc 135(33):12305–12312. doi:10.1021/ja405133k

Granzin J, Wilden U, Choe HW, Labahn J, Krafft B, Buldt G (1998) X-ray crystal structure of arrestin from bovine rod outer segments. Nature 391(6670):918–921. doi:10.1038/36147

Granzin J, Cousin A, Weirauch M, Schlesinger R, Büldt G, Batra-Safferling R (2012) Crystal structure of p44, a constitutively active splice variant of visual arrestin. J Mol Biol 416: 611–618

Granzin J, Stadler A, Cousin A, Schlesinger R, Batra-Safferling R (2015) Structural evidence for the role of polar core residue Arg175 in arrestin activation. Sci Rep 5:15808. doi:10.1038/srep15808

Gurevich VV, Benovic JL (1993) Visual arrestin interaction with rhodopsin. Sequential multisite binding ensures strict selectivity toward light-activated phosphorylated rhodopsin. J Biol Chem 268(16):11628–11638

Gurevich VV, Benovic JL (1995) Visual arrestin binding to rhodopsin. Diverse functional roles of positively charged residues within the phosphorylation-recognition region of arrestin. J Biol Chem 270(11):6010–6016

Gurevich VV, Pals-Rylaarsdam R, Benovic JL, Hosey MM, Onorato JJ (1997) Agonist-receptor-arrestin, an alternative ternary complex with high agonist affinity. J Biol Chem 272(46):28849–28852

Gurevich VV, Hanson SM, Song X, Vishnivetskiy SA, Gurevich EV (2011) The functional cycle of visual arrestins in photoreceptor cells. Prog Retinal Eye Res 30(6):405–430. doi:10.1016/j.preteyeres.2011.07.002

Han M, Gurevich VV, Vishnivetskiy SA, Sigler PB, Schubert C (2001) Crystal structure of beta-arrestin at 1.9 Å: possible mechanism of receptor binding and membrane translocation. Structure 9(9):869–880

Hanson SM, Gurevich VV (2006) The differential engagement of arrestin surface charges by the various functional forms of the receptor. J Biol Chem 281(6):3458–3462. doi:10.1074/jbc.M512148200

Hanson SM, Francis DJ, Vishnivetskiy SA, Kolobova EA, Hubbell WL, Klug CS, Gurevich VV (2006) Differential interaction of spin-labeled arrestin with inactive and active phospho-rhodopsin. Proc Natl Acad Sci U S A 103(13):4900–4905. doi:10.1073/pnas.0600733103

Hanson SM, Dawson ES, Francis DJ, Van Eps N, Klug CS, Hubbell WL, Meiler J, Gurevich VV (2008) A model for the solution structure of the rod arrestin tetramer. Structure 16(6):924–934. doi:10.1016/j.str.2008.03.006

Hessel E, Herrmann A, Muller P, Schnetkamp PP, Hofmann KP (2000) The transbilayer distribution of phospholipids in disc membranes is a dynamic equilibrium evidence for rapid flip and flop movement. Eur J Biochem 267(5):1473–1483

Hessel E, Muller P, Herrmann A, Hofmann KP (2001) Light-induced reorganization of phospholipids in rod disc membranes. J Biol Chem 276(4):2538–2543. doi:10.1074/jbc.M009061200

Hirsch JA, Schubert C, Gurevich VV, Sigler PB (1999) The 2.8 Å crystal structure of visual arrestin: a model for arrestin's regulation. Cell 97(2):257–269

Kang Y, Zhou XE, Gao X, He Y, Liu W, Ishchenko A, Barty A, White TA, Yefanov O, Han GW, Xu Q, de Waal PW, Ke J, Tan MH, Zhang C, Moeller A, West GM, Pascal BD, Van Eps N, Caro LN, Vishnivetskiy SA, Lee RJ, Suino-Powell KM, Gu X, Pal K, Ma J, Zhi X, Boutet S, Williams GJ, Messerschmidt M, Gati C, Zatsepin NA, Wang D, James D, Basu S, Roy-Chowdhury S, Conrad CE, Coe J, Liu H, Lisova S, Kupitz C, Grotjohann I, Fromme R, Jiang Y, Tan M, Yang H, Li J, Wang M, Zheng Z, Li D, Howe N, Zhao Y, Standfuss J, Diederichs K, Dong Y, Potter CS, Carragher B, Caffrey M, Jiang H, Chapman HN, Spence JC,

Fromme P, Weierstall U, Ernst OP, Katritch V, Gurevich VV, Griffin PR, Hubbell WL, Stevens RC, Cherezov V, Melcher K, Xu HE (2015) Crystal structure of rhodopsin bound to arrestin by femtosecond X-ray laser. Nature 523(7562):561–567. doi:10.1038/nature14656

Kim M, Vishnivetskiy SA, Van Eps N, Alexander NS, Cleghorn WM, Zhan X, Hanson SM, Morizumi T, Ernst OP, Meiler J, Gurevich VV, Hubbell WL (2012) Conformation of receptor-bound visual arrestin. Proc Natl Acad Sci U S A 109(45):18407–18412. doi:10.1073/pnas.1216304109

Kim YJ, Hofmann KP, Ernst OP, Scheerer P, Choe HW, Sommer ME (2013) Crystal structure of pre-activated arrestin p44. Nature 497(7447):142–146. doi:10.1038/nature12133

Kirchberg K, Kim TY, Moller M, Skegro D, Dasara Raju G, Granzin J, Buldt G, Schlesinger R, Alexiev U (2011) Conformational dynamics of helix 8 in the GPCR rhodopsin controls arrestin activation in the desensitization process. Proc Natl Acad Sci U S A 108(46):18690–18695. doi:10.1073/pnas.1015461108

Lally CC, Bauer B, Selent J, Sommer ME (2017) C-edge loops of arrestin function as a membrane anchor. Nat Commun 8:14258. doi:10.1038/ncomms14258

Lee KA, Nawrot M, Garwin GG, Saari JC, Hurley JB (2010) Relationships among visual cycle retinoids, rhodopsin phosphorylation, and phototransduction in mouse eyes during light and dark adaptation. Biochemistry 49(11):2454–2463. doi:10.1021/bi1001085

Milano SK, Kim YM, Stefano FP, Benovic JL, Brenner C (2006) Nonvisual arrestin oligomerization and cellular localization are regulated by inositol hexakisphosphate binding. J Biol Chem 281(14):9812–9823. doi:10.1074/jbc.M512703200

Nobles KN, Xiao K, Ahn S, Shukla AK, Lam CM, Rajagopal S, Strachan RT, Huang TY, Bressler EA, Hara MR, Shenoy SK, Gygi SP, Lefkowitz RJ (2011) Distinct phosphorylation sites on the beta(2)-adrenergic receptor establish a barcode that encodes differential functions of beta-arrestin. Sci Signal 4(185):ra51. doi:10.1126/scisignal.2001707

Nuber S, Zabel U, Lorenz K, Nuber A, Milligan G, Tobin AB, Lohse MJ, Hoffmann C (2016) Beta-arrestin biosensors reveal a rapid, receptor-dependent activation/deactivation cycle. Nature 531(7596):661–664. doi:10.1038/nature17198

Oakley RH, Laporte SA, Holt JA, Barak LS, Caron MG (2001) Molecular determinants underlying the formation of stable intracellular G protein-coupled receptor-beta-arrestin complexes after receptor endocytosis*. J Biol Chem 276(22):19452–19460. doi:10.1074/jbc.M101450200

Ostermaier MK, Peterhans C, Jaussi R, Deupi X, Standfuss J (2014a) Functional map of arrestin-1 at single amino acid resolution. Proc Natl Acad Sci U S A. doi:10.1073/pnas.1319402111

Ostermaier MK, Schertler GF, Standfuss J (2014b) Molecular mechanism of phosphorylation-dependent arrestin activation. Curr Opin Struct Biol 29C:143–151. doi:10.1016/j.sbi.2014.07.006

Palczewski K, Buczylko J, Imami NR, McDowell JH, Hargrave PA (1991a) Role of the carboxyl-terminal region of arrestin in binding to phosphorylated rhodopsin. J Biol Chem 266(23):15334–15339

Palczewski K, Pulvermüller A, Buczylko J, Gutmann C, Hofmann KP (1991b) Binding of inositol phosphates to arrestin. FEBS Lett 295(1–3):195–199

Palczewski K, Pulvermüller A, Buczylko J, Hofmann KP (1991c) Phosphorylated rhodopsin and heparin induce similar conformational changes in arrestin. J Biol Chem 266(28):18649–18654

Palczewski K, Buczylko J, Ohguro H, Annan RS, Carr SA, Crabb JW, Kaplan MW, Johnson RS, Walsh KA (1994) Characterization of a truncated form of arrestin isolated from bovine rod outer segments. Protein Sci 3(2):314–324

Peterhans C, Lally CC, Ostermaier MK, Sommer ME, Standfuss J (2016) Functional map of arrestin binding to phosphorylated opsin, with and without agonist. Sci Rep 6:28686. doi:10.1038/srep28686

Pulvermüller A, Maretzki D, Rudnicka-Nawrot M, Smith WC, Palczewski K, Hofmann KP (1997) Functional differences in the interaction of arrestin and its splice variant, p44, with rhodopsin. Biochemistry 36(30):9253–9260. doi:10.1021/bi970772g

Reiter E, Ahn S, Shukla AK, Lefkowitz RJ (2012) Molecular mechanism of beta-arrestin-biased agonism at seven-transmembrane receptors. Ann Rev Pharmacol Toxicol 52:179–197. doi:10.1146/annurev.pharmtox.010909.105800

Ren XR, Reiter E, Ahn S, Kim J, Chen W, Lefkowitz RJ (2005) Different G protein-coupled receptor kinases govern G protein and beta-arrestin-mediated signaling of V2 vasopressin receptor. Proc Natl Acad Sci U S A 102(5):1448–1453. doi:10.1073/pnas.0409534102

Schleicher A, Kühn H, Hofmann KP (1989) Kinetics, binding constant, and activation energy of the 48-kDa protein-rhodopsin complex by extra-metarhodopsin II. Biochemistry 28(4):1770–1775

Schröder K, Pulvermüller A, Hofmann KP (2002) Arrestin and its splice variant $Arr^{1-370A}(p^{44})$. Mechanism and biological role of their interaction with rhodopsin. J Biol Chem 277(46):43987–43996. doi:10.1074/jbc.M206211200

Shenoy SK, Modi AS, Shukla AK, Xiao K, Berthouze M, Ahn S, Wilkinson KD, Miller WE, Lefkowitz RJ (2009) Beta-arrestin-dependent signaling and trafficking of 7-transmembrane receptors is reciprocally regulated by the deubiquitinase USP33 and the E3 ligase Mdm2. Proc Natl Acad Sci U S A 106(16):6650–6655. doi:10.1073/pnas.0901083106

Shi GW, Chen J, Concepcion F, Motamedchaboki K, Marjoram P, Langen R (2005) Light causes phosphorylation of nonactivated visual pigments in intact mouse rod photoreceptor cells. J Biol Chem 280(50):41184–41191. doi:10.1074/jbc.M506935200 (M506935200)

Shukla AK, Manglik A, Kruse AC, Xiao K, Reis RI, Tseng WC, Staus DP, Hilger D, Uysal S, Huang LY, Paduch M, Tripathi-Shukla P, Koide A, Koide S, Weis WI, Kossiakoff AA, Kobilka BK, Lefkowitz RJ (2013) Structure of active beta-arrestin-1 bound to a G-protein-coupled receptor phosphopeptide. Nature 497(7447):137–141. doi:10.1038/nature12120

Shukla AK, Westfield GH, Xiao K, Reis RI, Huang LY, Tripathi-Shukla P, Qian J, Li S, Blanc A, Oleskie AN, Dosey AM, Su M, Liang CR, Gu LL, Shan JM, Chen X, Hanna R, Choi M, Yao XJ, Klink BU, Kahsai AW, Sidhu SS, Koide S, Penczek PA, Kossiakoff AA, Woods VL Jr, Kobilka BK, Skiniotis G, Lefkowitz RJ (2014) Visualization of arrestin recruitment by a G-protein-coupled receptor. Nature 512(7513):218–222. doi:10.1038/nature13430

Sommer ME, Smith WC, Farrens DL (2006) Dynamics of arrestin-rhodopsin interactions: acidic phospholipids enable binding of arrestin to purified rhodopsin in detergent. J Biol Chem 281(14):9407–9417. doi:10.1074/jbc.M510037200

Sommer ME, Hofmann KP, Heck M (2012) Distinct loops in arrestin differentially regulate ligand binding within the GPCR opsin. Nat Commun 3:995. doi:10.1038/ncomms2000

Sommer ME, Hofmann KP, Heck M (2014) Not just signal shutoff: the protective role of arrestin-1 in rod cells. Handb Exp Pharmacol 219:101–116. doi:10.1007/978-3-642-41199-1_5

Szczepek M, Beyriere F, Hofmann KP, Elgeti M, Kazmin R, Rose A, Bartl FJ, von Stetten D, Heck M, Sommer ME, Hildebrand PW, Scheerer P (2014) Crystal structure of a common GPCR-binding interface for G protein and arrestin. Nat Commun 5:4801. doi:10.1038/ncomms5801

Tobin AB, Butcher AJ, Kong KC (2008) Location, location, location… site-specific GPCR phosphorylation offers a mechanism for cell-type-specific signalling. Trends Pharmacol Sci 29(8):413–420. doi:10.1016/j.tips.2008.05.006

Tohgo A, Choy EW, Gesty-Palmer D, Pierce KL, Laporte S, Oakley RH, Caron MG, Lefkowitz RJ, Luttrell LM (2003) The stability of the G protein-coupled receptor-beta-arrestin interaction determines the mechanism and functional consequence of ERK activation. J Biol Chem 278(8):6258–6267. doi:10.1074/jbc.M212231200

Vishnivetskiy SA, Schubert C, Climaco GC, Gurevich YV, Velez MG, Gurevich VV (2000) An additional phosphate-binding element in arrestin molecule. Implications for the mechanism of arrestin activation. J Biol Chem 275(52):41049–41057. doi:10.1074/jbc.M007159200

Vishnivetskiy SA, Raman D, Wei J, Kennedy MJ, Hurley JB, Gurevich VV (2007) Regulation of arrestin binding by rhodopsin phosphorylation level. J Biol Chem 282(44):32075–32083. doi:10.1074/jbc.M706057200

Vishnivetskiy SA, Francis D, Van Eps N, Kim M, Hanson SM, Klug CS, Hubbell WL, Gurevich VV (2010) The role of arrestin alpha-helix I in receptor binding. J Mol Biol 395 (1):42–54. doi:10.1016/j.jmb.2009.10.058

Vishnivetskiy SA, Baameur F, Findley KR, Gurevich VV (2013) Critical role of central 139-loop in stability and binding selectivity of arrestin-1. J Biol Chem. doi:10.1074/jbc.M113.450031

Wilson CJ, Copeland RA (1997) Spectroscopic characterization of arrestin interactions with competitive ligands: study of heparin and phytic acid binding. J Protein Chem 16(8):755–763

Zhuang T, Vishnivetskiy SA, Gurevich VV, Sanders CR (2010) Elucidation of inositol hexaphosphate and heparin interaction sites and conformational changes in arrestin-1 by solution nuclear magnetic resonance. Biochemistry 49(49):10473–10485. doi:10.1021/bi101596g

Zhuang T, Chen Q, Cho MK, Vishnivetskiy SA, Iverson TM, Gurevich VV, Sanders CR (2013) Involvement of distinct arrestin-1 elements in binding to different functional forms of rhodopsin. Proc Natl Acad Sci U S A 110(3):942–947. doi:10.1073/pnas.1215176110

Zindel D, Butcher AJ, Al-Sabah S, Lanzerstorfer P, Weghuber J, Tobin AB, Bunemann M, Krasel C (2014) Engineered hyperphosphorylation of the beta$_2$-adrenoceptor prolongs arrestin-3 binding and induces arrestin internalization. Mol Pharmacol. doi:10.1124/mol.114.095422

Chapter 9
Localization of Conformational Dynamics of Arrestins by HDX-MS

Ji Young Park, Hee Ryung Kim and Ka Young Chung

Abstract Arrestins are a family of multifunctional adapter proteins that were originally discovered by their ability to desensitize G protein-coupled receptors (GPCRs). Besides desensitization of GPCRs, arrestins regulate various signaling molecules, such as mitogen-activated protein kinase (MAPK) signaling pathway proteins and ubiquitination pathway proteins. To have such diverse functions, arrestins are structurally dynamic. In this chapter, we will discuss the structural dynamics of arrestins revealed by hydrogen/deuterium exchange mass spectrometry (HDX-MS).

Keywords Arrestin · Structural dynamics · H/D exchange · Mass spectrometry · GPCR · Conformational change

What Can HDX-MS Reveal About Protein Conformation?

Proteins have pivotal biochemical functions including signal transduction, enzyme catalysis, and DNA regulation. Protein functions are closely related to their structures, and therefore many biochemical and biophysical techniques have been adopted for protein structure analysis. X-ray crystallography, nuclear magnetic resonance (NMR), and cryoelectron microscopy (cryoEM) are the methods of choice for protein structure analysis because they provide high-resolution 3D structures. However, these techniques have limitations as they are often not suitable

J.Y. Park · H.R. Kim · K.Y. Chung (✉)
School of Pharmacy, Sungkyunkwan University, Suwon 16419, Republic of Korea
e-mail: kychung2@skku.edu

J.Y. Park
e-mail: vasity@naver.com

H.R. Kim
e-mail: heeryung.jan.kim@gmail.com

V.V. Gurevich (ed.), *The Structural Basis of Arrestin Functions*,
DOI 10.1007/978-3-319-57553-7_9

for highly dynamic or flexible proteins, require large quantities of highly homogenous or highly concentrated protein samples and harsh analysis conditions (e.g. crystallization and freezing conditions for X-ray crystallography and cryoEM, respectively), and have protein size restrictions. Other biophysical and biochemical techniques, such as electron paramagnetic resonance (EPR) and fluorescence resonance energy transfer (FRET), are used to overcome these limitations. However, these techniques also have disadvantages because proteins must be modified with probes.

Mass spectrometry (MS) has proven to be a powerful tool for protein analysis. MS has been widely used for protein identification, which has led to the advance of proteomics. Recently, MS has been actively adopted for protein structure analysis since Katta and Chait showed that MS can monitor amide hydrogen/deuterium exchange (HDX) (Katta and Chait 1991). HDX coupled with MS is called HDX-MS (Fig. 9.1), and the theory of HDX-MS is well-described in several review articles (Konermann et al. 2008; Marcsisin and Engen 2010). Briefly, backbone

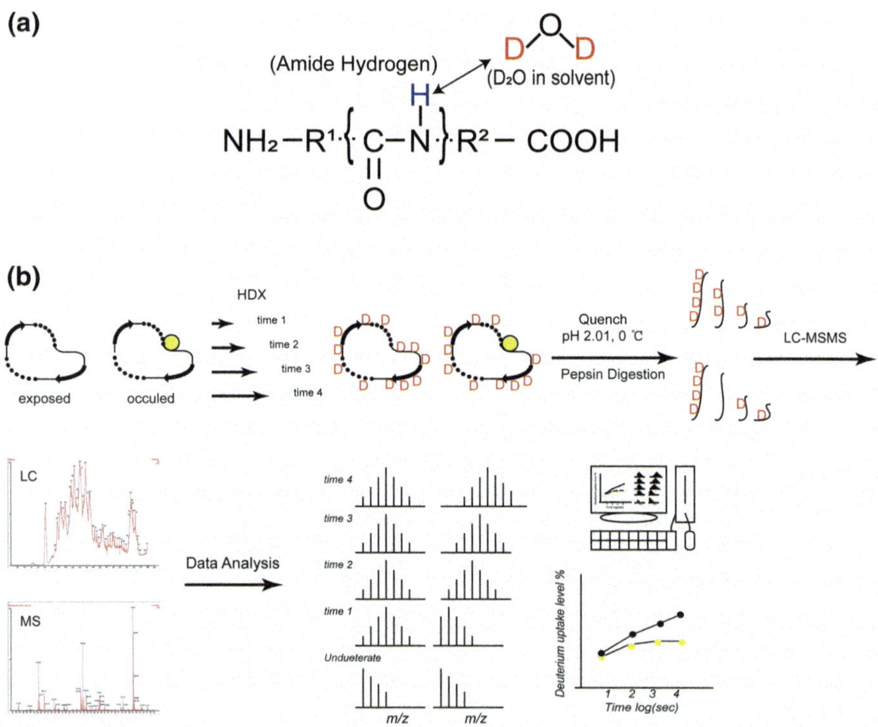

Fig. 9.1 Principles of HDX-MS. **a** Scheme of exchange between hydrogen and deuterium in the backbone amide hydrogen. Amide hydrogen is indicated in *blue* and deuterium in *red*. **b** Overall protocol of HDX-MS. Effect of a small molecule (*yellow circle*) binding on HDX is shown as an example. Secondary structures of proteins are illustrated in *bold arrows* (β-strand) and *dotted lines* (α-helix). The exchanged deuterium is in *red capital letter D*

amide hydrogens in a protein constantly exchange with hydrogens in H_2O of the solvent. If the solvent is prepared with deuterated water (D_2O), the amide hydrogens are exchanged to deuterium resulting in heavier proteins (Fig. 9.1a) (Konermann et al. 2008). The deuterium uptake rate is affected by pH, temperature, solvent accessibility, and hydrogen bonding, therefore, the deuterium uptake rate provides insights into protein conformation when pH and temperature are constant. Highly flexible and exposed regions undergo rapid HDX, whereas structured or occluded regions undergo slow HDX (Fig. 9.1b). The number of exchanged deuterium atoms is monitored by MS as deuterium is approximately 1 Da heavier than hydrogen.

HDX-MS has many advantages, although it does not determine the 3D structures, nor does it provide atomic resolution structures (Marcsisin and Engen 2010; Xiao et al. 2015). HDX-MS requires a relatively small amount of protein (nanomole) at a relatively low concentration (micromolar). A protein is kept in an aquatic solution and does not need to be modified. There are no size limitations, and flexible regions can be analyzed along with structured regions. Therefore, HDX-MS is a good complementary technique for protein conformation analysis.

Proteins are highly dynamic biomolecules, and many studies have used HDX-MS to monitor the dynamic conformations of various proteins (Fig. 9.2) (Marcsisin and Engen 2010; Iacob and Engen 2012). HDX-MS can provide information about the flexibility of specific protein regions, the interfaces of protein-small molecule or protein-protein interactions, the allosteric conformational change upon small molecule or protein binding, the conformational changes during

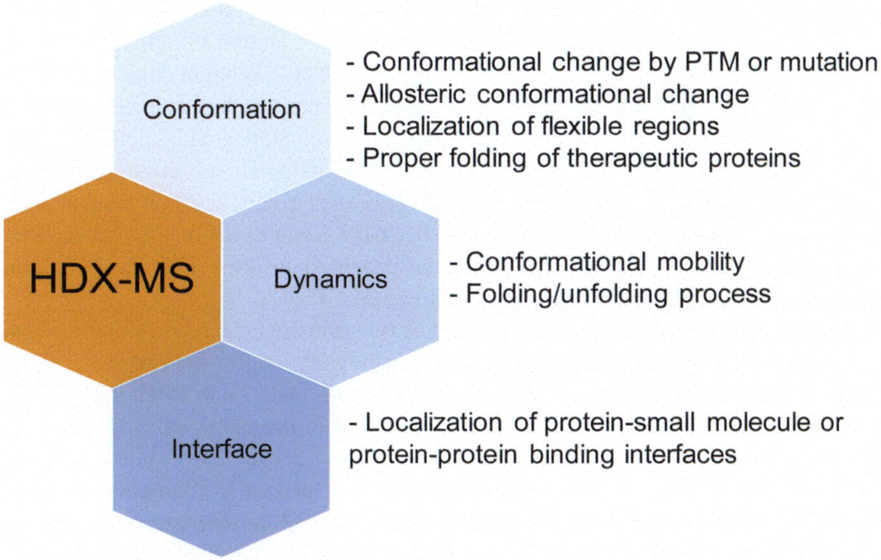

Fig. 9.2 Application of HDX-MS for protein structure analysis

folding/unfolding process, and the conformational changes due to mutations or post-translational modifications. This chapter will discuss the structural mechanism of arrestins as analyzed by HDX-MS.

Introduction to Arrestin Structure

Arrestins are a family of cytosolic proteins with four homologous isoforms (arrestin 1–4). Arrestin-1 (or visual arrestin) and arrestin-4 (or cone arrestin) are exclusively expressed in the visual system, whereas arrestin-2 (or β-arrestin-1) and arrestin-3 (or β-arrestin-2) are ubiquitously expressed (Lohse and Hoffmann 2014). Arrestins were originally discovered to desensitize GPCR signaling by interacting with ligand-activated phosphorylated GPCRs and/or by internalizing GPCRs after interacting with trafficking proteins such as clathrin, AP-2, and N-ethylmaleimide-sensitive fusion protein (Lohse and Hoffmann 2014; Park et al. 2016; Ferguson et al. 1996; Goodman et al. 1996; Laporte et al. 1999; McDonald et al. 1999). Subsequent studies discovered that activated arrestins interact not only with GPCRs and trafficking proteins, but also with many other signaling proteins including mitogen-activated protein kinases (MAPKs), which trigger non-classical GPCR signaling pathways (i.e., the G protein-independent signaling pathway) (DeFea 2011; Luttrell and Miller 2013; Shenoy and Lefkowitz 2011; Smith and Rajagopal 2016). Arrestin-mediated signaling has become an important GPCR signaling pathway as many studies have reported that G protein-mediated and arrestin-mediated signaling pathways produce different physiological or pathological outcomes. Therefore, understanding the mechanism of GPCR-G protein or GPCR-arrestin selectivity is needed for functionally selective GPCR-targeting in drug development (Shukla et al. 2014a; Violin et al. 2014; Whalen et al. 2011). For this reason, the structural mechanism of arrestin functions has been extensively studied using various techniques, including X-ray crystallography, NMR, EPR, FRET, BRET, and HDX-MS (Granzin et al. 1998, 2015; Hirsch et al. 1999; Han et al. 2001; Carter et al. 2005; Zhan et al. 2011; Kim et al. 2012, 2013; Shukla et al. 2013, 2014; Zhuang et al. 2013; Zhuo et al. 2014; Kang et al. 2015; Nuber et al. 2016; Lee et al. 2016). These studies have provided insights into the structural mechanism of GPCR-mediated arrestin activation.

All four arrestin isoforms are composed of two domains (i.e., the N-domain and the C-domain) with similar seven-strand β-sandwich conformations (Fig. 9.3a) (Granzin et al. 1998; Hirsch et al. 1999; Han et al. 2001; Zhan et al. 2011). The interaction between N- and C-domains is stabilized in the basal state by the polar core and the three-element interaction (Ostermaier et al. 2014; Gurevich and Gurevich 2006a). The polar core (Fig. 9.3a, red) is primarily maintained by salt bridges between Arg169/Arg393 and Asp290/Asp297 at the domain interface (the sequence numbering is based on the human β-arrestin-1 unless otherwise specified). The three-element interaction connects β-strand I and α-helix I in the N-domain with β-strand XX in the C-tail (Fig. 9.3a, blue). Previous biochemical and

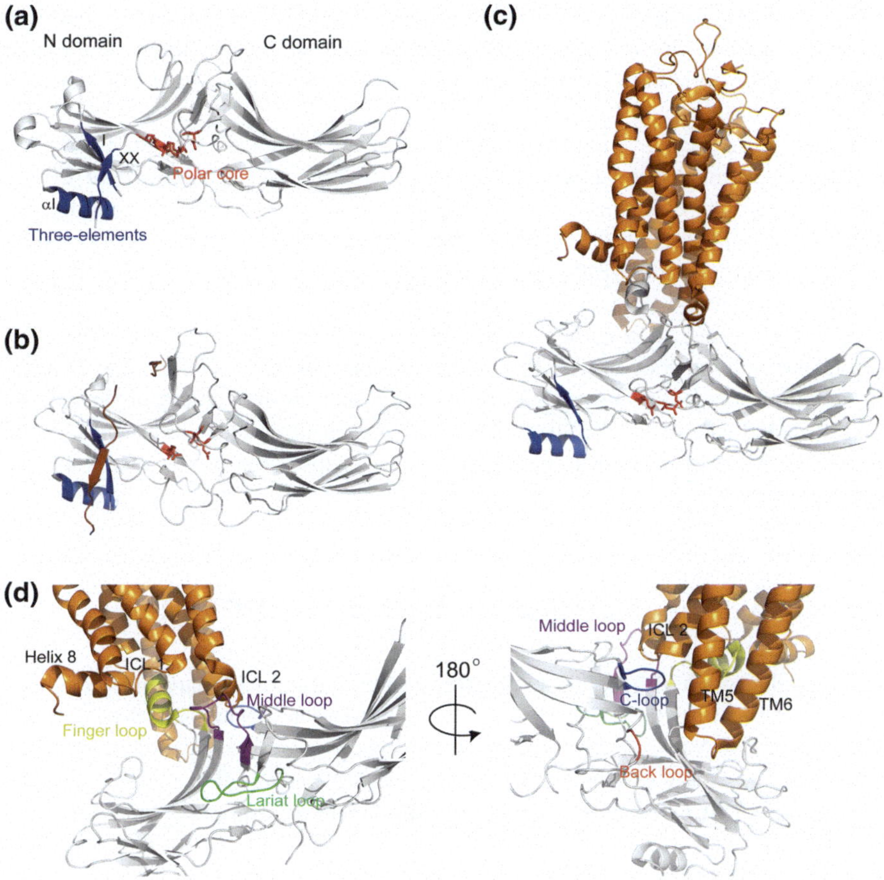

Fig. 9.3 High resolution X-ray crystal structures of arrestins and their structural features. **a** Overall structure of arrestins. X-ray crystal structure of β-arrestin-1 (PDB: 1G4R) is shown as an example of overall structure of arrestin. The polar core is shown in *red sticks*, and the three-element in *blue*. **b** X-ray crystal structure of V2Rpp-bound β-arrestin-1 (PDB: 4JQI). V2Rpp is shown in *brown*. Note that β-strand XX in **a** is missing and V2Rpp replaced the region. The polar core and the three-element interaction are shown as in (**a**). **c** X-ray crystal structure of rhodopsin-arrestin-1 complex (PDB: 4ZWJ). Rhodopsin is shown in *orange* and arrestin is shown in *gray*. The polar core and the three-element interaction are shown as in (**a**). **d** Interface between arrestin-1 and rhodopsin

biophysical studies proposed that disruption of the polar core or the three-element induces pre-activated arrestins, which interact with GPCRs in a phosphorylation-independent manner (Ostermaier et al. 2014; Gurevich and Gurevich 2013). Thus, these regions were considered to be "the phosphate sensor". This has been proven by the X-ray crystal structure of β-arrestin-1 interacting with the phosphorylated vasopressin receptor C-terminal peptide (V2Rpp) (Fig. 9.3b) (Shukla et al. 2013). In this structure, both the polar core and the three-element interaction are disrupted;

V2Rpp replaces the β-strand XX by forming a charge-charge interaction between phosphorylated resides in V2Rpp and lysines in the β-strand I of the three-element interaction; the polar core is broken probably as a result of the interaction between V2Rpp and the β-strand I.

Recently, the high-resolution 3D structure of the rhodopsin-arrestin-1 complex has been determined using a femtosecond X-ray laser (Fig. 9.3c) (Kang et al. 2015). Unfortunately, the C-terminal tail of rhodopsin was not visible in this rhodopsin-arrestin-1 complex (Fig. 9.3c). However, the structure revealed other binding interfaces between an arrestin and a GPCR; loops near the N- and C-domain interface of arrestin-1 contact the cytoplasmic side of rhodopsin (Fig. 9.3c, d). The finger loop formed an α-helix, which was inserted into the rhodopsin core contacting intracellular loop 1 (ICL1), transmembrane domain 7 (TM7), and helix 8. The middle loop moved toward the finger loop, and the movement of the middle loop provided room for ICL2 to be located between the middle loop and C-loop. Finally, the back loop contacts with TM5. These loops are now considered as "the activation sensor".

Arrestin Activation Mechanism Revealed by HDX-MS

Although the two high-resolution structures described above greatly improved our understanding of the GPCR-mediated arrestin activation mechanism, there are still limitations. The X-ray crystal structures only show a static snapshot of proteins. Most of all, the X-ray structure of the β-arrestin1-V2Rpp complex contains only the C-terminal tail of a receptor, whereas the rhodopsin-arrestin-1 complex structure does not show the C-terminal tail of rhodopsin. HDX-MS has been successfully used to overcome these limitations (Shukla et al. 2014b; Kang et al. 2015). Moreover, HDX-MS can show not only the changes in conformation, but also changes in conformational mobility.

HDX-MS has been adopted to study the conformational consequences of pre-activation (e.g. disruption of phosphate sensors by mutations) (Carter et al. 2005; Kim et al. 2015). Our group used HDX-MS to analyze three well-studied pre-activated mutants of β-arrestin-1 (e.g. R169E, 3A, and p44) (Kim et al. 2015). The R169E mutation disrupts the polar core by reversing the charge of Arg in position 169. The 3A and p44 mutations disrupt the three-element region by interfering with the interaction of the β-strand XX or truncating it, respectively. Our study demonstrated that pre-activation of β-arrestin-1 induces conformational changes not only near the phosphate sensors (β-strand I through III, α-helix I, and the lariat loop), but also in other regions (the middle loop, β-strand IV/V loop, and β-strand XII through XVI) (Figs. 9.4a and 9.5b). This report supports the hypothesis that phosphates of a receptor binding to the phosphate sensor induce conformational changes in other regions and suggests that these regions might be the activation sensor, which may lead to further interaction between the cytosolic

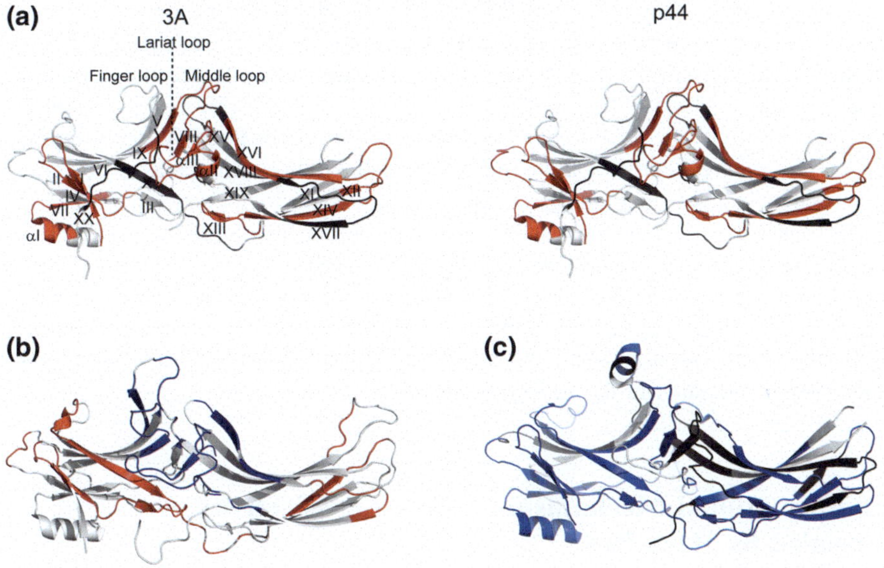

Fig. 9.4 Conformational changes of activated arrestins revealed by HDX-MS. **a** HDX profile change upon pre-activation mutation on β-arrestin-1. Regions with increased deuterium uptake level upon 3A or p44 mutation-mediated pre-activation compared to basal state are color-coded as *red* on the X-ray crystal structure of β-arrestin-1 (PDB: 1JSY). *Black* indicates regions without identified peptic peptides, and *gray* indicates the regions with no difference in deuterium uptake levels upon pre-activation. **b** HDX-MS profile change of β-arrestin-1 upon interaction with V2Rpp. The HDX-MS profile change is color-coded on the X-ray crystal structure of V2Rpp-bound β-arrestin-1 (PDB: 4JQI). **c** HDX-MS profile change of arrestin-1 upon interaction with rhodopsin. The HDX-Ms profile change is color-coded on the X-ray crystal structure of rhodopsin-bound arrestin-1 (PDB: 4ZWJ). For **b**, **c**, *red* indicates regions with increased deuterium uptake levels and *blue* indicates regions with decreased uptake levels upon binding to V2Rpp or rhodopsin. *Black* and *gray* indicate as in (**a**)

face of a GPCR and the activation sensor of an arrestin (Kim et al. 2012, 2013; Shukla et al. 2013; Vishnivetskiy et al. 2013).

The HDX-MS study by the Lefkowitz group more clearly predicted the activation sensor before the rhodopsin-arrestin-1 crystal structure was solved (Shukla et al. 2014b). When the HDX profile of β-arrestin-1 in the basal state is compared with that of β-arrestin-1 bound to the β_2-adrenergic receptor-V2Rpp chimera (β_2AR/V2Rpp), the loops in the N- and C-domain interface (i.e., the finger loop, middle loop, and lariat loop) showed reduced deuterium uptake indicating that these regions are probably the receptor interacting interfaces (Fig. 9.4b). Later, a study by the Xu group using both X-ray crystallography and HDX-MS analysis of the rhodopsin-arrestin-1 complex confirmed the interfaces between a GPCR and an arrestin, and provided a more complete understanding of the GPCR-mediated arrestin activation (Kang et al. 2015) (Fig. 9.4c).

However, some arrestin regions showed different HDX-MS profile changes upon GPCR binding depending on the GPCR-arrestin pair, as evidenced by the Lefkowitz and Xu groups (Fig. 9.4b, c). In the β_2AR/V2Rpp-β-arrestin-1 complex, some regions including β-strands II through III, β-strand V, and β-strands X through XI undergo increased deuterium uptake upon β_2AR/V2Rpp binding (Fig. 9.4b), whereas most of the regions in arrestin-1 showed decreased deuterium uptake upon rhodopsin binding in the rhodopsin-arrestin-1 complex (Fig. 9.4c). This observation leads to a question: what will be the functional consequences of the different conformational changes between arrestin-1 and β-arrestin-1 upon binding to different GPCRs? An HDX-MS study would provide a hint for understanding the arrestin selectivity either for GPCRs or other downstream signaling molecules, and we will elaborate more on HDX-MS-revealed conformational selectivity of arrestins in the following section.

Conformational Mechanism of Arrestin Selectivity Revealed by HDX-MS

β-arrestin-1 and β-arrestin-2 are 75% identical by amino acid sequence and 85% similar when accounting for conservative substitutions (Lohse and Hoffmann 2014). Knock-out mouse and other biochemical studies suggest that β-arrestins are functionally redundant (Kohout et al. 2001). Nonetheless, a few studies showed that they are different in their affinity for GPCRs, expression patterns, and functional effects (Smith and Rajagopal 2016; Oakley et al. 2000). GPCRs are classified into two groups based on their affinity for arrestins: class A GPCRs bind β-arrestin-2 with higher affinity than β-arrestin-1, whereas class B GPCRs bind β-arrestin-1 and β-arrestin-2 with similar high affinities (Lohse and Hoffmann 2014; Oakley et al. 2000; Tohgo et al. 2003). This implies that β-arrestin-2 has wider receptor selectivity than β-arrestin-1. Besides receptor selectivity, β-arrestin-1 and β-arrestin-2 have different cellular functions, which are well-summarized in a recent review (Srivastava et al. 2015).

A few studies using X-ray crystallography, limited tryptic proteolysis, and DEER spectroscopy have analyzed conformational differences between β-arrestin-1 and β-arrestin-2 (Han et al. 2001; Zhan et al. 2011; Zhuo et al. 2014; Nobles et al. 2007). However, conformational differences between β-arrestin-1 and β-arrestin-2 in the basal and active states were not investigated in these studies. A HDX-MS study from our group successfully showed the dynamic conformational differences between β-arrestin-1 and β-arrestin-2 (Fig. 9.5) (Yun et al. 2015).

In the basal state, β-arrestin-2 underwent higher deuterium uptake in β-strands II through IV whereas β-arrestin-1 underwent higher deuterium uptake in the middle loop (Fig. 9.5a). These findings indicate β-strands II through IV of β-arrestin-2 are more dynamic in the basal state and its middle loop is less dynamic compared to β-arrestin-1. Residues responsible for maintaining the polar core are located in

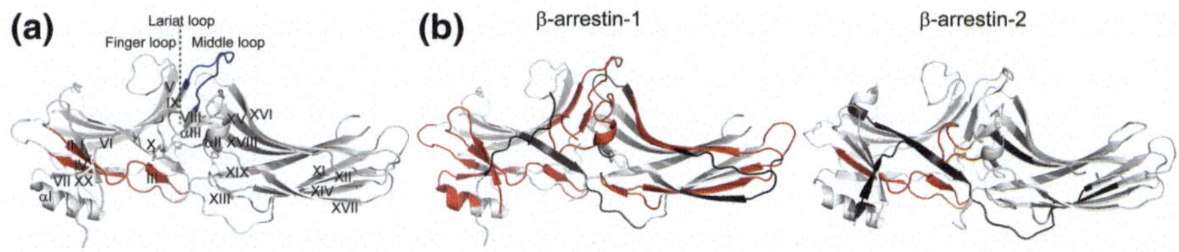

Fig. 9.5 HDX profile comparison of β-arrestin-1 and β-arrestin-2. **a** Comparison in the basal state. Regions with different HDX profile between β-arrestin-1 and β-arrestin-2 are color-coded on crystal structure of β-arrestin-1 (PDB: 1JSY). Regions with lower HDX levels in β-arrestin-2 compared to β-arrestin-1 are shown in blue, and regions with higher HDX levels in β-arrestin-2 compared to β-arrestin-1 are shown in *red*. **b** Comparison in the R169E mutation-induced pre-activated state. HDX profile changes upon pre-activation mutation (R169E) on β-arrestin-1 (*right*) and β-arrestin-2 (*left*) are color-coded on X-ray crystal structure β-arrestin-1 (PDB: 1JSY) and β-arrestin-2 (PDB: 3P2D). *Red* indicates the regions with increased HDX levels upon pre-activation compared to the basal state, *black* indicates regions without identified peptic peptides, and *gray* indicates the regions with no difference in deuterium uptake levels upon pre-activation

β-strands II through IV (e.g. D26 in β-arrestin1 and D27 in β-arrestin-2), and thus it implies the flexibility in these regions may lead to disruption of the polar core more easily, consequently resulting in higher binding affinity between β-arrestin-2 and receptors. The middle loop is especially interesting because the movement of this loop is involved in the interaction of the activation sensors of arrestin-1 with the cytoplasmic surface of rhodopsin and it is known to prevent arrestin-1 from binding to non-preferred forms of rhodopsin (Vishnivetskiy et al. 2013). Therefore, the differences in conformational dynamics of the middle loop may be one of the factors that determine receptor selectivity of β-arrestin-1 and β-arrestin-2.

Pre-activation induced different conformational changes between β-arrestin-1 and β-arrestin-2. Pre-activation by the R169E mutation caused many regions of β-arrestin-1 to uptake higher deuterium levels compared to the wild type; these regions include β-strands I through IV, α-helix I, β-strand VII, the middle loop, XI/XII loop, β-strands XIII through XVI, and the lariat loop (Fig. 9.4a). β-strands II through IV and the lariat loop of β-arrestin-2 also showed statistically increased deuterium uptake after pre-activation. Overall, this HDX-MS study proved that β-strands II through IV and the lariat loop became more dynamic in both β-arrestin-1 and β-arrestin-2 in pre-activated compared to basal states, however β-strand I, α-helix I, β-strand VII, the middle loop, XI/XII loop, and β-strands XIII through XVI became more dynamic only in β-arrestin-1 (Fig. 9.5b). β-strands V through VI and β-strands XV through XVI in arrestin-1 and β-arrestin-1 were reported to play key roles in receptor preference (Vishnivetskiy et al. 2004). Therefore, the regions that became more dynamic only in β-arrestin-1 may also be another conformational factor that determines the differential GPCR selectivity between β-arrestin-1 and β-arrestin-2.

The different conformational dynamics between β-arrestin-1 and β-arrestin-2 in both basal and active states may result in different cellular functions other than

GPCR selectivity. Arrestins are known to interact with more than 30 protein partners, but building the interface map between them has been a slow process (Gurevich and Gurevich 2006b; Gurevich et al. 2008). The differential binding preference for other cellular molecules may be due to the differential conformational dynamics between β-arrestin-1 and β-arrestin—(for example, JNK3 preferentially interacts with β-arrestin-2 rather than β-arrestin-1) (Gurevich and Gurevich 2006b; Gurevich et al. 2008). However, since the interfaces between arrestin and its partners have not been fully mapped, it is hard to delineate what these structural differences imply in terms of their functional specialization. Therefore, completing this map will allow us to resolve the structural mechanism of arrestin's cellular functions.

Perspectives

As we have discussed in this chapter, HDX-MS has been a great help in the quest for understanding the conformational mechanism of arrestins. However, there are still more questions to be explored by HDX-MS. First, the studies that analyzed receptor-arrestin complexes by HDX-MS failed to obtain an informative HDX profile of the receptors. Although a few studies could get HDX-MS data from receptors (West et al. 2011; Duc et al. 2015), it is still challenging to analyze the conformation of membrane proteins by HDX-MS, especially when membrane proteins are analyzed with other signaling proteins (unpublished, personal experience). A more precise understanding of GPCR-mediated arrestin activation would be possible upon generating conformational information regarding GPCRs. Therefore, it will be interesting to overcome this limitation and analyze the HDX profile of receptors upon arrestin binding.

Second, the current hypothesis regarding the order of events during the GPCR-arrestin interaction is that binding of the GPCR phosphorylated residues to an arrestin (i.e., the phosphate sensor) occurs first, followed by insertion of the arrestin (i.e., the activation sensor) at the cytoplasmic face of the GPCR (Shukla et al. 2014b). This hypothesis, however, has not been experimentally confirmed. Crystal structures or HDX-MS studies of arrestins have delineated both active and inactive conformations, but crystallography cannot readily investigate intermediate states. On the other hand, HDX-MS has been used to investigate the order of events during conformational change mainly in the protein folding and denaturing process (Marcsisin and Engen 2010). It is possible that HDX-MS can also be useful in analyzing the order of events during the GPCR-arrestin interaction.

Third, a few studies have suggested a "signaling barcode" model for GPCR-arrestin signaling, e.g. binding of the differently phosphorylated GPCR C-terminal tail to arrestins induces different conformational changes (Nobles et al. 2011; Yang et al. 2015). HDX-MS can be a useful tool to understand the dynamic conformational changes induced by differently phosphorylated GPCR binding. Lastly, HDX-MS can be used for analysis of the binding interfaces between relatively weak or dynamic binding partners. There are more than 30 arrestin-binding partners including adaptor proteins and MAP kinases, but the binding interfaces are not well understood. Therefore, HDX-MS can be a complementary tool for other biochemical and biophysical techniques utilized in arrestin binding interface analysis.

References

Carter JM, Gurevich VV, Prossnitz ER, Engen JR (2005) Conformational differences between arrestin2 and pre-activated mutants as revealed by hydrogen exchange mass spectrometry. J Mol Biol 351(4):865–878. doi:10.1016/j.jmb.2005.06.048 S0022-2836(05)00720-5 [pii]

DeFea KA (2011) Beta-arrestins as regulators of signal termination and transduction: how do they determine what to scaffold? Cell Signal 23(4):621–629. doi:10.1016/j.cellsig.2010.10.004 S0898-6568(10)00291-3 [pii]

Duc NM, Du Y, Thorsen TS, Lee SY, Zhang C, Kato H, Kobilka BK, Chung KY (2015) Effective application of bicelles for conformational analysis of G protein-coupled receptors by hydrogen/deuterium exchange mass spectrometry. J Am Soc Mass Spectrom 26(5):808–817. doi:10.1007/s13361-015-1083-4

Ferguson SS, Downey WE 3rd, Colapietro AM, Barak LS, Menard L, Caron MG (1996) Role of beta-arrestin in mediating agonist-promoted G protein-coupled receptor internalization. Science 271(5247):363–366

Goodman OB Jr, Krupnick JG, Santini F, Gurevich VV, Penn RB, Gagnon AW, Keen JH, Benovic JL (1996) Beta-arrestin acts as a clathrin adaptor in endocytosis of the beta2-adrenergic receptor. Nature 383(6599):447–450. doi:10.1038/383447a0

Granzin J, Wilden U, Choe HW, Labahn J, Krafft B, Buldt G (1998) X-ray crystal structure of arrestin from bovine rod outer segments. Nature 391(6670):918–921. doi:10.1038/36147

Granzin J, Stadler A, Cousin A, Schlesinger R, Batra-Safferling R (2015) Structural evidence for the role of polar core residue Arg175 in arrestin activation. Sci Rep 5:15808. doi:10.1038/srep15808 srep15808 [pii]

Gurevich VV, Gurevich EV (2006a) The structural basis of arrestin-mediated regulation of G-protein-coupled receptors. Pharmacol Ther 110(3):465–502. doi:10.1016/j.pharmthera.2005.09.008 S0163-7258(05)00211-1 [pii]

Gurevich EV, Gurevich VV (2006b) Arrestins: ubiquitous regulators of cellular signaling pathways. Genome Biol 7(9):236. doi:10.1186/gb-2006-7-9-236 gb-2006-7-9-236 [pii]

Gurevich VV, Gurevich EV (2013) Structural determinants of arrestin functions. Prog Mol Biol Transl Sci 118:57–92. doi:10.1016/B978-0-12-394440-5.00003-6 B978-0-12-394440-5.00003-6 [pii]

Gurevich VV, Gurevich EV, Cleghorn WM (2008) Arrestins as multi-functional signaling adaptors. Handb Exp Pharmacol 186:15–37. doi:10.1007/978-3-540-72843-6_2

Han M, Gurevich VV, Vishnivetskiy SA, Sigler PB, Schubert C (2001) Crystal structure of beta-arrestin at 1.9 Å: possible mechanism of receptor binding and membrane translocation. Structure 9(9):869–880 S096921260100644X [pii]

Hirsch JA, Schubert C, Gurevich VV, Sigler PB (1999) The 2.8 Å crystal structure of visual arrestin: a model for arrestin's regulation. Cell 97(2):257–269 S0092-8674(00)80735-7 [pii]

Iacob RE, Engen JR (2012) Hydrogen exchange mass spectrometry: are we out of the quicksand? J Am Soc Mass Spectrom 23(6):1003–1010. doi:10.1007/s13361-012-0377-z

Kang Y, Zhou XE, Gao X, He Y, Liu W, Ishchenko A, Barty A, White TA, Yefanov O, Han GW, Xu Q, de Waal PW, Ke J, Tan MH, Zhang C, Moeller A, West GM, Pascal BD, Van Eps N, Caro LN, Vishnivetskiy SA, Lee RJ, Suino-Powell KM, Gu X, Pal K, Ma J, Zhi X, Boutet S, Williams GJ, Messerschmidt M, Gati C, Zatsepin NA, Wang D, James D, Basu S, Roy-Chowdhury S, Conrad CE, Coe J, Liu H, Lisova S, Kupitz C, Grotjohann I, Fromme R,

Jiang Y, Tan M, Yang H, Li J, Wang M, Zheng Z, Li D, Howe N, Zhao Y, Standfuss J, Diederichs K, Dong Y, Potter CS, Carragher B, Caffrey M, Jiang H, Chapman HN, Spence JC, Fromme P, Weierstall U, Ernst OP, Katritch V, Gurevich VV, Griffin PR, Hubbell WL, Stevens RC, Cherezov V, Melcher K, Xu HE (2015) Crystal structure of rhodopsin bound to arrestin by femtosecond X-ray laser. Nature 523(7562):561–567. doi:10.1038/nature14656 nature14656 [pii]

Katta V, Chait BT (1991) Conformational changes in proteins probed by hydrogen-exchange electrospray-ionization mass spectrometry. Rapid Commun Mass Spectrom 5(4):214–217. doi:10.1002/rcm.1290050415

Kim M, Vishnivetskiy SA, Van Eps N, Alexander NS, Cleghorn WM, Zhan X, Hanson SM, Morizumi T, Ernst OP, Meiler J, Gurevich VV, Hubbell WL (2012) Conformation of receptor-bound visual arrestin. Proc Natl Acad Sci USA 109(45):18407–18412. doi:10.1073/pnas.1216304109 1216304109 [pii]

Kim YJ, Hofmann KP, Ernst OP, Scheerer P, Choe HW, Sommer ME (2013) Crystal structure of pre-activated arrestin p44. Nature 497(7447):142–146. doi:10.1038/nature12133 nature12133 [pii]

Kim DK, Yun Y, Kim HR, Seo MD, Chung KY (2015) Different conformational dynamics of various active states of beta-arrestin1 analyzed by hydrogen/deuterium exchange mass spectrometry. J Struct Biol 190(2):250–259. doi:10.1016/j.jsb.2015.04.006 S1047-8477(15)00078-7 [pii]

Kohout TA, Lin FS, Perry SJ, Conner DA, Lefkowitz RJ (2001) Beta-arrestin 1 and 2 differentially regulate heptahelical receptor signaling and trafficking. Proc Natl Acad Sci USA 98(4):1601–1606. doi:10.1073/pnas.041608198 041608198 [pii]

Konermann L, Tong X, Pan Y (2008) Protein structure and dynamics studied by mass spectrometry: H/D exchange, hydroxyl radical labeling, and related approaches. J Mass Spectrom 43(8):1021–1036. doi:10.1002/jms.1435

Laporte SA, Oakley RH, Zhang J, Holt JA, Ferguson SS, Caron MG, Barak LS (1999) The beta2-adrenergic receptor/betaarrestin complex recruits the clathrin adaptor AP-2 during endocytosis. Proc Natl Acad Sci USA 96(7):3712–3717

Lee MH, Appleton KM, Strungs EG, Kwon JY, Morinelli TA, Peterson YK, Laporte SA, Luttrell LM (2016) The conformational signature of beta-arrestin2 predicts its trafficking and signalling functions. Nature 531(7596):665–668. doi:10.1038/nature17154 nature17154 [pii]

Lohse MJ, Hoffmann C (2014) Arrestin interactions with G protein-coupled receptors. Handb Exp Pharmacol 219:15–56. doi:10.1007/978-3-642-41199-1_2

Luttrell LM, Miller WE (2013) Arrestins as regulators of kinases and phosphatases. Prog Mol Biol Transl Sci 118:115–147. doi:10.1016/B978-0-12-394440-5.00005-X B978-0-12-394440-5.00005-X [pii]

Marcsisin SR, Engen JR (2010) Hydrogen exchange mass spectrometry: what is it and what can it tell us? Anal Bioanal Chem 397(3):967–972. doi:10.1007/s00216-010-3556-4

McDonald PH, Cote NL, Lin FT, Premont RT, Pitcher JA, Lefkowitz RJ (1999) Identification of NSF as a beta-arrestin1-binding protein. Implications for beta2-adrenergic receptor regulation. J Biol Chem 274(16):10677–10680

Nobles KN, Guan Z, Xiao K, Oas TG, Lefkowitz RJ (2007) The active conformation of beta-arrestin1: direct evidence for the phosphate sensor in the N-domain and conformational differences in the active states of beta-arrestins1 and -2. J Biol Chem 282(29):21370–21381. doi:10.1074/jbc.M611483200 M611483200 [pii]

Nobles KN, Xiao K, Ahn S, Shukla AK, Lam CM, Rajagopal S, Strachan RT, Huang TY, Bressler EA, Hara MR, Shenoy SK, Gygi SP, Lefkowitz RJ (2011) Distinct phosphorylation sites on the beta(2)-adrenergic receptor establish a barcode that encodes differential functions of beta-arrestin. Sci Signal 4(185):ra51. doi:10.1126/scisignal.2001707 4/185/ra51 [pii]

Nuber S, Zabel U, Lorenz K, Nuber A, Milligan G, Tobin AB, Lohse MJ, Hoffmann C (2016) Beta-arrestin biosensors reveal a rapid, receptor-dependent activation/deactivation cycle. Nature 531(7596):661–664. doi:10.1038/nature17198 nature17198 [pii]

Oakley RH, Laporte SA, Holt JA, Caron MG, Barak LS (2000) Differential affinities of visual arrestin, beta arrestin1, and beta arrestin2 for G protein-coupled receptors delineate two major classes of receptors. J Biol Chem 275(22):17201–17210. doi:10.1074/jbc.M910348199 M910348199 [pii]

Ostermaier MK, Schertler GF, Standfuss J (2014) Molecular mechanism of phosphorylation-dependent arrestin activation. Curr Opin Struct Biol 29:143–151. doi:10.1016/j.sbi.2014.07.006 S0959-440X(14)00078-5 [pii]

Park JY, Lee SY, Kim HR, Seo MD, Chung KY (2016) Structural mechanism of GPCR-arrestin interaction: recent breakthroughs. Arch Pharm Res 39(3):293–301. doi:10.1007/s12272-016-0712-1 10.1007/s12272-016-0712-1 [pii]

Shenoy SK, Lefkowitz RJ (2011) Beta-arrestin-mediated receptor trafficking and signal transduction. Trends Pharmacol Sci 32(9):521–533. doi:10.1016/j.tips.2011.05.002 S0165-6147(11)00080-0 [pii]

Shukla AK, Manglik A, Kruse AC, Xiao K, Reis RI, Tseng WC, Staus DP, Hilger D, Uysal S, Huang LY, Paduch M, Tripathi-Shukla P, Koide A, Koide S, Weis WI, Kossiakoff AA, Kobilka BK, Lefkowitz RJ (2013) Structure of active beta-arrestin-1 bound to a G-protein-coupled receptor phosphopeptide. Nature 497(7447):137–141. doi:10.1038/nature12120 nature12120 [pii]

Shukla AK, Singh G, Ghosh E (2014a) Emerging structural insights into biased GPCR signaling. Trends Biochem Sci 39(12):594–602. doi:10.1016/j.tibs.2014.10.001 S0968-0004(14)00183-2 [pii]

Shukla AK, Westfield GH, Xiao K, Reis RI, Huang LY, Tripathi-Shukla P, Qian J, Li S, Blanc A, Oleskie AN, Dosey AM, Su M, Liang CR, Gu LL, Shan JM, Chen X, Hanna R, Choi M, Yao XJ, Klink BU, Kahsai AW, Sidhu SS, Koide S, Penczek PA, Kossiakoff AA, Woods VL Jr, Kobilka BK, Skiniotis G, Lefkowitz RJ (2014b) Visualization of arrestin recruitment by a G-protein-coupled receptor. Nature 512(7513):218–222. doi:10.1038/nature13430 nature13430 [pii]

Smith JS, Rajagopal S (2016) The beta-arrestins: multifunctional regulators of G protein-coupled receptors. J Biol Chem 291(17):8969–8977. doi:10.1074/jbc.R115.713313 R115.713313 [pii]

Srivastava A, Gupta B, Gupta C, Shukla AK (2015) Emerging functional divergence of beta-arrestin isoforms in GPCR function. Trends Endocrinol Metab 26(11):628–642. doi:10.1016/j.tem.2015.09.001 S1043-2760(15)00175-7 [pii]

Tohgo A, Choy EW, Gesty-Palmer D, Pierce KL, Laporte S, Oakley RH, Caron MG, Lefkowitz RJ, Luttrell LM (2003) The stability of the G protein-coupled receptor-beta-arrestin interaction determines the mechanism and functional consequence of ERK activation. J Biol Chem 278(8):6258–6267. doi:10.1074/jbc.M212231200 M212231200 [pii]

Violin JD, Crombie AL, Soergel DG, Lark MW (2014) Biased ligands at G-protein-coupled receptors: promise and progress. Trends Pharmacol Sci 35(7):308–316. doi:10.1016/j.tips.2014.04.007 S0165-6147(14)00069-8 [pii]

Vishnivetskiy SA, Hosey MM, Benovic JL, Gurevich VV (2004) Mapping the arrestin-receptor interface. Structural elements responsible for receptor specificity of arrestin proteins. J Biol Chem 279(2):1262–1268. doi:10.1074/jbc.M308834200 M308834200 [pii]

Vishnivetskiy SA, Baameur F, Findley KR, Gurevich VV (2013) Critical role of the central 139-loop in stability and binding selectivity of arrestin-1. J Biol Chem 288(17):11741–11750. doi:10.1074/jbc.M113.450031 M113.450031 [pii]

West GM, Chien EY, Katritch V, Gatchalian J, Chalmers MJ, Stevens RC, Griffin PR (2011) Ligand-dependent perturbation of the conformational ensemble for the GPCR beta2 adrenergic receptor revealed by HDX. Structure 19(10):1424–1432. doi:10.1016/j.str.2011.08.001 S0969-2126(11)00278-4 [pii]

Whalen EJ, Rajagopal S, Lefkowitz RJ (2011) Therapeutic potential of beta-arrestin- and G protein-biased agonists. Trends Mol Med 17(3):126–139. doi:10.1016/j.molmed.2010.11.004 S1471-4914(10)00182-6 [pii]

Xiao K, Chung J, Wall A (2015) The power of mass spectrometry in structural characterization of GPCR signaling. J Recept Signal Transduct Res 35(3):213–219. doi:10.3109/10799893.2015.1072979

Yang F, Yu X, Liu C, Qu CX, Gong Z, Liu HD, Li FH, Wang HM, He DF, Yi F, Song C, Tian CL, Xiao KH, Wang JY, Sun JP (2015) Phospho-selective mechanisms of arrestin conformations and functions revealed by unnatural amino acid incorporation and (19)F-NMR. Nat Commun 6:8202. doi:10.1038/ncomms9202 ncomms9202 [pii]

Yun Y, Kim DK, Seo MD, Kim KM, Chung KY (2015) Different conformational dynamics of beta-arrestin1 and beta-arrestin2 analyzed by hydrogen/deuterium exchange mass spectrometry. Biochem Biophys Res Commun 457(1):50–57. doi:10.1016/j.bbrc.2014.12.079 S0006-291X(14)02261-X [pii]

Zhan X, Gimenez LE, Gurevich VV, Spiller BW (2011) Crystal structure of arrestin-3 reveals the basis of the difference in receptor binding between two non-visual subtypes. J Mol Biol 406 (3):467–478. doi:10.1016/j.jmb.2010.12.034 S0022-2836(10)01366-5 [pii]

Zhuang T, Chen Q, Cho MK, Vishnivetskiy SA, Iverson TM, Gurevich VV, Sanders CR (2013) Involvement of distinct arrestin-1 elements in binding to different functional forms of rhodopsin. Proc Natl Acad Sci USA 110(3):942–947. doi:10.1073/pnas.1215176110 1215176110 [pii]

Zhuo Y, Vishnivetskiy SA, Zhan X, Gurevich VV, Klug CS (2014) Identification of receptor binding-induced conformational changes in non-visual arrestins. J Biol Chem 289(30):20991–21002. doi:10.1074/jbc.M114.560680 M114.560680 [pii]

Chapter 10
GPCR Footprint on Arrestins and Manipulation of Receptor Specificity

Sergey A. Vishnivetskiy, Wayne L. Hubbell, Candice S. Klug and Vsevolod V. Gurevich

Abstract A wide variety of methods were used to identify where exactly do GPCRs bind arrestins. Experiments implicated large part of the concave sides of the two arrestin domains and identified surprisingly few residues on this extensive surface that determine which receptor does each arrestin prefer. Manipulation of these “receptor-discriminator” residues in one of the non-visual subtypes, arrestin-3, yielded non-visual arrestins with greatly enhanced receptor specificity. Targeting particular receptors with arrestin mutants with special functional characteristics, such as the ability to bind unphosphorylated GPCRs, activate or fail to activate select pathways of arrestin-mediated signaling opens the prospect of creating “designer” arrestins to direct the signaling from GPCRs of our choice to the pathways we want.

Keywords Arrestin · GPCR · Receptor binding · Receptor specificity · Signaling bias

Arrestin was first discovered in the visual system for its ability to bind light-activated phosphorylated rhodopsin (Kuhn et al. 1984) (P-Rh*) and preclude (arrest) further G protein activation (Wilden et al. 1986). This raised a question of where exactly does the receptor bind arrestin. This issue was addressed long before

---

S.A. Vishnivetskiy · V.V. Gurevich (✉)
Department of Pharmacology, Vanderbilt University, 2200 Pierce Avenue, Preston Research Building, Rm. 454, Nashville, TN 3723, USA
e-mail: vsevolod.gurevich@vanderbilt.edu

W.L. Hubbell
Department of Ophthalmology, Jules Stein Eye Institute, University of California, Los Angeles, Los Angeles, CA 90095, USA

C.S. Klug
Department of Biophysics, Medical College of Wisconsin, Milwaukee, WI 53226, USA

V.V. Gurevich (ed.), *The Structural Basis of Arrestin Functions*,
DOI 10.1007/978-3-319-57553-7_10

any structural information became available. Progressive truncations of visual arrestin-1[1] removing increasing parts of it starting from the C-terminus (Gurevich and Benovic 1992; Gurevich and Benovic 1993) showed that both the N- and C-terminal halves of the 404-residue molecule have rhodopsin-binding elements. The N-terminal 1-191 piece bound P-Rh* best, but its binding to this form of rhodopsin was roughly equal to the sum of the binding to inactive phosphorylated (P-Rh) and active unphosphorylated (Rh*) forms, in sharp contrast to the full-length protein, where the binding to P-Rh* was many times greater than the sum of P-Rh and Rh* binding. These data suggested that the C-terminal half also contains important rhodopsin-binding elements (Gurevich and Benovic 1992; Gurevich and Benovic 1993). However, limited resolution of truncation mutagenesis did not allow the identification of arrestin residues engaged by the rhodopsin more precisely. The first non-visual subtype cloned, termed β-arrestin at the time (current systematic name arrestin-2) showed clear preference for non-visual GPCRs over rhodopsin, in contrast to the visual arrestin-1 (Lohse et al. 1992; Lohse et al. 1990). Functional characterization of the first arrestin-1/2 chimeras implicated the central part of the molecule in receptor interactions, but did not improve the precision of the identification of receptor-binding elements (Gurevich and Benovic 1993).

The next attempt used H/D exchange in arrestin, based on a logical assumption that the H/D exchange would be slowed by rhodopsin in parts of arrestin that it binds (Ohguro et al. 1994). This work identified many receptor-binding elements. Individual phosphate-binding and other residues likely to contact the receptor in arrestin-1 were also identified by mutagenesis (Gurevich and Benovic 1995, 1997). As soon as the first crystal structures of arrestin-1 revealed that arrestin-1 is a two-domain molecule (Granzin et al. 1998; Hirsch et al. 1999), localization of the residues implicated by the H/D exchange and mutagenesis suggested that receptor footprint localizes to the concave sides of both domains (Hirsch et al. 1999). This conclusion was further supported by targeted mutagenesis (Hanson et al. 2006a, b; Vishnivetskiy et al. 2000) and extensive site-directed spin labeling/EPR studies (Hanson et al. 2006a, b), which indicated that the finger loop (Hirsch et al. 1999) (see Fig. 1.1 in Chap. 1) in the central crest of the receptor-binding side of arrestin is engaged by P-Rh, and even stronger by P-Rh* (Hanson et al. 2006a, b). Interestingly, the same surface with essentially the same elements and residues in arrestin-1 and non-visual arrestin-2 were implicated in receptor binding using chimera construction, mutagenesis, and EPR (Vishnivetskiy et al. 2004, 2011).

It is important to note that all four vertebrate arrestin subtypes, visual arrestin-1 (Granzin et al. 1998; Hirsch et al. 1999), cone-specific arrestin-4 (Sutton et al. 2005), as well as non-visual arrestin-2 (Han et al. 2001; Milano et al. 2002) and arrestin-3 (Zhan et al. 2011), have very similar structures, where essentially the

[1]Here we use the systematic names of arrestin proteins: arrestin-1 (historic names S-antigen, 48 kDa protein, visual or rod arrestin), arrestin-2 (β-arrestin or β-arrestin1), arrestin-3 (β-arrestin2 or hTHY-ARRX), and arrestin-4 (cone or X-arrestin; for unclear reasons its gene is called "arrestin 3" in the HUGO database).

same intra-molecular interactions stabilize their basal conformations. As could be expected, the ability of all arrestins to undergo transition into high-affinity receptor-binding state is associated with disruption of these interactions, which depends on the residues forming the "clasps", rather than receptor-binding elements. All arrestins can be "pre-activated" by mutations disrupting the polar core and the three-element interaction that anchors the C-tail to the body of the N-domain (Sutton et al. 2005; Vishnivetskiy et al. 1999; Celver et al. 2002; Gurevich 1998; Kovoor et al. 1999) (see Fig. 1.1 in Chap. 1). The conformational changes accompanying arrestin activation and the actual shape of the "active" receptor-bound arrestins are discussed in Chaps. 1 and 11–14.

While the surface area of arrestins implicated in the receptor binding is rather extensive, surprisingly few residues determine arrestin preference for particular GPCRs. The exchange of two elements, comprising residues 49–90 and 237–268 in bovine arrestin-1, which correspond to residues 45–86 and 231–262 in bovine arrestin-2, completely switches receptor preference from rhodopsin, characteristic for arrestin-1, to M2 muscarinic receptor, that arrestin-1 does not bind well, whereas arrestin-2 binds with high affinity (Vishnivetskiy et al. 2004) (Fig. 10.1). Considering fairly high homology among arrestins (Gurevich and Gurevich 2006), these data suggest that no more than 10–12 surface residues determine receptor specificity. Indeed, replacing these key residues in arrestin-2 with their arrestin-1 homologues yields an arrestin that binds P-Rh* as well as WT arrestin-1 (Vishnivetskiy et al. 2011). Importantly, the residues that determine receptor specificity appear to be key drivers of the arrestin-GPCR interaction: replacement of ten of these residues along with the two phosphate-binding lysines in β-strand I (see Chap. 6) with alanines (KNC mutations) yielded versions of arrestin-1, -2, and -3

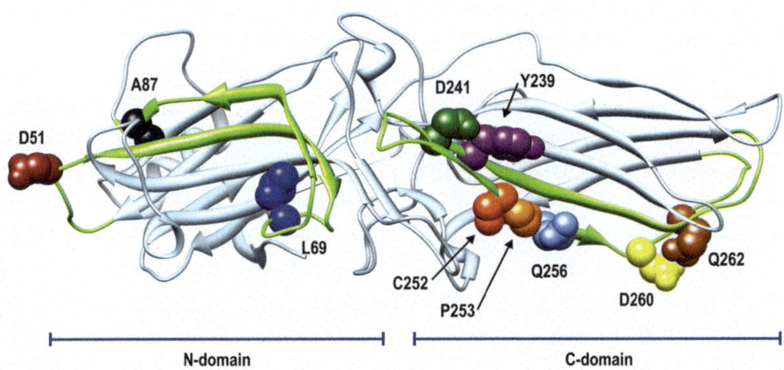

Fig. 10.1 Receptor-discriminator residues in arrestin-3. The structure of basal arrestin-3 (PDB 3P2D) (Zhan et al. 2011) is shown. The regions that determine receptor specificity identified using arrestin-1/2 chimeras (Vishnivetskiy et al. 2004) are shown in green. Individual residues the mutations of which dramatically change receptor preference of arrestin-3 (Gimenez et al. 2012b, Gimenez et al. 2014) are shown as CPK models

that do not bind receptors (Vishnivetskiy et al. 2011). These mutants turned out to be useful negative controls in experiments where the binding of arrestins and different GPCRs was measured in living cells using BRET between luciferase-tagged receptors and Venus-tagged arrestins (Gimenez et al. 2012a, b, 2014).

These results paved the way to an attempt to make inherently promiscuous non-visual arrestins that bind numerous GPCRs (Gimenez et al. 2012b; Barak et al. 1997) more specific for particular receptors. Interestingly, it turned out that one of the non-exposed residues plays critical role in GPCR specificity of arrestins. In the arrestin-1, which is quite specific for rhodopsin, Val90 localized between the two β-strand sheets in the "sandwich" of the N-domain, interacts with several hydrophobic residues (Val45, Val57, Val59, Phe118), which apparently stabilizes the β-sandwich (Hirsch et al. 1999). Its potential partners are conserved in arrestin-2 (Han et al. 2001) and arrestin-3 (Zhan et al. 2011), but this residue is a serine or alanine, respectively, in these non-visual subtypes with broad receptor specificity (Barak et al. 1997; Gurevich et al. 1995). In fact, the replacement of Val90 in arrestin-1 with a serine (the homolog in arrestin-2 is Ser86) increases arrestin-1 binding to non-visual receptors more than many other mutations (Han et al. 2001), suggesting that structurally rigid N-domain predisposes an arrestin to be specific for particular GPCRs. Thus, the construction of receptor-specific variants was performed in the context of arrestin-3-Ala87Val base mutant (Gimenez et al. 2012b). Naturally, placing all 20 possible residues into 10 positions would yield a huge number of combinations that cannot be tested experimentally. However, the analysis of arrestin evolution suggests that very few residues were in the key positions in the last ~600 million years (Gurevich and Gurevich 2006). Placing in key positions only residues that are found there in existing arrestins reduces the number of combinations that need to be tested to manageable. The first experiments involving manipulation of 11 surface residues yielded mutants that showed up to 5-fold discrimination among four GPCRs: β2-adrenergic (β2AR), M2 muscarinic, D1 and D2 dopamine receptors (Gimenez et al. 2012b) (Fig. 10.1), in contrast to the parental arrestin-3 that binds all these receptors similarly. The combination of mutations changing receptor preference in the same direction yielded arrestin-3 variant (Asp260Lys + Gln262Pro) with >60-fold preference of other GPCRs over β2AR (Gimenez et al. 2012b), confirming the potential of this approach. The mutations tested in this study similarly affected basal (agonist-independent) arrestin binding to these receptors and agonist-induced increase in the binding (Gimenez et al. 2012b). The next series of arrestin-3 mutants had two-residue substitutions and insertions in some of these receptor-discriminator elements, mimicking the sequence of arrestins from invertebrate species (*C. elegans*, *Drosophila*, and the bee *Apis mellifera*) (Gimenez et al. 2014). Some of these new mutations differentially affected pre-docking to inactive Y1 and Y2 neuropeptide Y receptors and agonist-induced increase in binding, indicating that these two phenomena are separable. One mutation, Tyr239Thr, increased the preference for Y1 over Y2 receptor >5-fold (Gimenez et al. 2014). Interestingly, virtually none of the mutations in either series significantly affected arrestin-3 binding to D1 dopamine and Y1 neuropeptide Y receptors (Gimenez et al. 2012b, 2014), suggesting that additional arrestin elements must play key role in these interactions.

The identification of additional targets for mutagenesis to enhance receptor specificity was greatly facilitated by the structure of the complex of mouse arrestin-1 with rhodopsin (Kang et al. 2015). This structure revealed four main patches of the arrestin-receptor interface, which include some elements that were not identified as receptor discriminators by arrestin-1/2 chimera analysis (Vishnivetskiy et al. 2004, 2011). The central finger loop, which was identified as key receptor-binding element by EPR (Hanson et al. 2006a, b; Vishnivetskiy et al. 2011) and chimera construction (Vishnivetskiy et al. 2004), was found to become helical and insert into the cavity that opens between the cytoplasmic sides of the trans-membrane rhodopsin helices (TM) upon activation (Farrens et al. 1996). The arrestin finger loop interacts with the C-terminus of TM7, the N-terminus of helix 8, and rhodopsin intracellular loop (ICL) 1 (Kang et al. 2015). The second patch includes arrestin middle loop [(Shukla et al. 2013); also termed 139-loop in arrestin-1 (Kim et al. 2012; Vishnivetskiy et al. 2013)] (around residue Val140), the C-loop [around Tyr 251; this element was identified in chimeras (Vishnivetskiy et al. 2004, 2011)] that bind ICL2 of rhodopsin, and arrestin back loop (Arg319 and Thr320) that binds the C-terminus of rhodopsin TM5. The third patch includes arrestin β-strand following the finger loop (residues 79–86) interacting with rhodopsin TM5, TM6, and ICL3. The fourth patch includes the interaction of arrestin β-strand I with the rhodopsin C-terminus. This interaction was predicted based on the presence of phosphate-binding lysines in the arrestin β-strand I (Vishnivetskiy et al. 2000) and localization of all phosphorylation sites in the rhodopsin C-terminus (Vishnivetskiy et al. 2007; Azevedo et al. 2015). This latter patch likely involves receptor-attached phosphates, and therefore is unlikely to play a significant role in receptor specificity. Thus, the structure of the complex (Kang et al. 2015) identified new targets that need to be explored in a quest to construct non-visual arrestins with high receptor specificity.

To summarize, the comparison of the receptor-G protein complexes (Rasmussen et al. 2007; Carpenter et al. 2016) and the receptor-arrestin complex (Kang et al. 2015) reveals the structural basis of the competition between G proteins and arrestins for the active receptor described earlier (Wilden et al. 1986; Krupnick et al. 1997; Wilden 1995): both classes of proteins interact with the inter-helical cavity on the intracellular side and many of the same cytoplasmic elements of the receptor. Biochemical, biophysical, and structural data mapped an extensive arrestin-receptor interface and identified arrestin residues within it responsible for the receptor preference of arrestin proteins. This paves the way to the construction of non-visual arrestins targeting particular GPCRs.

References

Azevedo AW, Doan T, Moaven H, Sokal I, Baameur F, Vishnivetskiy SA, Homan KT, Tesmer JJ, Gurevich VV, Chen J, Rieke F (2015) C-terminal threonines and serines play distinct roles in the desensitization of rhodopsin, a G protein-coupled receptor. Elife 4:05981

Barak LS, Ferguson SS, Zhang J, Caron MG (1997) A beta-arrestin/green fluorescent protein biosensor for detecting G protein-coupled receptor activation. J Biol Chem 272:27497–27500

Carpenter B, Nehmé R, Warne T, Leslie AG, Tate CG (2016) Structure of the adenosine A(2A) receptor bound to an engineered G protein. Nature 536:104–107

Celver J, Vishnivetskiy SA, Chavkin C, Gurevich VV (2002) Conservation of the phosphate-sensitive elements in the arrestin family of proteins. J Biol Chem 277:9043–9048

Farrens DL, Altenbach C, Yang K, Hubbell WL, Khorana HG (1996) Requirement of rigid-body motion of transmembrane helices for light activation of rhodopsin. Science 274:768–770

Gimenez LE, Kook S, Vishnivetskiy SA, Ahmed MR, Gurevich EV, Gurevich VV (2012a) Role of receptor-attached phosphates in binding of visual and non-visual arrestins to G protein-coupled receptors. J Biol Chem 287:9028–9040

Gimenez LE, Vishnivetskiy SA, Baameur F, Gurevich VV (2012b) Manipulation of very few receptor discriminator residues greatly enhances receptor specificity of non-visual arrestins. J Biol Chem 287:29495–29505

Gimenez LE, Babilon S, Wanka L, Beck-Sickinger AG, Gurevich VV (2014) Mutations in arrestin-3 differentially affect binding to neuropeptide Y receptor subtypes. Cell Signal 26:1523–1531

Granzin J, Wilden U, Choe HW, Labahn J, Krafft B, Buldt G (1998) X-ray crystal structure of arrestin from bovine rod outer segments. Nature 391:918–921

Gurevich VV (1998) The selectivity of visual arrestin for light-activated phosphorhodopsin is controlled by multiple nonredundant mechanisms. J Biol Chem 273:15501–15506

Gurevich VV, Benovic JL (1992) Cell-free expression of visual arrestin. Truncation mutagenesis identifies multiple domains involved in rhodopsin interaction. J Biol Chem 267:21919–21923

Gurevich VV, Benovic JL (1993) Visual arrestin interaction with rhodopsin: sequential multisite binding ensures strict selectivity towards light-activated phosphorylated rhodopsin. J Biol Chem 268:11628–11638

Gurevich VV, Benovic JL (1995) Visual arrestin binding to rhodopsin: diverse functional roles of positively charged residues within the phosphorylation-recignition region of arrestin. J Biol Chem 270:6010–6016

Gurevich VV, Benovic JL (1997) Mechanism of phosphorylation-recognition by visual arrestin and the transition of arrestin into a high affinity binding state. Mol Pharmacol 51:161–169

Gurevich EV, Gurevich VV (2006) Arrestins are ubiquitous regulators of cellular signaling pathways. Genome Biol 7:236

Gurevich VV, Dion SB, Onorato JJ, Ptasienski J, Kim CM, Sterne-Marr R, Hosey MM, Benovic JL (1995) Arrestin interaction with G protein-coupled receptors. Direct binding studies of wild type and mutant arrestins with rhodopsin, b2-adrenergic, and m2 muscarinic cholinergic receptors. J Biol Chem 270:720–731

Han M, Gurevich VV, Vishnivetskiy SA, Sigler PB, Schubert C (2001) Crystal structure of beta-arrestin at 1.9 Å: possible mechanism of receptor binding and membrane translocation. Structure 9:869–880

Hanson SM, Francis DJ, Vishnivetskiy SA, Klug CS, Gurevich VV (2006a) Visual arrestin binding to microtubules involves a distinct conformational change. J Biol Chem 281: 9765–9772

Hanson SM, Francis DJ, Vishnivetskiy SA, Kolobova EA, Hubbell WL, Klug CS, Gurevich VV (2006b) Differential interaction of spin-labeled arrestin with inactive and active phosphorhodopsin. Proc Natl Acad Sci USA 103:4900–4905

Hirsch JA, Schubert C, Gurevich VV, Sigler PB (1999) The 2.8 Å crystal structure of visual arrestin: a model for arrestin's regulation. Cell 97:257–269

Kang Y, Zhou XE, Gao X, He Y, Liu W, Ishchenko A, Barty A, White TA, Yefanov O, Han GW, Xu Q, de Waal PW, Ke J, Tan MHE, Zhang C, Moeller A, West GM, Van Eps N, Caro LN, Vishnivetskiy SA, Lee RJ, Suino-Powell KM, Gu X, Pal K, Ma J, Zhi X, Boutet S, Williams GJ, Messerschmidt M, Gati C, Zatsepin NA, Wang D, James D, Basu S, Roy-Chowdhuty S, Conrad S, Coe J, Liu H, Lisova S, Kupitz C, Grotjohann I, Fromme R, Jiang Y, Tan M, Yang H, Li J, Wang M, Zheng Z, Li D, Zhao Y, Standfuss J, Diederichs K, Dong Y, Potter CS, Carragher B, Caffrey M, Jiang H, Chapman HN, Spence JCH, Fromme P, Weierstall U, Ernst OP, Katritch V, Gurevich VV, Griffin PR, Hubbell WL, Stevens RC, Cherezov V, Melcher K, Xu HE (2015) Crystal structure of rhodopsin bound to arrestin determined by femtosecond X-ray laser. Nature 523:561–567

Kim M, Vishnivetskiy SA, Van Eps N, Alexander NS, Cleghorn WM, Zhan X, Hanson SM, Morizumi T, Ernst OP, Meiler J, Gurevich VV, Hubbell WL (2012) Conformation of receptor-bound visual arrestin. Proc Nat Acad Sci USA 109:18407–18412

Kovoor A, Celver J, Abdryashitov RI, Chavkin C, Gurevich VV (1999) Targeted construction of phosphorylation-independent b-arrestin mutants with constitutive activity in cells. J Biol Chem 274:6831–6834

Krupnick JG, Gurevich VV, Benovic JL (1997) Mechanism of quenching of phototransduction. Binding competition between arrestin and transducin for phosphorhodopsin. J Biol Chem 272:18125–18131

Kuhn H, Hall SW, Wilden U (1984) Light-induced binding of 48-kDa protein to photoreceptor membranes is highly enhanced by phosphorylation of rhodopsin. FEBS Lett 176:473–478

Lohse MJ, Benovic JL, Codina J, Caron MG, Lefkowitz RJ (1990) Beta-arrestin: a protein that regulates beta-adrenergic receptor function. Science 248:1547–1550

Lohse MJ, Andexinger S, Pitcher J, Trukawinski S, Codina J, Faure JP, Caron MG, Lefkowitz RJ (1992) Receptor-specific desensitization with purified proteins. Kinase dependence and receptor specificity of beta-arrestin and arrestin in the beta 2-adrenergic receptor and rhodopsin systems. J Biol Chem 267:8558–8564

Milano SK, Pace HC, Kim YM, Brenner C, Benovic JL (2002) Scaffolding functions of arrestin-2 revealed by crystal structure and mutagenesis. Biochemistry 41:3321–3328

Ohguro H, Palczewski K, Walsh KA, Johnson RS (1994) Topographic study of arrestin using differential chemical modifications and hydrogen/deuterium exchange. Protein Sci 3: 2428–2434

Rasmussen SG, Choi HJ, Rosenbaum DM, Kobilka TS, Thian FS, Edwards PC, Burghammer M, Ratnala VR, Sanishvili R, Fischetti RF, Schertler GF, Weis WI, Kobilka BK (2007) Crystal structure of the human beta2 adrenergic G-protein-coupled receptor. Nature 450:383–387

Shukla AK, Manglik A, Kruse AC, Xiao K, Reis RI, Tseng WC, Staus DP, Hilger D, Uysal S, Huang LY, Paduch M, Tripathi-Shukla P, Koide A, Koide S, Weis WI, Kossiakoff AA, Kobilka BK, Lefkowitz RJ (2013) Structure of active beta-arrestin-1 bound to a G-protein-coupled receptor phosphopeptide. Nature 497:137–141

Sutton RB, Vishnivetskiy SA, Robert J, Hanson SM, Raman D, Knox BE, Kono M, Navarro J, Gurevich VV (2005) Crystal structure of cone arrestin at 2.3 Å: evolution of receptor specificity. J Mol Biol 354:1069–1080

Vishnivetskiy SA, Paz CL, Schubert C, Hirsch JA, Sigler PB, Gurevich VV (1999) How does arrestin respond to the phosphorylated state of rhodopsin? J Biol Chem 274:11451–11454

Vishnivetskiy SA, Schubert C, Climaco GC, Gurevich YV, Velez M-G, Gurevich VV (2000) An additional phosphate-binding element in arrestin molecule: implications for the mechanism of arrestin activation. J Biol Chem 275:41049–41057

Vishnivetskiy SA, Hosey MM, Benovic JL, Gurevich VV (2004) Mapping the arrestin-receptor interface: structural elements responsible for receptor specificity of arrestin proteins. J Biol Chem 279:1262–1268

Vishnivetskiy SA, Raman D, Wei J, Kennedy MJ, Hurley JB, Gurevich VV (2007) Regulation of arrestin binding by rhodopsin phosphorylation level. J Biol Chem 282:32075–32083

Vishnivetskiy SA, Gimenez LE, Francis DJ, Hanson SM, Hubbell WL, Klug CS, Gurevich VV (2011) Few residues within an extensive binding interface drive receptor interaction and determine the specificity of arrestin proteins. J Biol Chem 286:24288–24299

Vishnivetskiy SA, Baameur F, Findley KR, Gurevich VV (2013) Critical role of the central 139-loop in stability and binding selectivity of arrestin-1. J Biol Chem 288:11741–11750

Wilden U (1995) Duration and amplitude of the light-induced cGMP hydrolysis in vertebrate photoreceptors are regulated by multiple phosphorylation of rhodopsin and by arrestin binding. Biochemistry 34:1446–1454

Wilden U, Hall SW, Kuhn H (1986) Phosphodiesterase activation by photoexcited rhodopsin is quenched when rhodopsin is phosphorylated and binds the intrinsic 48-kDa protein of rod outer segments. Proc Natl Acad Sci 83:1174–1178

Zhan X, Gimenez LE, Gurevich VV, Spiller BW (2011) Crystal structure of arrestin-3 reveals the basis of the difference in receptor binding between two non-visual arrestins. J Mol Biol 406:467–478

Part III
Arrestin Flexibility and Activation

Chapter 11
The Structure of the Polar Core Mutant R175E and Its Functional Implications

Renu Batra-Safferling and Joachim Granzin

Abstract Mutation of arginine 175 to glutamic acid (R175E), a central residue in the polar core and previously predicted as the 'phosphosensor', leads to a constitutively active arrestin that is able to terminate phototransduction by binding to non-phosphorylated, light-activated rhodopsin. Crystal structure of a R175E mutant arrestin at 2.7 Å resolution reveals significant differences compared to the basal state reported in full-length arrestin structures. Most striking differences are disruption of hydrogen bond network in the polar core, and three-element interaction (between β-strand I, α-helix I, and the C-tail), including disordering of several residues in the receptor-binding finger loop and the C-terminus (residues 361–404). Additionally, R175E structure shows a 7.5° rotation of the amino and carboxy-terminal domains relative to each other. Comparison of the crystal structures of basal arrestin and R175E mutant provides insights into the mechanism of arrestin activation, where the latter likely represents an intermediate activation state prior to formation of the high-affinity complex with the G protein-coupled receptor.

Keywords Arrestin · Phosphorylation-independent · Photoreceptor · Polar core · Rhodopsin

Introduction

Mutagenesis in the polar core region results in phosphorylation-independent arrestins

---

R. Batra-Safferling (✉) · J. Granzin
Institute of Complex Systems (ICS-6), Structural Biochemistry, Forschungszentrum Jülich, 52425 Jülich, Germany
e-mail: r.batra-safferling@fz-juelich.de

V.V. Gurevich (ed.), *The Structural Basis of Arrestin Functions*,
DOI 10.1007/978-3-319-57553-7_11

Selective binding of visual arrestin-1 to light-activated, phosphorylated rhodopsin in rod photoreceptors led the scientists to look for the putative phosphate sensors in arrestin (Gurevich and Benovic 1993; Wilden et al. 1986). The first logical explanation proposed was that for high-affinity binding, arrestin interacts with rhodopsin-attached phosphates via multiple positive charges. No structural information was available on any of the arrestin family members at that time. Sequence analysis led to the identification of a linear stretch of a short basic region (residues 163–176) including six positive charges (Gurevich and Benovic 1995). Site-directed mutagenesis of these residues was performed in bovine arrestin-1, where the expected result was that mutagenesis of the phosphate sensor residue will result in 'activation' of arrestin-1, increasing its binding to light-activated rhodopsin (Rh*). As reported previously and discussed in detail in Chap. 5: 'Phosphate sensor and construction of phosphorylation-independent arrestins' of this book, substitution of five positive charges (K163S, K166S, K167S, R171Q, and K176S) resulted in decreased binding to light-activated, phosphorylated rhodopsin (P-Rh*) with no effect on Rh* interaction (Gurevich and Benovic 1995). Mutation of the sixth residue R175 into neutral or negatively charged residue, however, showed a significant increase in Rh* binding that was equivalent to the wild type arrestin-1 binding to P-Rh*. This finding thus matched well the predicted result of activation of the phosphate sensor, a role taken by R175 in arrestin-1. In the later studies, R175 was substituted with every possible residue (Gurevich and Benovic 1997). The resulting mutants with negatively charged residues in this position showed the highest increase in Rh* binding. In contrast, the positively charged residue Lys showed no activation of arrestin-1. Based on these results, the simplest model proposed was that in the basal state, R175 is engaged in an intramolecular interaction with a negatively charged residue. Disruption of this interaction is induced by receptor-attached phosphates, signaling the arrestin molecule to 'dock' on the receptor (Gurevich et al. 2011). The initial extensive mutagenesis studies confirmed first, the functional significance of the role of R175 in its direct binding to the negatively charged phosphate group from the phosphorylated receptor; and second, the assumption that disruption of R175 interaction with an internal negatively charged residue underlies activation of arrestin-1, enabling its binding to the receptor (Gray-Keller et al. 1997; Gurevich and Benovic 1997). Residue R175 was thus proposed to function as a phosphorylation-sensitive trigger.

Crystal structures of native and recombinant bovine rod arrestins were published soon after, showing an all β-strand two-domain structure where R175 is the central residue of the polar core that forms the most distinct feature of arrestin conformation (Granzin et al. 1998; Hirsch et al. 1999) (Also see Chap. 2: 'The first structures of arrestin-1'). Sandwiched between the N- and the C-domains, the polar core comprises charged residues from the N-domain (D30, R175), the interfacial gate loop of the C-domain (D296, D303), and the C-tail (R382) (Fig. 11.1). These residues show high degree of sequence and structural conservation among visual and non-visual members of the arrestin family, as evident in the later published crystal structures of cone arrestins (Sutton et al. 2005) and in β-arrestin (Han et al. 2001; Milano et al. 2002). Subsequent exhaustive mutagenesis reports identified

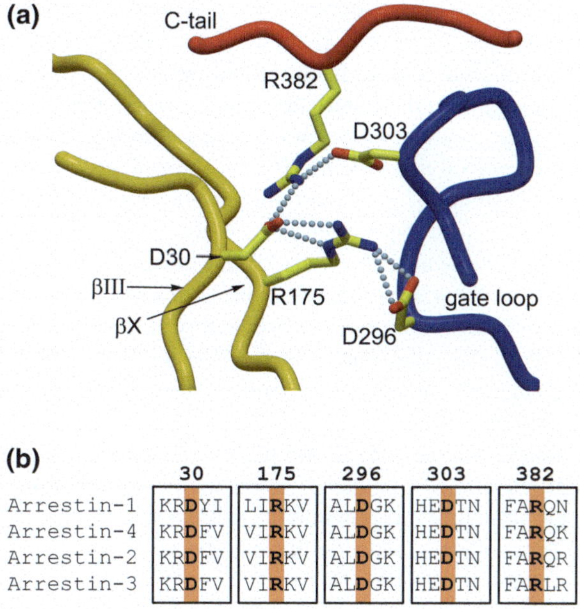

Fig. 11.1 The polar core region in arrestins. **a** A network of ionic interactions among five conserved charged residues that stabilizes the basal conformation of arrestin: Positively charged R175 (N-domain, *yellow*), R382 (C-tail, *red*) and negatively charged residues D30 (N-domain), D296 and D303 (C-domain, *blue*). The residues are shown as stick models, colored by element: carbon, *yellow*; nitrogen, *blue*; oxygen, *red*. **b** Primary sequence alignment of the polar core residues of four vertebrate arrestins (Arrestin-1, P08168 from *Bos taurus*; Arrestin-4, Q9PTE7 from *Ambystoma tigrinum*; Arrestin-2, P17870 from *Bos taurus*; and Arrestin-3, P32120 from *Bos taurus*). The polar core residues are highlighted in *orange* and the numbering on top refers to that of arrestin-1. Sequence alignments were performed using BLAST (http://www.uniprot.org)

other polar core residues in arrestin (D30, R175, D296, D303, R382) which when substituted result in constitutively active mutant arrestins that bind the light-activated receptors independently of their phosphorylation (Celver et al. 2002; Kovoor et al. 1999; Pan et al. 2003; Vishnivetskiy et al. 1999). The crystal structure revealed a central salt bridge between R175 and D296 in the polar core, confirming the existence of previously proposed internal negatively charged partner of R175. Charge reversal of either residue yielded activated arrestins, whereas the combination of charge reversals (R175E + D296R) restored selectivity towards P-Rh* (Vishnivetskiy et al. 1999). Collectively, these results demonstrate that disruption of the salt bridge between R175 and D296 in the polar core via receptor-attached phosphates converts arrestin from inactive state visualized in the crystal structures to the high-affinity binding state.

Other than the polar core, basal state in arrestin is stabilized via so-called three-element interaction between β-strand I, α-helix I (in the N-domain), and the C-tail (Gurevich and Gurevich. 2004; Palczewski et al. 1991). In the inactive state, the C-tail is positioned on the N-domain in an extended conformation, closely interacting with residues from the β-strand I, α-helix I and the polar core (Fig. 11.2). Here, the C-tail residue R382 participates in the polar core hydrogen bond network, anchoring it in the polar core. Single amino acid substitutions of the residues mediating three-element interaction also result in active arrestins that show dramatically lowered selectivity towards P-Rh* (Vishnivetskiy et al. 2000). The two lysines in the β-strand I, K14 and K15 were shown to be necessary for high-affinity binding of wild type arrestin to P-Rh*, but not for the binding of activated arrestin mutants, where the polar or the three-element interaction is disrupted already. It was thus proposed that these lysines first interact with the phosphate groups of receptor protein, subsequently directing it to the phosphosensor residue R175 in the polar core (Gurevich et al. 2011, 2014; Vishnivetskiy et al. 2000). The lysines are, however, not required for the activated arrestins where the R175E is already exposed.

Polar core residues including R175 are highly conserved in the arrestin family (Fig. 11.1b). High resolution structure of activated mutants such as R175E is thus expected to provide structural insights into the mechanism of arrestin activation that can also be applied to other arrestin family members.

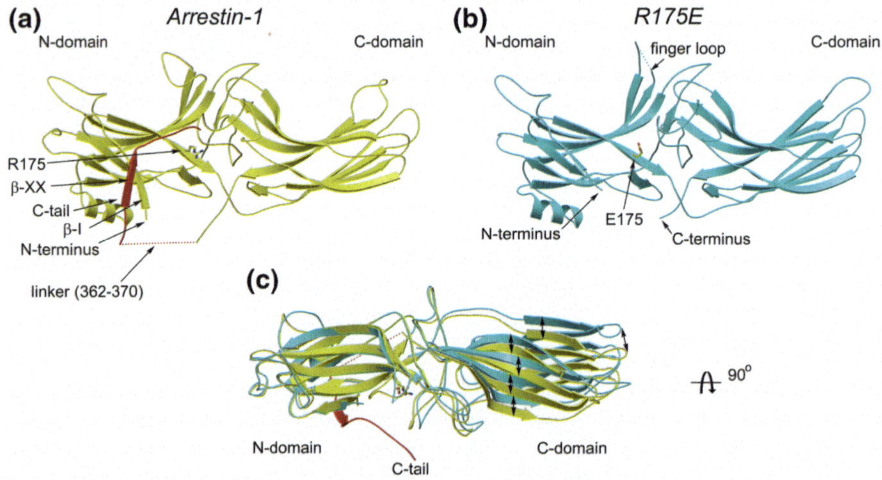

Fig. 11.2 Comparison of overall conformations between arrestin-1 and R175E. **a** Ribbon representation of arrestin-1 (*PDB ID* 3UGX, molecule A) where C-tail is highlighted in *red* and the missing residues 360–370 are shown as *dotted line*, **b** R175E (*PDB ID* 4ZRG), position of the finger loop is indicated (**c**) superposition of N-domains of arrestin-1 (*yellow*) and R175E (*cyan*) shows a ~7.5° rotation in C domain of the mutant arrestin. *Double-headed arrows* indicate the rotation shift in the individual β-strands. Compared to panels **a** and **b**, the view here is rotated by 90° along the horizontal axis with the reader's view upon receptor-binding concave surface. Arginine and mutated residue glutamic acid at position 175 in the polar core are shown as stick models

Crystal Structure of R175E Mutant of Arrestin-1

Crystal structure of R17E mutant of visual arrestin-1 was recently reported (Granzin et al. 2015), where recombinant protein was generated using site-directed mutagenesis by introducing a single point mutation on wild type arrestin-1 background. Overexpression of mutant protein was carried out in *Saccharomyces cerevisiae* F11 α strain. Purified R175E showed binding to different forms of rhodopsin P-Rh*, P-Rh and Rh*, confirming its activated state. These results are in agreement with the mutagenesis reports published previously (Gray-Keller et al. 1997; Vishnivetskiy et al. 1999). Crystallization setups often require highly concentrated protein. In contrast to the wild-type arrestin-1 that could be concentrated up to ~40 mg/mL, the R175E protein showed tendency to aggregate at higher concentrations and could only be concentrated to 13 mg/mL that was used for the crystallization setups. Additionally, conditions under which the crystals grew differ significantly for the two proteins (*Arrestin-1*: 40 mM PIPES, pH 7.2; 0.6 M KCl; 22% ethylene glycol; 6% polyethylene glycol 6000; 13% polyethylene glycol 200; 4% polyethylene glycol 1000, and *R175E*: 100 mM Tris, pH 7.5, 8–18% (w/v) polyethylene glycol 4000), indicating a possible conformational rearrangement (Granzin et al. 2012, 2015).

The mutant R175E crystallized in space group $P2_12_12_1$ with one molecule per asymmetric unit. The structure was determined by molecular replacement using the crystal structure of bovine arrestin-1 expressed in *S. cerevisiae* (*PDB ID* 3UGX) as a search model, and was refined to 2.7 Å resolution. In the so far published arrestin-1 crystal structures, there are four molecules per asymmetric unit. The only other arrestin to crystallize in the same orthorhombic space group is the cone arrestin (Sutton et al. 2005). Hydrodynamic radii calculated using size-exclusion chromatography and small angle X-ray scattering (SAXS) suggested that R175E mutant is monomeric in solution (see original reference (Granzin et al. 2015) for a more detailed account on SAXS studies of arrestin-1, R175E and p44—a monomeric splice variant of arrestin-1) and in contrast to the wild type rod arrestin, R175E does not show self-association.

Overall Structure

The structure of R175E revealed the typical arrestin fold consisting of N- and C-domains, each composed of a seven-stranded β-sandwich (Fig. 11.2b). The monomer model of R175E comprises residues 11–360, but lacks 1–10 and 361–404 residues of the N-terminus and C-terminus, respectively (Fig. 11.2). Additionally, no electron density could be detected for the finger loop residues 69–75, predicted to be disordered. The presence of a single helix on the convex surface distinguishes N-domain from the C-domain. For the reasons described in the previous section, mutation of R175 to E in the polar core region in the fulcrum of the domain interface is expected to result in 'active' conformation revealing the associated conformational changes. The superposition of arrestin-1 and R175E structures resulted in a poor overlay where the N- and the C-domains of R175E undergo a ~7.5° rotation relative to each other (Fig. 11.2c).

Polar Core Region

A hydrogen bond network involving residues from both N- and C-domains in the polar core is highly conserved in the arrestin family. Strikingly, this network collapses and no interactions seen in the basal states could be observed in R175E crystal structure (Figs. 11.1 and 11.3). It was thus surprising that the overall positioning (Cα) of the individual residues comprising the polar core is similar in R175E and the basal state of arrestin-1 with an rmsd of ~1.24 Å. This is due to the fact that the residues are partially localized on the β-strands (D30, R/E175) and are stabilized by backbone hydrogen interactions such as E175-N…O-R29 and E175-O…N-Y31 involving the Cα of the mutated residue E175. The side chains with missing hydrogen bond interactions show an increase in the B-factor by 50%.

Loops in the Central Crest Region: Gate Loop, Finger Loop and Middle Loop

Several of the polar core residues interacting with R175 in the basal state of arrestin are localized on the gate loop (D296, D303 and N305). The side chain of E175 is exposed in R175E structure, facing away from the center (Fig. 11.3b). Several residues of the gate loop likewise show rotamers different from the basal state, facing away from center. Lack of hydrogen bond interactions in the polar core likely results in an outward shift of the gate loop, opening the polar core cleft slightly (Cα R/E 175E–Cα D303 distance is 10.6 and 12.2 Å in arrestin-1 and R175E, respectively) (Fig. 11.3c).

Loops on the concave side on receptor-interface show high flexibility in both arrestin-1 and arrestin-1_P-Rh* complex, and are in distinct conformations in monomers of arrestin-1 crystal tetramer (Granzin et al. 1998; Hirsch et al. 1999). Several of these loops are involved in receptor binding (Gurevich and Gurevich 2004; Kang et al. 2015; Ostermaier et al. 2014b). Specifically, these include the finger loop (residues 68–79), loop 139 or middle loop (residues 133–142), and lariat loop (283–305, including the gate loop residues 296–305). In contrast, loops 157 and 344 on the edge of N- and C-domains correspond to the 'plastic' regions that show conformational variability in the reported crystal structures of arrestins (Kim et al. 2012). In R175E, these loops show conformations similar to that in the parental wild type arrestin-1 (Fig. 11.2).

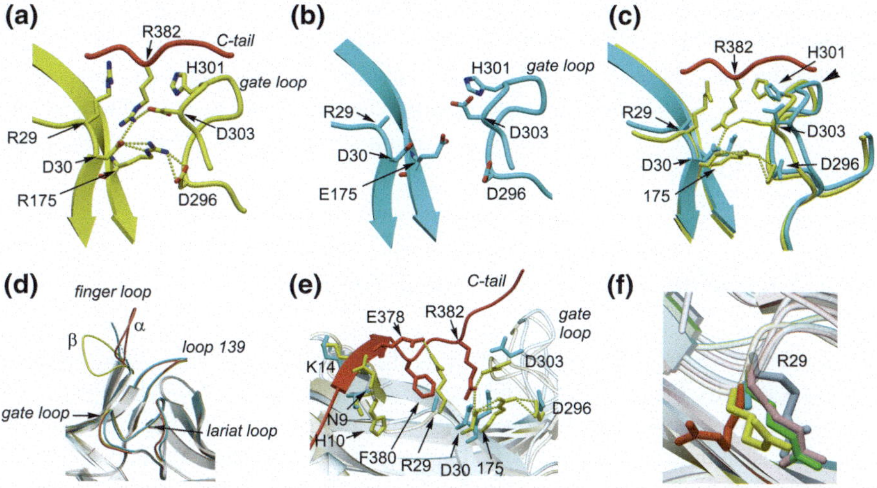

Fig. 11.3 Structural differences between arrestin-1 (*PDB ID* 3UGX, molecule A) and R175E. **a** The hydrogen bond network in arrestin-1 around R175 connecting residues (shown as a stick model) from polar core, gate loop and the C-tail. Hydrogen bonds are shown as *dotted lines* (for clarity, H-bonds shown are ≤ 3.2 Å), **b** R175E polar core region shows a complete collapse of the hydrogen bond network, **c** superposition of arrestin-1 and R175E showing differences in the polar core region. Gate loop in R175E shows an outward shift (indicated by an *arrowhead*) and side chains of the residues in this loop show different rotamers. Additionally, E175 is exposed, and the side chain of R29 is truncated beyond Cβ position due to missing electron density. **d** Superposition of arrestin-1 (*yellow*) and R175E (*cyan*) showing the finger loop, loop 139 and the lariat loop. Both, α (extended, as in molecule B in *PDB ID* 3UGX, *orange*) and β (bent, as in molecule A in *PDB ID* 3UGX, *yellow*) conformations are seen in arrestin-1. In R175E, no electron density could be traced for residues 69 to 75 of the loop that is likely disordered (shown as *dotted line*). **e** Anchoring of C-tail, and three-element interaction in arrestin-1 (*yellow*) and R175E (*cyan*). Hydrogen-bond interactions between residues from the polar core region, the N-terminus and the C-tail are present only in arrestin-1. Also, note the anti-parallel positioning of the C-tail where R382 is the terminal residue in arrestin-1 crystal structure to participate in the polar core network. Disruption of polar core in R175E releases R382, a major constraint that once released leads to enhanced C-tail flexibility. Energetically, this in turn can initiate the disruption of further H-bond interactions (between the C-tail residues 380, 379, 378 with the central residue R29), causing the R29 side chain to disorder. **f** Superposition of different arrestin structures in basal and active states showing side chain flexibility of residue R29 (Color code: *PDB ID's* 4ZRG *cyan*, 3UGX *yellow*, 3UGU *orange*, 4J2Q pink, 4JQI *light blue*, 4ZWJ *green*). Side chain of R29 in R175E crystal structure is truncated beyond Cβ position due to missing electron density

Both biochemical and structural data have established a functionally significant role of the finger loop in binding to the receptor, where the loop occupies a special position in the center of the N- and C-domains (Fig. 11.3d) (Gurevich and Gurevich 2004; Hanson et al. 2006; Kang et al. 2015; Kim et al. 2012; Zhuang et al. 2013). Residues of this loop adopt two different conformations in the heterodimers of the arrestin-1 basal state structure (Hirsch et al. 1999). It is extended in the α-conformer, whereas in the β-conformer, it takes up a bent conformation folding in towards the concave side of the N-domain. In R175E, residues 69–75 are disordered, which further supports the previously suggested conformational flexibility in the loop (Fig. 11.3d). Although there is no direct contact between the finger loop and the polar core (Fig. 11.2), conformational flexibility in finger loop is an attribute related to the activation-dependent changes such as disruption of the hydrogen bond network in the polar core (Granzin et al. 2012; Kim et al. 2013; Shukla et al. 2013).

C-tail Anchoring and Three-Element Interaction

Yet another basal state feature of arrestin is the three element interaction between β-strand I, α-helix I (in the N-domain), and the C-tail (Fig. 11.2a), where the latter is anchored to the polar core. In general, the majority of the C-terminal residues are flexible, as seen in previously published arrestin structures where residues 362–370 and their equivalents could not be resolved (Granzin et al. 1998; Han et al. 2001; Hirsch et al. 1999; Milano et al. 2002; Sutton et al. 2005). These residues form the linker region connecting the C-domain and the C-tail (Fig. 11.2a). In the basal state, the only ordered residues are 371–386 in the C-tail, where its interaction with the polar core blocks the binding of phosphorylated receptor. Specifically, these interactions are between residues of the C-tail (most conserved are R382, F380, E378), the polar core (D30, D303, R175), the arginine switch (R29) and the N-terminal β-strand I (here the C-tail forms an anti-parallel β-strand XX) (Fig. 11.3e). In R175E, the C-terminal residues 361–404 that include the linker region and the C-tail residues could not be traced due to discontinuous density, which indicates that the region is even more flexible. Disordered C-tail residues have been also observed in the active state of visual and non-visual arrestins (Kang et al. 2015; Shukla et al. 2013).

Release of the C-tail is the key step in arrestin activation that precedes high affinity binding of the phosphorylated C-terminus of the receptor (Gurevich and Gurevich 2004; Ostermaier et al. 2014b). In the crystal structure of arrestin-1, R382 in the C-tail shows hydrogen bonding with residues R29, D30, D303 and L173 in the vicinity of the polar core region (Fig. 11.3e). Of these residues, R29 is hydrogen bonded to E378, E379, F380 and R382, an interaction network that is largely responsible for anchoring of the C-tail in the basal state. Interestingly, the side chain of R29 is disordered in R175E, as it could not be traced beyond the Cβ position due to missing electron density (Fig. 11.3f). The side chain of R29 shows conformational variability in the previously reported arrestin structures (Fig. 11.3f). In the structure of the truncated splice variant p44 arrestin, R29 points 'in' the polar core where it interacts with D30 and R175, taking place of the missing C-tail residue R382 (Granzin et al. 2012). In contrast to variability in the side chain of R29, the backbone hydrogen bond R29-O…N-E175 is conserved in all arrestin structures including R175E suggesting importance of the close proximity of two residues in arrestin function. In the recently published crystal structure of β-arrestin-1-peptide complex, the equivalent residue of R29 (R25 in β-arrestin) is in direct contact with the GPCR phosphopeptide ligand confirming this region as the binding site for the phosphorylated receptor (Shukla et al. 2013). Several C-terminal amino acids (362–393, β-arrestin-1 numbering) are disordered in the complex structure, which are likely flexible as the binding site is occupied by the phosphopeptide. In another functional study, extensive mutagenesis was performed on arrestin-1 where the authors mutated each residue to alanine and compared binding of all 403 mutants to the GPCR rhodopsin (Ostermaier et al. 2014a). Interestingly, mutant R29A showed the weakest binding to P-Rh*. The authors thus suggested that the interaction between the phosphate groups of rhodopsin C-terminus and residue R29 in arrestin-1 controls the C-tail exchange mechanism, where the C-tail release exposes several charged residues for binding of the phosphorylated receptor (Ostermaier et al. 2014a).

In the R175E mutant, collapse of the hydrogen bond network in the polar core results in release of the C-tail, so that R382 is no longer anchored in the polar core. Consequently, disruption of the hydrogen bond interactions between the C-tail residues and residue R29 results in an increase in the side chain flexibility of the latter (Fig. 11.3e). To test that the C-terminal residues are flexible in the crystal structure and not absent due to proteolysis during crystallization, small angle X-ray scattering (SAXS) analysis technique was applied that provides accurate information on the molecular weight, oligomeric state and low-resolution shape envelope of proteins in solution. For comparison, the full-length arrestin-1, the mutant R175E and the truncated splice variant p44 were tested under identical conditions (Granzin et al. 2015). The molecular mass determined from the Porod volume and ab initio bead modeling confirmed that both R175E and p44 are monomeric, whereas the arrestin-1 is dimeric in solution. Both these monomeric proteins also contain one monomer per asymmetric unit in the crystals. The respective ab initio shape reconstructions of arrestin-1 fit a dimer and that of p44 and R175E a monomer each. The presence of flexible C-tail is verified in R175E ab initio shape envelope, which is more extended in the region of the flexible C-tail compared to that of p44 (Fig. 11.4).

Differences Between R175E and the 'Active' Conformations

Compared to the basal state, crystal structure of R175E shows a ~7.5° rotation of the N- and C-domains relative to each other around the center (Fig. 11.1c). Movement of domains was previously suggested in molecular modeling experiments (Modzelewska et al. 2006) and in truncation experiments in the interdomain hinge region (Vishnivetskiy et al. 2002). Recently published arrestin structures show 'pre-active' conformations, β-arrestin bound to a GPCR phosphopeptide (Shukla et al. 2013), and p44 splice variant of arrestin-1 (Kim et al. 2013). Most striking feature common in both the structures is a ~20° rotation between the N- and C-domains. Relative movement of the domains supports the previously proposed 'clam-shell' model where these transitions lead to 'high-affinity' state arrestin

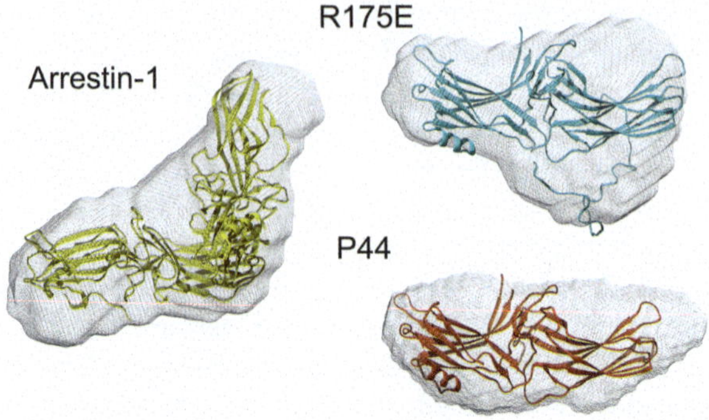

Fig. 11.4 Characterization in solution by small angle X-ray scattering (SAXS). Ab initio shape reconstruction models (*grey*, shown in mesh) determined by SAXS. Ribbon structures of the respective crystal structures: Arrestin-1: *Yellow*, dimer with molecules A and D (*PDB ID* 3UGX); R175E: *cyan* (*PDB ID* 4ZRG), and p44: *orange* (*PDB ID*: 3UGU). Orientation of R175E and p44 are as in Fig. 11.1. The dimer of arrestin-1 is positioned such that the bottom molecule A is oriented as in Fig. 11.2c

that in turn can bind the receptor. In contrast, other reports suggested no change in relative positions of N- and C-domains working against the 'clam-shell' model. For example, using site-directed spin labeling and double electron-electron resonance, Kim et al. measured long-range ($\sim$17–60 Å) intramolecular distances in arrestin-1 in solution in the presence of P-Rh* where the relative position of the N- and C-domains remains largely unchanged (Kim et al. 2012). Likewise, no domain movement could be seen in non-visual arrestins (Zhuo et al. 2014).

Most important clue regarding role of the domain movement in arrestin activation mechanism is provided by the crystal structure of rhodopsin-arrestin complex (Kang et al. 2015) that reveals a $\sim$20° rotation between the N- and C-domains, similar to that seen in the above-mentioned two 'pre-active' arrestins (Kim et al. 2013; Shukla et al. 2013). In the arrestin-receptor complex structure, the domain movement opens a cleft in arrestin to accommodate the short helix formed by the second intracellular loop of rhodopsin (Fig. 11.5). The physiological relevance of the structural model of the complex and the activation state of both proteins was validated by several methods such as double electron-electron resonance, hydrogen-deuterium exchange mass spectrometry, cell-based rhodopsin-arrestin interaction assays and site-specific disulfide cross-linking experiments. Even though the domain movement in R175E is smaller ($\sim$7.5° compared to $\sim$20° in 'pre-active' states), it is a common attribute seen for the activated arrestins.

Primary interactions at the arrestin-receptor interface involve three loops on the arrestin side, namely, the finger, the gate and the middle loops (loop 139) (Fig. 11.5). Loop 139, also called the middle loop, has been previously proposed to play a role in stability and binding selectivity of arrestin-1 (Vishnivetskiy et al. 2013). Large conformational changes in this loop were reported in arrestin-1 using double electron-electron resonance (DEER) technique (Kim et al. 2012).

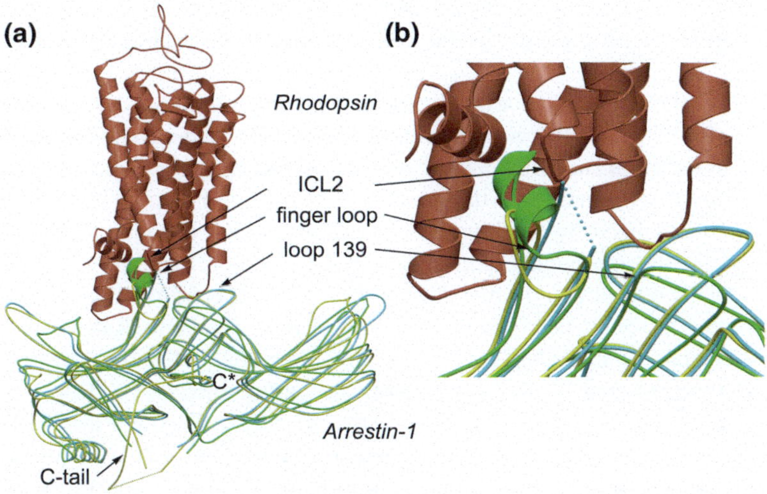

Fig. 11.5 Structural differences between the basal state arrestin-1, R175E and rhodopsin-arrestin-1 complex. Superposition of the three crystal structures (Color code: *PDB ID's* 4ZRG *cyan*, 3UGX *yellow*, 4ZWJ *green*) with a zoomed-in view of the interface shown in *right panel*. The finger loop, loop 139 (middle loop) and the second intracellular loop (ICL2) of rhodopsin at the interface are indicated; C* shows the C-tail terminus in basal state arrestin-1

In contrast, the corresponding loop was suggested to be highly flexible in β-arrestin in both the basal and active states (Zhuo et al. 2014). Additionally, functional mapping of arrestin-1 at single amino acid resolution showed no significant binding effects to Rh* when residues of this loop were mutated to alanine (Ostermaier et al. 2014a). Crystal structures of the pre-activated arrestins and the rhodopsin-arrestin complex show similar middle loop conformation, which is different from the basal state. Position of this loop at the receptor interface in the complex further confirms its role in receptor binding. As shown in Fig. 11.5, in order to bind rhodopsin, the loop seen in the basal state needs to move up to avoid the steric clashes. In the complex, the middle and the C-loops (residue Y251 region, central loop in the C-domain) move apart to form a cleft that accommodates the ICL2 loop of rhodopsin, which adopts a short helix conformation upon activation (Kang et al. 2015). Another primary interaction at the arrestin-rhodopsin interface involves the finger loop that adapts α-helical conformation (Feuerstein et al. 2009). This short helix embedded in the receptor interacts extensively with the C-terminus of TM7, N-terminus of helix 8 as well as loop ICL1 of rhodopsin. The gate loop conformation of 'activated' arrestin differs significantly from the basal state. It was previously suggested that residues of the gate loop (296–305) are exposed for interaction with the activated receptor upon release of the C-tail (Gurevich and Gurevich 2004). In the complex, the gate loop residues show no direct contact with rhodopsin (Kang et al. 2015). However, it is possible that conformational changes in gate loop are relayed to the domain movement that in turn is correlated to the conformational changes in the finger and middle loops. Structural evidence of a step-by-step progress of events in this series of conformational changes is not yet available.

In R175E, the C-tail is disordered exposing the lariat loop residues, and yet the overall conformation of the gate loop and middle loop remains similar to that of arrestin-1 in the basal state. In contrast, residues of the finger loop are disordered in R175E. A possible explanation is that the polar core disruption caused by the mutation of R175 is recognized by the finger loop residues which become highly flexible 'searching' for the appropriate binding site in the receptor. Flexibility in the finger loop is probably relevant for the transformation to a helical conformation when bound to the receptor, as seen in the rhodopsin-arrestin complex structure (Kang et al. 2015). Notably, crystals of the other two 'pre-active' arrestins p44 and β-arrestin were obtained in the presence of the ligands opsin and phosphopeptide V2Rpp-synthetic antibody fragment Fab30, respectively (Kim et al. 2013; Shukla et al. 2013). The finger loop in both these structures is in extended conformation, where the residues are engaged in crystal lattice contacts. Crystallization of R175E was performed without any ligand. The 20° rotational change in domains as well as significantly different loop conformations seen for the three above-mentioned central crest loop regions in case of 'activated' arrestins is likely induced by the presence of ligand protein/peptide, as revealed in the crystal structure of β-arrestin–peptide complex where the bound peptide forms an antiparallel β-strand replacing the C-tail (Shukla et al. 2013); and validated in the rhodopsin-arrestin complex (Kang et al. 2015).

Functional Implications of R175E Conformation: Basal, Pre-active or Active?

Previous mutagenesis studies have demonstrated that similar to R175E, other mutations of residues in the polar core, C-tail and those involved in the three-element interaction result in pre-active arrestins that bind light-activated rhodopsin irrespective of its phosphorylation state (Vishnivetskiy et al. 1999). The most likely explanation common to these mutations is the release of stereochemical constraints responsible for keeping arrestin in its basal state. For the highly conserved salt bridge interaction between R175 and D296, it was proposed that neutralization or charge reversal of the residues result in mutants that mimic the effect of negatively charged phosphate groups from the receptor, allowing them to bypass the need for receptor phosphorylation (Vishnivetskiy et al. 1999). Formation of this state would involve structural rearrangements in the polar core, C-tail and three-element interaction. Additionally, the authors speculated that a change in the position of the N- and C-domains relative to each other is necessary for arrestin to achieve the high-affinity binding state. The structure of R175E is the sole report of a polar core mutant that shows a completely disrupted hydrogen bond network in the polar core including the one salt bridge between D30 and R175, which is still formed in the 'active' conformations (Kang et al. 2015; Kim et al. 2013; Shukla et al. 2013). The destabilized polar core causes displacement of the C-tail, breaking the three-element interaction and a ~7.5° rotation between N- and C-domains. Release of the C-tail results in an increase in the overall positive electrostatic potential exposing the key phospho-sensor residues (Fig. 11.6), allowing it to bind the receptor irrespective of its phosphorylation state as has been previously reported for the truncated arrestins that lack the C-tail element (Granzin et al. 2012; Hanson and Gurevich 2006; Kim et al. 2013).

Extensive biochemical and structural studies over the past two decades support the C-tail exchange mechanism as the current model of arrestin activation (Gurevich and Gurevich 2004; Gurevich et al. 2014; Ostermaier et al. 2014b) (Fig. 11.7). Phosphorylated C-terminus of rhodopsin displaces the C-tail of arrestin, destabilizing the polar core, which is accompanied by structural rearrangements in the central crest loops in arrestin. This allows for a significant rotation of ~20° between N- and C-domains, opening a cleft between the middle and C-loop of arrestin where the ICL2 helix of rhodopsin can now bind, as evident in the arrestin-rhodopsin complex (Kang et al. 2015).

In contrast to the basal state arrestin, R175E already has a disrupted polar core that causes the C-tail release, and is thus 'reverse' of the first step of activation model where replacement of the C-tail by the phosphorylated residues of the receptor leads to destabilization of the polar core. In this regard, one could say that R175E is pre-activated. Additionally, the overall increase in positive surface potential in R175E is expected to facilitate receptor binding mediated via electrostatic interactions. The charge distribution in arrestin and rhodopsin is complementary to each other for their interactions (Kang et al. 2015; Kim et al. 2013).

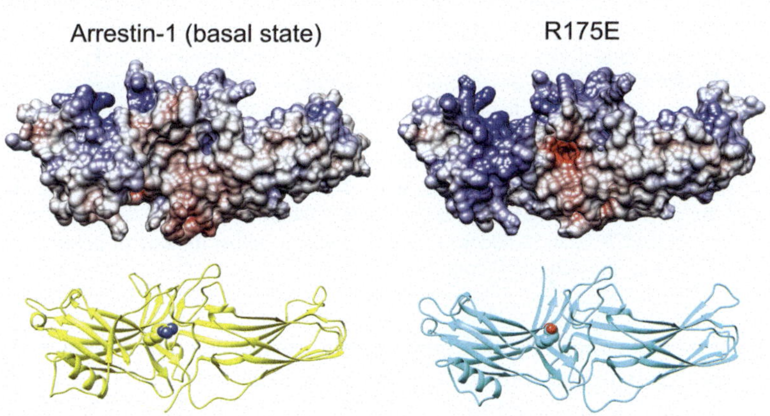

Fig. 11.6 Electrostatic surfaces of crystal structures of arrestin-1 and R175E. Comparison of the lower panels shows an increase in overall positive potential in R175E due to exposure of the key phosphor-sensor residues. Negatively and positively charged surface are colored *red* and *blue*, respectively. The *upper panels* are ribbon representations to show the molecule orientation with the residue 175 in the center (Compared to Fig. 11.2, the view is rotated by approximately 45° along the horizontal axis away from the reader)

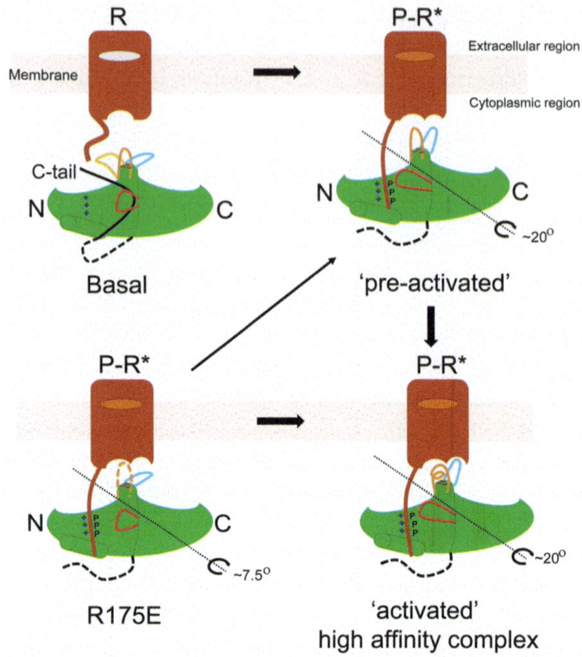

Fig. 11.7 Structural basis of arrestin activation. A cartoon model showing interaction between arrestin-1 and activated rhodopsin. Four conformations of arrestin-1 are depicted namely basal, pre-activated, activated as in the arrestin-rhodopsin complex and R175E. Basal state of arrestin-1 is the inactive conformation, primarily stabilized by a close network of interactions between residues in the polar core, N-terminus and the C-tail. Activation of rhodopsin (P-R*) allows binding of phosphorylated C-terminus to the N-domain of arrestin-1, disrupting the polar core network and releasing the C-tail. The resulting 'pre-activated' arrestin-1 state is typically characterized by structural differences in the central crest loop conformations and a ~20° rotation between the N- and C-domains. Consequently, opening of the loop 139 (middle loop) and C-loop leads to accommodation of ICL2 helix of rhodopsin, forming the high-affinity activated state of arrestin-1. Associated to fully-activated state of arrestin-1 is the α-helical conformation of the finger loop at the interface. Conformation of the polar core mutant R175E shows a fully-disrupted polar core, flexible finger loop, a 'free' C-tail, and ~7.5° rotation between the N- and C-domains, representing an 'intermediate' conformation prior to formation of the 'fully-activated' state. Once in contact with the GPCR ligand (rhodopsin), this can happen either directly or in steps involving changes first as in the 'pre-activated' conformation

The R175E mutant structure shows significantly smaller domain movement (~7.5° rotation) compared to the 'pre-active' structures. Additionally, no major conformational changes are observed in the middle, gate and C-loops in R175E. These rearrangements in arrestin are likely induced once it comes in contact with the receptor molecule followed by formation of the 'high-affinity' receptor-bound state as seen in the recent rhodopsin-arrestin and β-arrestin phosphopeptide complexes (Kang et al. 2015; Shukla et al. 2013). Taken together, the structural features of R175E and its differences from the basal and 'active' states, with disrupted polar core, three-element interaction, and with a 'free' C-tail likely represents an intermediate activation state prior to formation of the high-affinity arrestin-receptor complex (Fig. 11.7). For other constitutively active mutants of arrestin studied previously where the mutations are located in the polar core, the three-element or the C-tail regions, it can be speculated that the conformational rearrangements and the activation mechanism in the mutants would be similar to R175E.

References

Celver J, Vishnivetskiy SA, Chavkin C, Gurevich VV (2002) Conservation of the phosphate-sensitive elements in the arrestin family of proteins. J Biol Chem 277:9043–9048
Feuerstein SE, Pulvermuller A, Hartmann R, Granzin J, Stoldt M, Henklein P, Ernst OP, Heck M, Willbold D, Koenig BW (2009) Helix formation in arrestin accompanies recognition of photoactivated Rhodopsin. Biochem US 48:10733–10742
Granzin J, Wilden U, Choe HW, Labahn J, Krafft B, Buldt G (1998) X-ray crystal structure of arrestin from bovine rod outer segments. Nature 391:918–921
Granzin J, Cousin A, Weirauch M, Schlesinger R, Buldt G, Batra-Safferling R (2012) Crystal structure of p44, a constitutively active splice variant of visual arrestin. J Mol Biol 416:611–618
Granzin J, Stadler A, Cousin A, Schlesinger R, Batra-Safferling R (2015) Structural evidence for the role of polar core residue Arg175 in arrestin activation. Sci Rep 5:15808
Gray-Keller MP, Detwiler PB, Benovic JL, Gurevich VV (1997) Arrestin with a single amino acid substitution quenches light-activated rhodopsin in a phosphorylation-independent fashion. Biochem US 36:7058–7063
Gurevich VV, Benovic JL (1993) Visual arrestin interaction with rhodopsin—sequential multisite binding ensures strict selectivity toward light-activated phosphorylated rhodopsin. J Biol Chem 268:11628–11638
Gurevich VV, Benovic JL (1995) Visual arrestin binding to rhodopsin—diverse functional roles of positively charged residues within the phosphorylation-recognition region of arrestin. J Biol Chem 270:6010–6016
Gurevich VV, Benovic JL (1997) Mechanism of phosphorylation-recognition by visual arrestin and the transition of arrestin into a high affinity binding state. Mol Pharmacol 51:161–169
Gurevich VV, Gurevich EV (2004) The molecular acrobatics of arrestin activation. Trends Pharmacol Sci 25:105–111
Gurevich VV, Hanson SM, Song XF, Vishnivetskiy SA, Gurevich EV (2011) The functional cycle of visual arrestins in photoreceptor cells. Prog Retin Eye Res 30:405–430
Gurevich VV, Song X, Vishnivetskiy SA, Gurevich EV (2014) Enhanced phosphorylation-independent arrestins and gene therapy. Handb Exp Pharmacol 219:133–152
Han M, Gurevich VV, Vishnivetskiy SA, Sigler PB, Schubert C (2001) Crystal structure of beta-arrestin at 1.9 angstrom: Possible mechanism of receptor binding and membrane translocation. Structure 9:869–880
Hanson SM, Gurevich VV (2006) The differential engagement of arrestin surface charges by the various functional forms of the receptor. J Biol Chem 281:3458–3462
Hanson SM, Francis DJ, Vishnivetskiy SA, Kolobova EA, Hubbell WL, Klug CS, Gurevich VV (2006) Differential interaction of spin-labeled arrestin with inactive and active phosphorhodopsin. Proc Natl Acad Sci USA 103:4900–4905
Hirsch JA, Schubert C, Gurevich VV, Sigler PB (1999) The 2.8 angstrom crystal structure of visual arrestin: a model for arrestin's regulation. Cell 97:257–269
Kang Y, Zhou XE, Gao X, He Y, Liu W, Ishchenko A, Barty A, White TA, Yefanov O, Han GW, Xu Q, de Waal PW, Ke J, Tan MH, Zhang C, Moeller A, West GM, Pascal BD, Van Eps N,

Caro LN, Vishnivetskiy SA, Lee RJ, Suino-Powell KM, Gu X, Pal K, Ma J, Zhi X, Boutet S, Williams GJ, Messerschmidt M, Gati C, Zatsepin NA, Wang D, James D, Basu S, Roy-Chowdhury S, Conrad CE, Coe J, Liu H, Lisova S, Kupitz C, Grotjohann I, Fromme R, Jiang Y, Tan M, Yang H, Li J, Wang M, Zheng Z, Li D, Howe N, Zhao Y, Standfuss J, Diederichs K, Dong Y, Potter CS, Carragher B, Caffrey M, Jiang H, Chapman HN, Spence JC, Fromme P, Weierstall U, Ernst OP, Katritch V, Gurevich VV, Griffin PR, Hubbell WL, Stevens RC, Cherezov V, Melcher K, Xu HE (2015) Crystal structure of rhodopsin bound to arrestin by femtosecond X-ray laser. Nature 523:561–567

Kim M, Vishnivetskiy SA, Van Eps N, Alexander NS, Cleghorn WM, Zhan XZ, Hanson SM, Morizumi T, Ernst OP, Meiler J, Gurevich VV, Hubbell WL (2012) Conformation of receptor-bound visual arrestin. Proc Natl Acad Sci USA 109:18407–18412

Kim YJ, Hofmann KP, Ernst OP, Scheerer P, Choe HW, Sommer ME (2013) Crystal structure of pre-activated arrestin p44. Nature 497:142–146

Kovoor A, Celver J, Abdryashitov RI, Chavkin C, Gurevich VV (1999) Targeted construction of phosphorylation-independent beta-arrestin mutants with constitutive activity in cells. J Biol Chem 274:6831–6834

Milano SK, Pace HC, Kim YM, Brenner C, Benovic JL (2002) Scaffolding functions of arrestin-2 revealed by crystal structure and mutagenesis. Biochem US 41:3321–3328

Modzelewska A, Filipek S, Palczewski K, Park PS (2006) Arrestin interaction with rhodopsin: conceptual models. Cell Biochem Biophys 46:1–15

Ostermaier MK, Peterhans C, Jaussi R, Deupi X, Standfuss J (2014a) Functional map of arrestin-1 at single amino acid resolution. Proc Natl Acad Sci USA 111:1825–1830

Ostermaier MK, Schertler GFX, Standfuss J (2014b) Molecular mechanism of phosphorylation-dependent arrestin activation. Curr Opin Struct Biol 29:143–151

Palczewski K, Pulvermuller A, Buczylko J, Hofmann KP (1991) Phosphorylated rhodopsin and heparin induce similar conformational-changes in arrestin. J Biol Chem 266:18649–18654

Pan L, Gurevich EV, Gurevich VV (2003) The nature of the arrestin x receptor complex determines the ultimate fate of the internalized receptor. J Biol Chem 278:11623–11632

Shukla AK, Manglik A, Kruse AC, Xiao K, Reis RI, Tseng WC, Staus DP, Hilger D, Uysal S, Huang LY, Paduch M, Tripathi-Shukla P, Koide A, Koide S, Weis WI, Kossiakoff AA, Kobilka BK, Lefkowitz RJ (2013) Structure of active beta-arrestin-1 bound to a G-protein-coupled receptor phosphopeptide. Nature 497:137–141

Sutton RB, Vishnivetskiy SA, Robert J, Hanson SM, Raman D, Knox BE, Kono M, Navarro J, Gurevich VV (2005) Crystal structure of cone arrestin at 2.3A: evolution of receptor specificity. J Mol Biol 354:1069–1080

Vishnivetskiy SK, Paz CL, Schubert C, Hirsch JA, Sigler PB, Gurevich VV (1999) How does arrestin respond to the phosphorylated state of rhodopsin? J Biol Chem 274:11451–11454

Vishnivetskiy SA, Schubert C, Climaco GC, Gurevich YV, Velez MG, Gurevich VV (2000) An additional phosphate-binding element in arrestin molecule—implications for the mechanism of arrestin activation. J Biol Chem 275:41049–41057

Vishnivetskiy SA, Hirsch JA, Velez MG, Gurevich YV, Gurevich VV (2002) Transition of arrestin into the active receptor-binding state requires an extended interdomain hinge. J Biol Chem 277:43961–43967

Vishnivetskiy SA, Baameur F, Findley KR, Gurevich VV (2013) Critical role of the central 139-loop in stability and binding selectivity of arrestin-1. J Biol Chem 288:11741–11750

Wilden U, Hall SW, Kuhn H (1986) Phosphodiesterase activation by photoexcited rhodopsin is quenched when rhodopsin is phosphorylated and binds the intrinsic 48-kDa protein of rod outer segments. Proc Natl Acad Sci USA 83:1174–1178

Zhuang TD, Chen QY, Cho MK, Vishnivetskiy SA, Iverson TM, Gurevich VV, Sanders CR (2013) Involvement of distinct arrestin-1 elements in binding to different functional forms of rhodopsin. Proc Natl Acad Sci USA 110:942–947

Zhuo Y, Vishnivetskiy SA, Zhan XZ, Gurevich VV, Klug CS (2014) Identification of receptor binding-induced conformational changes in non-visual arrestins. J Biol Chem 289:20991–21002

Chapter 12
Active Conformations of Arrestins: Expected and Unexpected Changes

Abhishek Bandyopadhyay, Ya Zhuo, Wayne L. Hubbell, Candice S. Klug, Vsevolod V. Gurevich and Oliver P. Ernst

Abstract The idea that the conformation of the receptor-bound arrestin is dramatically different from that of a free arrestin in solution emerged soon after the discovery of the first member of the family, visual arrestin-1. However, most of the evidence for this emerged in the last few years from studies using biophysical methods and X-ray crystallography. The similarity of the conformational changes in different arrestin subtypes detected by various methods suggests that these rearrangements accompany arrestin "activation", i.e., the transition into the high-affinity receptor-binding state. The key changes include the release of the arrestin C-terminal tail, the twisting of the two arrestin domains relative to each other, and the changes in several flexible loops in the central "crest" on the receptor-binding surface and elsewhere in the molecule. From a signaling standpoint, the changes on the arrestin side not involved in receptor contact appear more important, as they allow the non-receptor partners to discriminate between free and receptor-bound

A. Bandyopadhyay · O.P. Ernst (✉)
Department of Biochemistry, University of Toronto,
1 King's College Circle, Toronto, ON M5S 1A8, Canada
e-mail: oliver.ernst@utoronto.ca

Y. Zhuo · C.S. Klug (✉)
Department of Biophysics, Medical College of Wisconsin,
8701 Watertown Plank Road, Milwaukee, WI 53226, USA
e-mail: candice@mcw.edu

W.L. Hubbell
University of California Los Angeles, Los Angeles, CA 90095, USA

V.V. Gurevich (✉)
Department of Pharmacology, Vanderbilt University, 2200 Pierce Avenue,
Preston Research Building, Rm. 454, Nashville, TN 37232, USA
e-mail: vsevolod.gurevich@vanderbilt.edu

O.P. Ernst
Department of Molecular Genetics, University of Toronto,
1 King's College Circle, Toronto, ON M5S 1A8, Canada

V.V. Gurevich (ed.), *The Structural Basis of Arrestin Functions*,
DOI 10.1007/978-3-319-57553-7_12

forms of arrestins. The identification of these changes facilitates elucidation of the structural basis of arrestin-mediated signaling, including mapping the binding sites for various signaling proteins.

Keywords Arrestin · Conformation · Protein–protein interactions · GPCRs · Cell signaling

Arrestin was discovered by Hermann Kuhn in 1978 as one of the proteins that binds rhodopsin in response to light activation (Kuhn 1978). In contrast to the other light-sensitive partners, arrestin (called 48-kDa protein at the time, also known as S-antigen) clearly preferred phosphorylated rhodopsin (Kuhn et al. 1984). It was then discovered that arrestin binding reduces transducin activation by light-activated rhodopsin (Wilden et al. 1986). This effect was later found to be due to direct competition of transducin and arrestin for the active rhodopsin (Krupnick et al. 1997; Wilden 1995). The paradigm of terminating receptor signaling by arrestin binding in a phosporylation-dependent manner proved applicable to other GPCRs (Benovic et al. 1987; Lohse et al. 1990, 1992). The study of the arrestin-rhodopsin interaction in 1989 revealed an unusually high Arrhenius activation energy, which suggested that arrestin undergoes a global conformational change in the interaction process (Schleicher et al. 1989). The first evidence suggested that this conformational change must involve the release of the arrestin C-terminus, which became much more accessible to trypsin upon rhodopsin binding (Palczewski et al. 1991b). While full-length arrestin-1 demonstrates a remarkable preference for active phosphorylated rhodopsin (P-Rh*) (Gurevich and Benovic 1992, 1993; Palczewski et al. 1991a), the elimination of the arrestin C-terminus by proteolysis (Palczewski et al. 1991a), mutagenesis (Gurevich and Benovic 1992, 1993), or alternative splicing in some species (Pulvermuller et al. 1997; Smith et al. 1994) greatly reduces its selectivity for P-Rh*, enhancing the binding to the other functional forms of rhodopsin, such as active unphosphorylated (Rh*), inactive phosphorylated (P-Rh), and inactive unphosphorylated (Rh). Thus, the arrestin C-terminus emerged as some kind of intra-molecular brake, preventing arrestin binding to all forms of rhodopsin except P-Rh*. In fact, this was the first indication that receptor binding can be facilitated by concomitant conformational changes in arrestin. Further mutagenesis studies suggested that in the basal state arrestin's C-terminus and N-terminus interact, and the disruption of this interaction by mutations from either side facilitates the binding to non-preferred functional forms of rhodopsin (Gurevich et al. 1994). These results suggested that P-Rh* has to disrupt this interaction to enable binding.

Another intra-molecular interaction that has to be disrupted by P-Rh* was discovered before structural information became available, and involved Arg175, apparently serving as the phosphate sensor in bovine arrestin-1 (Gurevich and Benovic 1995). First, neutralization of its charge by the Arg175Asn mutation greatly enhances arrestin binding to unphosphorylated active rhodopsin (Rh*) (Gurevich and Benovic 1995). Next, this arginine was replaced with all 19 possible residues. It turned out that the reversal of its charge by Arg175Glu or Arg175Asp mutations increases Rh* binding most, and that charge neutralization is also effective, whereas the most conservative substitution, Arg175Lys that preserves the charge, does not increase Rh* binding (Gurevich and Benovic 1997). Thus, it seemed reasonable to propose that in the basal state Arg175 interacts with a negatively charged partner within arrestin, and that charge reversal or neutralization breaks that interaction, essentially doing the job that receptor-attached phosphates are supposed to do, making receptor phosphorylation unnecessary for high-affinity arrestin binding (Gurevich and Benovic 1997).

The first crystal structures of arrestin-1 (Granzin et al. 1998; Hirsch et al. 1999) confirmed these hypotheses (Fig. 12.1). Both structures revealed that there are three negatively charged aspartates in the vicinity of Arg175. This arrangement (shown as sphere model in Fig. 12.1) also included one Arg residue from the C-terminus and was termed the "*polar core*" (Hirsch et al. 1999), as it was located in the middle of the molecule between the two arrestin domains, the N- and the C-domain (shown in lime green and deep teal, respectively, Fig. 12.1). The fact that these five charged residues are shielded, almost solvent-excluded, whereas in soluble proteins charged side chains are usually exposed, suggested the functional significance of the polar core. The salt bridge between Arg175 and Asp296 in the polar core was proposed to act as the main phosphate sensor in arrestin-1 (Hirsch et al. 1999; Vishnivetskiy et al. 1999) (see Chap. 6 for details on phosphate sensing). The high-resolution structure that captured part of the arrestin-1 C-terminus (shown in grey in Fig. 12.1) (Hirsch et al. 1999) also revealed that the C-terminus is indeed anchored to the N-terminus and also interacts with the only α-helix in arrestin, localized in its N-domain. This arrangement was termed the "*three-element interaction*" (Hirsch et al. 1999); the residues involved are highlighted in orange in Fig. 12.1. Just like the polar core, the three-element interaction is conserved in all arrestin crystal structures (Han et al. 2001; Hirsch et al. 1999; Milano et al. 2002; Sutton et al. 2005; Zhan et al. 2011), and relevant residues are present in the sequence of all arrestins starting from *Caenorhabditis elegans* (Gurevich and Gurevich 2006a), suggesting the functional importance of the three-element interaction. Indeed, mutagenesis studies showed that this is yet another intra-molecular interaction that can be destabilized by the receptor-attached phosphates upon P-Rh* binding (Vishnivetskiy et al. 2000). The three-element interaction is mediated largely by bulky hydrophobic side chains (Hirsch et al. 1999). Alanine substitution of participating residues in the C-terminus (Gurevich 1998), or other interacting elements (Vishnivetskiy et al. 2000) greatly enhances arrestin-1 binding to non-preferred forms of rhodopsin, such as Rh* and dark P-Rh, reducing its selectivity for P-Rh*. These data suggested that P-Rh* must destabilize this interaction along with the polar core to enable high-affinity binding of arrestin.

The two arrestin domains are connected by an "*inter-domain hinge*" region (shown in blue in Fig. 12.1), which is invariably 12 residues long in all arrestin subtypes (Han et al. 2001; Hirsch et al. 1999; Sutton et al. 2005; Zhan et al. 2011). It was found that increasing deletions within the inter-domain hinge in arrestin-1 (Vishnivetskiy et al. 2002) and in both non-visual arrestins (Hanson et al. 2007a)

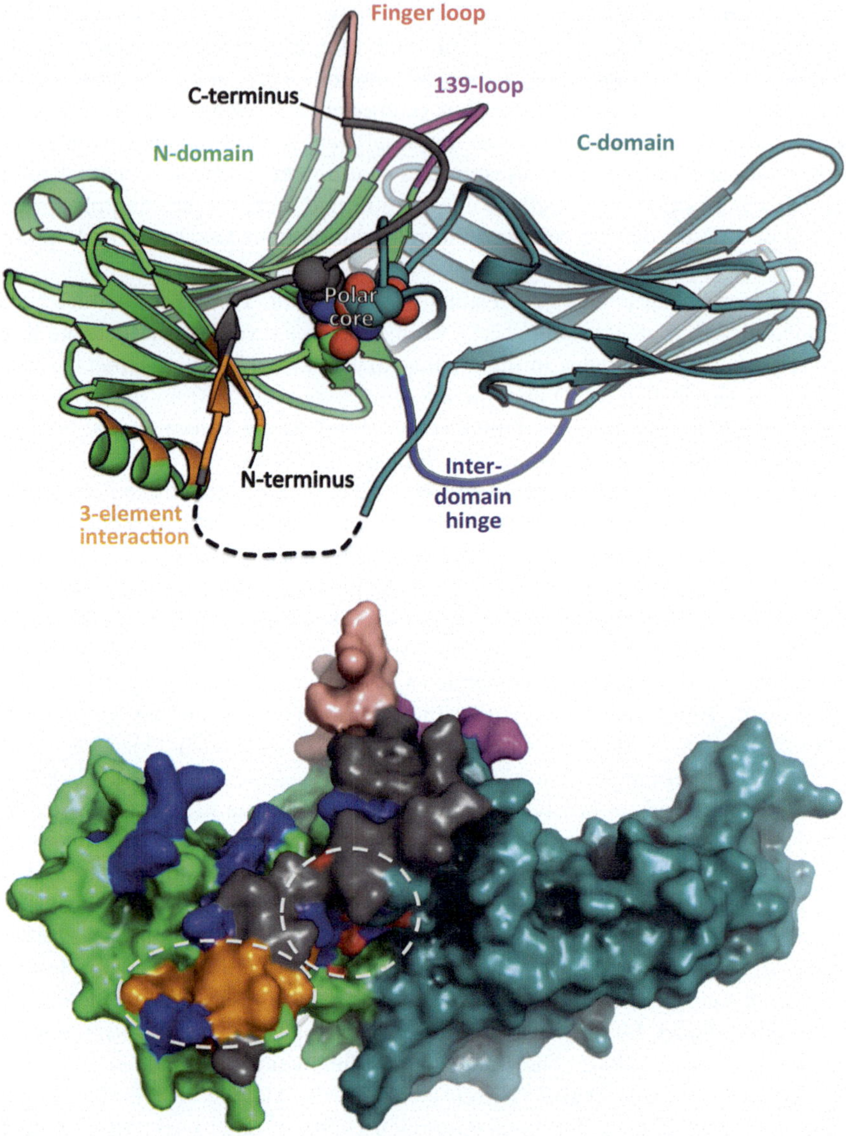

Fig. 12.1 *Top* Structure of bovine arrestin-1 in the basal conformation (PDB code: 1cf1). The N- and C-domains are colored *lime green* and *deep teal*, respectively. The main structural features revealed by the crystal structure (Hirsch et al. 1999) are highlighted. *Orange*, the *three-element interaction* involving the α-helix I, β-strand I and β-strand XX. The *finger loop* and the *139-loop* are coloured *salmon* and *magenta*, respectively. The *polar core* is represented as side chain *sphere* model. The region between β-strand XIX and β-strand XX is not resolved in the crystal structure and shown as a *dotted line*. *Bottom* Surface representation of arrestin-1. The coloring is as above. The phosphorylated receptor C-terminal tail approaches the phosphate sensing residues (shown in *blue*) to disrupt the three-element interaction and polar core to release arrestin's C-terminal tail. For phosphosensors deployed by P-Rh* and P-opsin (Peterhans et al. 2016)

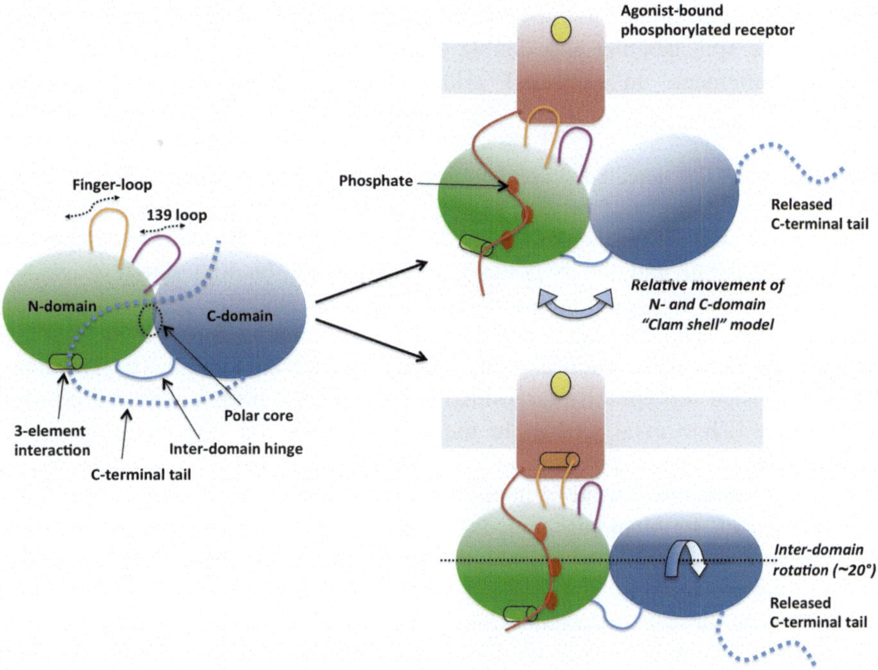

Fig. 12.2 *Left* Cartoon representation of the known basal conformation of arrestins. The coloring scheme is as follows: N- and C-domains are *green* and *slate*, finger loop *orange* and 139-loop *magenta*. The C-terminal tail is shown as a *blue dotted line* and the α-helix I in the N-domain is indicated as a *cylinder*. The three-element interaction, polar core and the inter-domain hinge are indicated. The flexibility in the finger loop and the 139-loop is indicated by the *double-headed arrow* above them. *Top right* Clam shell model. A model proposed based on deletion mutants of the hinge region of arrestin-1. In this model, the N- and the C-domains latch onto the receptor in a manner reminiscent of the closed shell of a clam. Such a structure has not been observed crystallographically or by DEER spectroscopy. *Bottom right* Model of the crystallographically observed inter-domain twist of activated (Kim et al. 2013; Shukla et al. 2013) or receptor-bound (Kang et al. 2015) arrestin. The receptor is colored in *red* and the agonist is indicated as a *yellow oval*. The inter-domain rotation of ~20° is indicated by an *arrow*. The receptor C-terminal tail with attached phosphates is shown in *red*. The released C-terminal tail of arrestin is indicated. The finger loop adopts a helical conformation upon receptor engagement (Feuerstein et al. 2009; Kang et al. 2015; Szczepek et al. 2014)

progressively decreases their binding to the receptor, so that when only five residues necessary to connect the domains are left, the binding is virtually abolished. These data indicated that an extra slack is needed for binding of the receptor, indirectly implying that a movement of the two arrestin domains relative to each other must be involved. The movement was proposed to be either a "*clam-like*" closing of the two domains on the receptor (Vishnivetskiy et al. 2002), or a "*twisting*" of the two domains relative to each other (Modzelewska et al. 2006) (Fig. 12.2).

The issue of receptor binding-induced conformational changes in arrestin was addressed by spin labeling and double electron–electron resonance (DEER) distance measurements in arrestin-1 (Hanson et al. 2006; Kim et al. 2012; Vishnivetskiy et al. 2010), followed by a similar study of both non-visual arrestins (Zhuo et al. 2014). The study of the visual arrestin-1 was the most comprehensive (Kim et al. 2012). It detected several conformational changes, some expected, some not. The most expected one was the release of the C-terminal tail, which was indicated by a significant increase in the inter-spin distance between spin labels on the C-terminal tail and the N-domain, upon receptor binding. Interestingly, in arrestin-1 the C-terminal tail seems to simply "flop around" after release, yielding a dramatic increase in the inter-spin distance distribution with contributions that extend from 20 to 50 Å (Hanson et al. 2006; Vishnivetskiy et al. 2010). However, in the non-visual subtypes the C-terminal tail appeared to have a preferred position after release (Zhuo et al. 2014). In the case of arrestin-2, the major inter-spin distance distribution was centered at 55 Å, suggesting the C-terminal tail assumes a specific conformation instead of flopping around (Zhuo et al. 2014). In arrestin-3, the dissociated C-terminal tail seemed to adopt at least three distinct positions since the inter-spin distance distribution contained three distance populations (31, 43, 54 Å) (Zhuo et al. 2014). The least expected change was a dramatic movement of the "*139-loop*" (also called *middle loop*) from its basal position overhanging the cavity of the C-domain towards the N-domain and to the side of the molecule (Kim et al. 2012). Inter-spin distance distributions involving position 139 all showed large changes upon receptor binding (Kim et al. 2012). In non-visual arrestins the "136-loop" is homologous to the 139-loop in arrestin-1. However, a limited movement of the 136-loop upon receptor binding was found in arrestin-2. By measuring the distances between spin labels on the 136-loop and labeled reference sites at several positions in both the N- and C-domains, the DEER data for arrestin-2 showed a smaller movement of the 136-loop toward the N-domain and away from the C-domain (Zhuo et al. 2014). Moreover, the arrestin-2 136-loop was shown to be highly flexible, as distance distributions between site 136 and multiple references sites in the absence and presence of the receptor were broad (Zhuo et al. 2014). In addition, another remarkable conformational change, which involves the dislocation of the "*finger loop*" (colored in salmon in Fig. 12.1), was shown to be conserved in arrestin-1,-2 and -3 upon receptor binding (Hanson et al. 2006; Kim et al. 2012; Zhuo et al. 2014). By measuring the inter-spin distance between the finger loop and a labeled reference site on the N-terminal β-sheet domain, the DEER distances showed a small increase upon receptor binding in arrestin-1 (3 Å), arrestin-2 (4 Å) and arrestin-3 (2 Å), indicating the finger loop is extending (Hanson et al. 2006; Kim et al. 2012; Zhuo et al. 2014). Last, no clam-shell-like inter-domain movement was found for arrestin-1 and arrestin-2 upon receptor binding, based on DEER distance measurements (Kim et al. 2012;

Zhuo et al. 2014). When spin labels were placed on the C-domain and the N-domain, the inter-domain distance between the pairs didn't show significant changes upon receptor binding (Kim et al. 2012; Zhuo et al. 2014).

Thus, limited proteolysis, DEER distance measurements, and mutagenesis identified several receptor binding-induced conformational changes in arrestins before any structural information on the active conformation became available. The data suggested that these changes are similar, although not identical, in all subtypes. The most definitive among these rearrangements was the disruption of the three-element interaction accompanied by the release of the C-terminal tail, demonstrated by all methods mentioned above. The destabilization of the polar core was also indicated, although the molecular details remained unclear. A large movement of the 139-loop was proven beyond reasonable doubt by extensive DEER distance measurements. The DEER data also argued against clam shell-like movement of the two domains, suggesting that some other kind of movement must occur. In addition, the movements of several flexible loops, including the finger loop (Hirsch et al. 1999) that was predicted to engage the receptor directly based on earlier spin labeling studies of arrestin-1 (Hanson et al. 2006; Sommer et al. 2005) and arrestin-2 (Vishnivetskiy et al. 2011), and of loops on the tips of both arrestin domains were clearly demonstrated (Kim et al. 2012; Zhuo et al. 2014). Loops at the edge of the C-domain are engaged in membrane interaction (see Chap. 8 for details).

The structures of p44 (a splice variant of arrestin-1) (Kim et al. 2013) and an active conformation of arrestin-2 (Shukla et al. 2013) resembling receptor bound-like conformation appeared soon after and confirmed many of the earlier findings. However, for crystallization purposes the proteins were modified from the native arrestin. Crystallized active arrestin-1 was in fact its short splice variant p44 (Kim et al. 2013), which is not found in all species (Smith et al. 1994) and was earlier crystallized in the conformation very similar to the basal conformation of full-length arrestin-1 (Granzin et al. 2012). p44 was crystallized in the presence of opsin which might have induced the active arrestin conformation during the crystallization process (Kim et al. 2013). For arrestin-2 a truncated form, earlier shown to be pre-activated (Kovoor et al. 1999) was used, whereas the receptor was substituted by a multi-phosphorylated C-terminal peptide of V2 vasopressin receptor. In addition, a conformation specific antibody fragment was used to stabilize this complex (Shukla et al. 2013). Neither of these two pre-activated arrestin structures included bound receptor, which left their relevance uncertain. As mentioned in the publication, in the active conformation of the arrestin-2 structure, part of the finger loop and the 136-loop participate in crystal lattice contacts, which makes an exact interpretation of their conformational changes compared to the basal state difficult (Shukla et al. 2013). However, the first (and so far the only) structure of the arrestin-receptor complex (Kang et al. 2015) supported the conclusions of the earlier studies (see Chap. 13 for details), although that structure has its own caveats. The limitations arise from the use of a constitutively active opsin mutant, which was not phosphorylated, so that a phosphorylation-independent arrestin-1 mutant had to be used, which was fused to the receptor C-terminus (Kang et al. 2015).

Nonetheless, a number of striking conformational changes in arrestin were observed in all cases, which can be considered obligatory for the active receptor-bound arrestin. The most striking change is the twisting of the two arrestin domains relative to each other by ~20°, observed in the arrestin-1 complex with rhodopsin (Kang et al. 2015), arrestin-2 complex with receptor phosphopeptide (Shukla et al. 2013), and one of the p44 structures (Kim et al. 2013). Considering that in the structure of phosphorylation-independent arrestin-1-Arg175Glu polar core mutant the domains rotate by ~7.5° (Granzin et al. 2015) (see Chap. 11 for details), domain rotation can be considered a proven part of the conformational rearrangements associated with receptor binding. Importantly, this rotation becomes possible only when both the polar core and the three-element interaction are destabilized, so that both main "clasps" holding arrestin in its basal state are loosened. Another change that can be considered proven occurs in the central finger loop (shown in salmon in Fig. 12.1). It straightens up in all structures and assumes an α-helical conformation in the arrestin-receptor complex crystal structure (Kang et al. 2015), as well as in the complex of the finger loop peptide with the active rhodopsin conformation as shown by X-ray crystallography (Szczepek et al. 2014) and solution NMR (Feuerstein et al. 2009). Although in the two crystal structures the orientation of the finger loop helix is somewhat different, in both cases the peptide inserts into the cytoplasmic cavity of the active rhodopsin conformation. The cytoplasmic opening of the receptor is a hallmark of GPCR activation (Altenbach et al. 2008; Choe et al. 2011; Kang et al. 2015; Rasmussen et al. 2011). Thus, the finger loop binds the structural element that only exists in the active GPCR conformation making the finger loop a good candidate for the role of the activation sensor proposed earlier along with phosphorylation sensing (see Chap. 6) to explain arrestin selectivity for the active phosphorylated receptor (Gurevich and Benovic 1993; Gurevich and Gurevich 2004) (however, see an alternative view in Chap. 7). A third change observed in all these structures was the movement of the 139-loop (Kim et al. 2012; Vishnivetskiy et al. 2013), also called the middle loop (shown in magenta in Fig. 12.1) (Kang et al. 2015; Shukla et al. 2013). In the arrestin basal state it half-covers the concave side of the C-domain, whereas in the active receptor-bound state it moves towards the N-domain and actually participates in the receptor binding (Kang et al. 2015) (see also Chap. 13). Interestingly though, the distance distributions between site 139 (or site 136 in arrestin-2 and -3) and each of the reference sites were very broad, which suggests that this loop is also highly flexible (Zhuo et al. 2014). This could be responsible for the interaction of arrestin-2 and arrestin-3 to the hundreds of receptors by fine-tuning its structure. Other loop movements were detected in some cases, but not in others, and might be associated with receptor binding or membrane interactions (see Chap. 8).

It is also worth noting that X-ray crystallography often yields atomic resolution structures of macromolecules and their complexes. Since almost every protein is flexible, crystal structures are extremely useful in designing experiments for probing the dynamic nature of protein–protein and protein-ligand interactions. While the idea that the arrestin-receptor complex likely exists in many shapes seemed purely speculative in 2006 (Gurevich and Gurevich 2006b), now there is a lot of evidence that this is the case (see Chaps. 7, 10, 16 and 17 and references therein). In this regard, DEER is uniquely advantageous and complementary to other structural methods because of its ability to report on multiple conformations of a protein or a protein–protein complex in the same experiment which is in contrast to other techniques that report on averages of multiple conformations (Hanson et al. 2007c). The DEER measurements of selected distances between rhodopsin and arrestin (residue 74 in human rhodopsin to residues 61, 140 and 241 in mouse arrestin-1, respectively) confirmed the crystal structure of the rhodopsin-arrestin complex; however distance distributions with multiple peaks also suggested the existence of other conformations not observed in the crystal structure [see Fig. 12.3 in (Kang et al. 2015)]. Although mammalian arrestins alone or bound to receptor likely sample multiple conformations, the general trend observed for receptor–arrestin complexes appears to suggest a common mode of binding so far, such as the engagement of the finger loop and relatively large inter-domain twisting to transition into the high affinity receptor-binding conformation.

An interesting addition to our growing understanding of receptor-arrestin interactions is the squid rhodopsin-arrestin complex. The squid visual system shows significant similarity with the mammalian visual system. The squid retina contains a photoreceptor, rhodopsin, a heterotrimeric G protein belonging to the G-alpha q family, an arrestin and a kinase termed squid rhodopsin kinase (SQRK) (Mayeenuddin and Mitchell 2003; Ryba et al. 1993; Swardfager and Mitchell 2007). In contrast to mammalian arrestins, squid arrestin forms in the rhabdomeric

```
Bovine Arrestin-1 1MkANKpAPNhvIFKKISRDKSVTIYLGKRDYIDHVERVEPVDGVVLVDPELVKGKRVYVSLTCAFRYGQEDIDVMGLSFR80
Bovine Arrestin-2 1M-GDK-GTR--VFKKASPNGKLTVYLGKRDFVDHIDLVEPVDGVVLVDPEYLKERRVYVTLTCAFRYGREDLDVLGLTFR76
Bovine Arrestin-3 1M-GEKpGTR--VFKKSSPNCKLTVYLGKRDFVDHLDKVDPVDGVVLVDPDYLKDRKVFVTLTCAFRYGREDLDVLGLSFR77
Bovine Arrestin-4 1M------SR--VFKKTCSNGKLSIYLGKRDFVDHVDMVEPIDGVVLVDPEYLKGRKMFVMLTCAFRYGHDDLDVIGLTFR72
Squid Arrestin    1MvTKK--AK--VYKKASPNGKLTTYLAKRDYYDHKEWQDNIDGVCVVDPDYLKNRKVFGLIVVAFRYGREDMDVMGVSFR76

Bovine Arrestin-1 81RDLYFSQVQVFPPVGASGA-T-TRLQESLIKKLGANTYPFLLTFPDYLPCSVMLQPAPQDVGKSCGVDFEIKAFATHSTd158
Bovine Arrestin-2 77KDLFVANVQSFPPAPEDKK-PLTRLQERLIKKLGEHAYPFTFEIPPNLPCSVTLQPGPEDTGKACGVDYEVKAFCAENL-154
Bovine Arrestin-3 78KDLFIANYQAFPPTPNPPR-PPTRLQERLLRKLGQHAHPFFFTIPQNLPCSVTLQPGPEDTGKACGVDFEIRAFCAKSL-155
Bovine Arrestin-4 73KDLYVQVQQVVPAESSSPRgSLTVLQERLLHKLGDNAYPFTLQMVVNLPCSVTLQPGPDDTGKACGVDFEVKSFCAENL-151
Squid Arrestin    77KDFAVKQMQIYPPLEENQR-PLTKLQAKLLNKLGENAVPFHYDLPTNTPDTVCIQPSEYDGGAPCGVDYQVTTYVSQNM-154

Bovine Arrestin-1 159veEDKIPKKSSVRLLIRKVQHAPR182
Bovine Arrestin-2 155--EEKIHKRNSVRLVIRKVQYAPE176
Bovine Arrestin-3 156--EEKSHKRNSVRLVIRKVQFAPE177
Bovine Arrestin-4 152--EEKVSKRDSVRLVIRKIQFAPL173
Squid Arrestin    155--DDKIHKRNSVSLSIRKLSYFEF176

Bovine Arrestin-1 292GIALDGKIKHEDTN305
Bovine Arrestin-2 286GLALDGKLKHEDTN299
Bovine Arrestin-3 287GLALDGKLKHEDTN300
Bovine Arrestin-4 283GLALDGKLKQGDTN296
Squid Arrestin    289GLALNGKVKYEDTM302
```

Fig. 12.3 Sequence alignment of bovine arrestin-1 to arrestin-4 and the arrestin from the atlantic squid *Loligo pealeii*. The known phosphate sensing residues are highlighted in *bold* and *blue*. The central arginine in the polar core is in *black*, *bold*, and *underlined*. The aspartate counter ion conserved for all the arrestins shown above is in *green*, *bold*, and *underlined*. The second counter ion is in *gold*, also *underlined* and *bold*. The corresponding asparagine in squid arrestin is in *purple* and *bold*. The residue numbers for the respective sequences are shown at the beginning and the end of each line of the alignment. It is clear that the phosphate sensing residues identified in the mammalian arresins are conserved in squid arrestin

membrane (equivalent to rod outer segments of a mammalian retina) a high affinity complex with rhodopsin in a light-dependent and receptor phosphorylation-independent manner (Robinson et al. 2015). As discussed above, for mammalian arrestins, disrupting the three-element interaction or the polar core through mutation or deletion (such as truncation of the C-terminal tail) generates arrestins with significantly reduced preference toward the phosphorylated state of the activated receptor that are able to bind non-phosphorylated receptors, yet none of them show a complete independence from receptor phosphorylation as found for the squid arrestin (Robinson et al. 2015). This fact is simply not due to any obvious difference in the primary sequence of squid arrestin compared to mammalian arrestin, since squid arrestin shows fairly high degree of sequence identity (about 40–44%) with arrestin-2 and arrestin-3 (β-arrestins) (Mayeenuddin and Mitchell 2003).

Especially puzzling is the fact that many of the known phosphate sensing residues in arrestin-1 (Peterhans et al. 2016) and arrestin-2 (Shukla et al. 2013) are actually conserved in squid arrestin (Fig. 12.3). Not surprisingly though, the main polar core residues are conserved in squid arrestin (Arg169/Asp26), which are Arg175 and Asp30 in bovine arrestin-1 and Arg169 and Asp26 in bovine arrestin-2 (Fig. 12.3). Whereas in mammalian arrestins, the central arginine has a second aspartate counterion, such as Asp296, Asp290, Asp291 and Asp287 in bovine arrestin-1, -2, -3 and -4, respectively, the corresponding position 293 in squid arrestin has an asparagine residue. This particular asparagine is unlikely to cause any significant changes in the structure of squid arrestin, since a similar mutation in arrestin-1 (Asp296Asn) did not render the arrestin 'pre-activated' and phosphorylation insensitive (Vishnivetskiy et al. 1999).

It is also worth noting that unlike bovine rhodopsin, squid rhodopsin does not lose the retinal after isomerization (Hubbard and St George 1958). This could be one factor for the relatively high stability of the complex. The C-terminal tail of squid rhodopsin does not contain the characteristic contiguous stretches of serines and threonines as seen in mammalian receptors, it possibly has only one phosphorylation site (by analogy to octopus rhodopsin) (Ohguro et al. 1998). Another unique feature of the squid visual system is that both rhodopsin and arrestin get phosphorylated by the same kinase, SQRK. Although rhodopsin phosphorylation is strictly dependent on light activation, arrestin phosphorylation is not. The phosphorylation sites on squid arrestin have been identified, they reside towards the very end of the C-terminus (Ser392 and Ser397) (Robinson et al. 2015). Interestingly, β-arrestins are also phosphorylated toward the end of the C-terminal tail (Ser412 in rat arrestin-2/β-arrestin-1, and Ser361 and Thr383 in rat arrestin-3/β-arrestin-2), and their phosphorylation is mediated by ERK1/2 (arrestin-2) or CKII (arrestin-3 at Thr383) (Lin et al. 1997, 2002). Like squid arrestin, C-terminal phosphorylation of non-visual arrestins does not depend on their association with receptors, and the phosphorylation sites are located almost at the very end in arrestin-2 and in a flexible region between β-strands XIX and XX in arrestin-3. These regions are not at all resolved in any of the crystallographic structures of arrestin-2 or -3. This suggests that the inherent flexibility in these regions allows them to engage with the kinase independently of their association with receptor. The phosphorylation of non-visual arrestins appears to impede clathrin interaction and therefore decrease receptor endocytosis via clathrin-coated pits (Gurevich and Benovic 1997; Lin et al. 2002). Unlike arrestin-2 and arrestin-3 (β-arrestins), the physiological consequence of phosphorylation of squid arrestin remains elusive (Robinson et al. 2015).

As far as squid arrestin's ability to form a high affinity complex with light activated rhodopsin is concerned, receptor or arrestin phosphorylation does not seem to be a deciding factor (Robinson 2015). The stoichiometry of the interaction at the native membrane could be effectively modelled as a 1:1 complex (Robinson 2015), a feature quite distinct from the bovine rhodopsin-arrestin system, where the stoichiometry of interaction in some cases appears to be dependent on the percentage of light-activated phosphorylated rhodopsins (Sommer et al. 2011), although high-affinity binding involves 1:1 interaction (Bayburt et al. 2011; Hanson et al. 2007b).

Structural and biophysical characterization of the squid rhodopsin-arrestin complex in native membranes is ongoing in our lab (Bandyopadhyay/Ernst). Using site-directed spin labeling of arrestin we found that squid arrestin engages the receptor upon light activation with the finger loop (Bandyopadhyay, Ernst et al., unpublished results). This is similar to the mammalian arrestin-1 (Hanson et al. 2006), however, for other regions, preliminary DEER data of arrestin in complex with rhodopsin show differences from mammalian arrestins, suggesting that the mode of binding in the squid system is not the same as in mammalian arrestin-receptor interactions. The squid rhodopsin crystal structures revealed an extended intracellular cytoplasmic region between trans-membrane helices 5 and 6 (Murakami and Kouyama 2008; Shimamura et al. 2008) that would suggest some variation from the binding mode seen in the bovine rhodopsin system (Kang et al. 2015). The precise conformation of receptor-bound squid arrestin is difficult to envision at the moment with limited information. Therefore, it is not possible to discount either a 'clam shell' or the 'domain twist' model for this interaction. It is also possible that this arrestin displays a yet unseen conformation upon receptor engagement.

Therefore, to fully understand arrestin-mediated signaling from GPCRs, we need additional receptor-arrestin structures of both phosphorylation dependent and independent types, where the receptors are activated by different ligands and phosphorylated in different positions for the phosphorylation sensitive complexes. It is perhaps safe to predict that the complexes will have different shapes, and quite likely each one would have sufficient flexibility to have both arrestin and receptor in more than one conformation. The differences might be subtle, likely detectable only by biophysical methods in solution. The ultimate task would be to figure out which conformation of the receptor-bound arrestin mediates which branches of signaling. The field is making the first steps in this direction (see Chaps. 9, 16 and 17). In the end, clear understanding of structural requirements of each arrestin-dependent signaling cascade will pave the way to the construction of "designer" arrestins directing the signaling from the receptors we want to the pathways of our choosing. Thus the 'unexpected' in the world of arrestin conformations is still wide for us to explore and more surprises likely await us.

Acknowledgements This work was supported by the NIH RO1 EY011500, GM109955, GM077561 (VVG), EY05216 and the Jules Stein Professorship Endowment (WLH), and Canada Excellence Research Chairs program (OPE). OPE is the Anne and Max Tanenbaum Chair at University of Toronto. We thank Dr. Ned Van Eps for comments on the manuscript.

References

Altenbach C, Kusnetzow AK, Ernst OP, Hofmann KP, Hubbell WL (2008) High-resolution distance mapping in rhodopsin reveals the pattern of helix movement due to activation. Proc Natl Acad Sci U S A 105:7439–7444

Bayburt TH, Vishnivetskiy SA, McLean M, Morizumi T, Huang C-C, Tesmer JJ, Ernst OP, Sligar SG, Gurevich VV (2011) Rhodopsin monomer is sufficient for normal rhodopsin kinase (GRK1) phosphorylation and arrestin-1 binding. J Biol Chem 286:1420–1428

Benovic JL, Kuhn H, Weyand I, Codina J, Caron MG, Lefkowitz RJ (1987) Functional desensitization of the isolated beta-adrenergic receptor by the beta-adrenergic receptor kinase: potential role of an analog of the retinal protein arrestin (48-kDa protein). Proc Natl Acad Sci U S A 84:8879–8882

Choe HW, Kim YJ, Park JH, Morizumi T, Pai EF, Krauss N, Hofmann KP, Scheerer P, Ernst OP (2011) Crystal structure of metarhodopsin II. Nature 471:651–655

Feuerstein SE, Pulvermuller A, Hartmann R, Granzin J, Stoldt M, Henklein P, Ernst OP, Heck M, Willbold D, Koenig BW (2009) Helix formation in arrestin accompanies recognition of photoactivated rhodopsin. Biochemistry 48:10733–10742

Granzin J, Wilden U, Choe HW, Labahn J, Krafft B, Buldt G (1998) X-ray crystal structure of arrestin from bovine rod outer segments. Nature 391:918–921

Granzin J, Cousin A, Weirauch M, Schlesinger R, Büldt G, Batra-Safferling R (2012) Crystal structure of p44, a constitutively active splice variant of visual arrestin. J Mol Biol 416: 611–618

Granzin J, Stadler A, Cousin A, Schlesinger R, Batra-Safferling R (2015) Structural evidence for the role of polar core residue Arg175 in arrestin activation. Sci Rep 5:15808

Gurevich VV (1998) The selectivity of visual arrestin for light-activated phosphorhodopsin is controlled by multiple nonredundant mechanisms. J Biol Chem 273:15501–15506

Gurevich VV, Benovic JL (1992) Cell-free expression of visual arrestin. Truncation mutagenesis identifies multiple domains involved in rhodopsin interaction. J Biol Chem 267:21919–21923

Gurevich VV, Benovic JL (1993) Visual arrestin interaction with rhodopsin: sequential multisite binding ensures strict selectivity towards light-activated phosphorylated rhodopsin. J Biol Chem 268:11628–11638

Gurevich VV, Benovic JL (1995) Visual arrestin binding to rhodopsin: diverse functional roles of positively charged residues within the phosphorylation-recognition region of arrestin. J Biol Chem 270:6010–6016

Gurevich VV, Benovic JL (1997) Mechanism of phosphorylation-recognition by visual arrestin and the transition of arrestin into a high affinity binding state. Mol Pharmacol 51:161–169

Gurevich VV, Gurevich EV (2004) The molecular acrobatics of arrestin activation. Trends Pharmacol Sci 25:105–111

Gurevich EV, Gurevich VV (2006a) Arrestins are ubiquitous regulators of cellular signaling pathways. Genome Biol 7:236

Gurevich VV, Gurevich EV (2006b) The structural basis of arrestin-mediated regulation of G protein-coupled receptors. Pharmacol Ther 110:465–502

Gurevich VV, Chen C-Y, Kim CM, Benovic JL (1994) Visual arrestin binding to rhodopsin: intramolecular interaction between the basic N-terminus and acidic C-terminus of arrestin may regulate binding selectivity. J Biol Chem 269:8721–8727

Han M, Gurevich VV, Vishnivetskiy SA, Sigler PB, Schubert C (2001) Crystal structure of beta-arrestin at 1.9 Å: possible mechanism of receptor binding and membrane translocation. Structure 9:869–880

Hanson SM, Francis DJ, Vishnivetskiy SA, Kolobova EA, Hubbell WL, Klug CS, Gurevich VV (2006) Differential interaction of spin-labeled arrestin with inactive and active phospho-rhodopsin. Proc Natl Acad Sci U S A 103:4900–4905

Hanson SM, Cleghorn WM, Francis DJ, Vishnivetskiy SA, Raman D, Song X, Nair KS, Slepak VZ, Klug CS, Gurevich VV (2007a) Arrestin mobilizes signaling proteins to the cytoskeleton and redirects their activity. J Mol Biol

Hanson SM, Gurevich EV, Vishnivetskiy SA, Ahmed MR, Song X, Gurevich VV (2007b) Each rhodopsin molecule binds its own arrestin. Proc Natl Acad Sci U S A 104:3125–3128

Hanson SM, Van Eps N, Francis DJ, Altenbach C, Vishnivetskiy SA, Arshavsky VY, Klug CS, Hubbell WL, Gurevich VV (2007c) Structure and function of the visual arrestin oligomer. EMBO J 26:1726–1736

Hirsch JA, Schubert C, Gurevich VV, Sigler PB (1999) The 2.8 Å crystal structure of visual arrestin: a model for arrestin's regulation. Cell 97:257–269

Hubbard R, St George RC (1958) The rhodopsin system of the squid. J Gen Physiol 41:501–528

Kang Y, Zhou XE, Gao X, He Y, Liu W, Ishchenko A, Barty A, White TA, Yefanov O, Han GW et al (2015) Crystal structure of rhodopsin bound to arrestin by femtosecond X-ray laser. Nature 523:561–567

Kim M, Vishnivetskiy SA, Van Eps N, Alexander NS, Cleghorn WM, Zhan X, Hanson SM, Morizumi T, Ernst OP, Meiler J et al (2012) Conformation of receptor-bound visual arrestin. Proc Natl Acad Sci U S A 109:18407–18412

Kim YJ, Hofmann KP, Ernst OP, Scheerer P, Choe HW, Sommer ME (2013) Crystal structure of pre-activated arrestin p44. Nature 497:142–146

Kovoor A, Celver J, Abdryashitov RI, Chavkin C, Gurevich VV (1999) Targeted construction of phosphorylation-independent b-arrestin mutants with constitutive activity in cells. J Biol Chem 274:6831–6834

Krupnick JG, Gurevich VV, Benovic JL (1997) Mechanism of quenching of phototransduction. Binding competition between arrestin and transducin for phosphorhodopsin. J Biol Chem 272:18125–18131

Kuhn H (1978) Light-regulated binding of rhodopsin kinase and other proteins to cattle photoreceptor membranes. Biochemistry 17:4389–4395

Kuhn H, Hall SW, Wilden U (1984) Light-induced binding of 48-kDa protein to photoreceptor membranes is highly enhanced by phosphorylation of rhodopsin. FEBS Lett 176:473–478

Lin FT, Krueger KM, Kendall HE, Daaka Y, Fredericks ZL, Pitcher JA, Lefkowitz RJ (1997) Clathrin-mediated endocytosis of the beta-adrenergic receptor is regulated by phosphorylation/dephosphorylation of beta-arrestin1. J Biol Chem 272:31051–31057

Lin FT, Chen W, Shenoy S, Cong M, Exum ST, Lefkowitz RJ (2002) Phosphorylation of beta-arrestin2 regulates its function in internalization of beta(2)-adrenergic receptors. Biochemistry 41:10692–10699

Lohse MJ, Benovic JL, Codina J, Caron MG, Lefkowitz RJ (1990) Beta-arrestin: a protein that regulates beta-adrenergic receptor function. Science 248:1547–1550

Lohse MJ, Andexinger S, Pitcher J, Trukawinski S, Codina J, Faure JP, Caron MG, Lefkowitz RJ (1992) Receptor-specific desensitization with purified proteins. Kinase dependence and receptor specificity of beta-arrestin and arrestin in the beta 2-adrenergic receptor and rhodopsin systems. J Biol Chem 267:8558–8564

Mayeenuddin LH, Mitchell J (2003) Squid visual arrestin: cDNA cloning and calcium-dependent phosphorylation by rhodopsin kinase (SQRK). J Neurochem 85:592–600

Milano SK, Pace HC, Kim YM, Brenner C, Benovic JL (2002) Scaffolding functions of arrestin-2 revealed by crystal structure and mutagenesis. Biochemistry 41:3321–3328

Modzelewska A, Filipek S, Palczewski K, Park PS (2006) Arrestin interaction with rhodopsin: conceptual models. Cell Biochem Biophys 46:1–15

Murakami M, Kouyama T (2008) Crystal structure of squid rhodopsin. Nature 453:363–367

Ohguro H, Yoshida N, Shindou H, Crabb JW, Palczewski K, Tsuda M (1998) Identification of a single phosphorylation site within octopus rhodopsin. Photochem Photobiol 68:824–828
Palczewski K, Buczyłko J, Imami NR, McDowell JH, Hargrave PA (1991a) Role of the carboxyl-terminal region of arrestin in binding to phosphorylated rhodopsin. J Biol Chem 266:15334–15339
Palczewski K, Pulvermuller A, Buczylko J, Hofmann KP (1991b) Phosphorylated rhodopsin and heparin induce similar conformational changes in arrestin. J Biol Chem 266:18649–18654
Peterhans C, Lally CC, Ostermaier MK, Sommer ME, Standfuss J (2016) Functional map of arrestin binding to phosphorylated opsin, with and without agonist. Sci Rep 6:28686
Pulvermuller A, Maretzki D, Rudnicka-Nawrot M, Smith WC, Palczewski K, Hofmann KP (1997) Functional differences in the interaction of arrestin and its splice variant, p44, with rhodopsin. Biochemistry 36:9253–9260
Rasmussen SG, DeVree BT, Zou Y, Kruse AC, Chung KY, Kobilka TS, Thian FS, Chae PS, Pardon E, Calinski D et al (2011) Crystal structure of the beta2 adrenergic receptor-Gs protein complex. Nature 477:549–555
Robinson KA (2015) Arrestin interactions with rhodopsin in the squid visual system. In: Graduate department of pharmacology and toxicology. University of Toronto
Robinson KA, Ou WL, Guan X, Sugamori KS, Bandyopadhyay A, Ernst OP, Mitchell J (2015) The effect of phosphorylation on arrestin-rhodopsin interaction in the squid visual system. J Neurochem 135:1129–1139
Ryba NJ, Findlay JB, Reid JD (1993) The molecular cloning of the squid (*Loligo forbesi*) visual Gq-alpha subunit and its expression in *Saccharomyces cerevisiae*. Biochem J 292(Pt 2): 333–341
Schleicher A, Kuhn H, Hofmann KP (1989) Kinetics, binding constant, and activation energy of the 48-kDa protein-rhodopsin complex by extra-metarhodopsin II. Biochemistry 28: 1770–1775
Shimamura T, Hiraki K, Takahashi N, Hori T, Ago H, Masuda K, Takio K, Ishiguro M, Miyano M (2008) Crystal structure of squid rhodopsin with intracellularly extended cytoplasmic region. J Biol Chem 283:17753–17756
Shukla AK, Manglik A, Kruse AC, Xiao K, Reis RI, Tseng WC, Staus DP, Hilger D, Uysal S, Huang LY et al (2013) Structure of active beta-arrestin-1 bound to a G-protein-coupled receptor phosphopeptide. Nature 497:137–141
Smith WC, Milam AH, Dugger D, Arendt A, Hargrave PA, Palczewski K (1994) A splice variant of arrestin. Molecular cloning and localization in bovine retina. J Biol Chem 269:15407–15410
Sommer ME, Smith WC, Farrens DL (2005) Dynamics of arrestin-rhodopsin interactions: arrestin and retinal release are directly linked events. J Biol Chem 280:6861–6871
Sommer ME, Hofmann KP, Heck M (2011) Arrestin-rhodopsin binding stoichiometry in isolated rod outer segment membranes depends on the percentage of activated receptors. J Biol Chem 286:7359–7369
Sutton RB, Vishnivetskiy SA, Robert J, Hanson SM, Raman D, Knox BE, Kono M, Navarro J, Gurevich VV (2005) Crystal structure of cone arrestin at 2.3 Å: evolution of receptor specificity. J Mol Biol 354:1069–1080
Swardfager W, Mitchell J (2007) Purification of visual arrestin from squid photoreceptors and characterization of arrestin interaction with rhodopsin and rhodopsin kinase. J Neurochem 101:223–231
Szczepek M, Beyriere F, Hofmann KP, Elgeti M, Kazmin R, Rose A, Bartl FJ, von Stetten D, Heck M, Sommer ME et al (2014) Crystal structure of a common GPCR-binding interface for G protein and arrestin. Nat Commun 5:4801
Vishnivetskiy SA, Paz CL, Schubert C, Hirsch JA, Sigler PB, Gurevich VV (1999) How does arrestin respond to the phosphorylated state of rhodopsin? J Biol Chem 274:11451–11454
Vishnivetskiy SA, Schubert C, Climaco GC, Gurevich YV, Velez M-G, Gurevich VV (2000) An additional phosphate-binding element in arrestin molecule: implications for the mechanism of arrestin activation. J Biol Chem 275:41049–41057

Vishnivetskiy SA, Hirsch JA, Velez M-G, Gurevich YV, Gurevich VV (2002) Transition of arrestin in the active receptor-binding state requires an extended interdomain hinge. J Biol Chem 277:43961–43968

Vishnivetskiy SA, Francis DJ, Van Eps N, Kim M, Hanson SM, Klug CS, Hubbell WL, Gurevich VV (2010) The role of arrestin alpha-helix I in receptor binding. J Mol Biol 395:42–54

Vishnivetskiy SA, Gimenez LE, Francis DJ, Hanson SM, Hubbell WL, Klug CS, Gurevich VV (2011) Few residues within an extensive binding interface drive receptor interaction and determine the specificity of arrestin proteins. J Biol Chem 286:24288–24299

Vishnivetskiy SA, Baameur F, Findley KR, Gurevich VV (2013) Critical role of the central 139-loop in stability and binding selectivity of arrestin-1. J Biol Chem 288:11741–11750

Wilden U (1995) Duration and amplitude of the light-induced cGMP hydrolysis in vertebrate photoreceptors are regulated by multiple phosphorylation of rhodopsin and by arrestin binding. Biochemistry 34:1446–1454

Wilden U, Hall SW, Kühn H (1986) Phosphodiesterase activation by photoexcited rhodopsin is quenched when rhodopsin is phosphorylated and binds the intrinsic 48-kDa protein of rod outer segments. Proc Natl Acad Sci U S A 83:1174–1178

Zhan X, Gimenez LE, Gurevich VV, Spiller BW (2011) Crystal structure of arrestin-3 reveals the basis of the difference in receptor binding between two non-visual arrestins. J Mol Biol 406:467–478

Zhuo Y, Vishnivetskiy SA, Zhan X, Gurevich VV, Klug CS (2014) Identification of receptor binding-induced conformational changes in non-visual arrestins. J Biol Chem 289:20991–21002

Chapter 13
The Arrestin-Receptor Complex: Exciting Answers and New Questions

Yanyong Kang, Karsten Melcher, Vsevolod V. Gurevich and H. Eric Xu

Abstract To better understand the molecular mechanism of arrestin-mediated signaling, detailed structural information on the arrestin-receptor complex is necessary. Biochemical studies provided some information about how arrestins are recruited by active receptors. The X-ray laser crystal structure of the rhodopsin–arrestin complex reveals unique structural features, which include the asymmetric binding of arrestin to rhodopsin. Arrestin adopts the active conformation, with a $\sim 20°$ rotation between the N- and C-domains of the molecule, which opens up a cleft in arrestin to accommodate a short helix formed by the second intracellular loop of rhodopsin. Rhodopsin–arrestin complex gives important insights into how G protein–coupled receptor signaling is terminated by arrestin and reveals structural basis of the mechanism of arrestin-biased signaling.

Keywords Crystal structure · GPCR · Rhodopsin · Arrestin · Biased signaling

Arrestins are responsible for the desensitization, internalization and G protein-independent signaling of G protein-coupled receptors (GPCRs) in an agonist-dependent manner (DeWire et al. 2007). Recent crystallographic (Kim et al. 2013; Shukla et al. 2013), mutational (reviewed in Gurevich and Gurevich 2004, 2006, 2014), and biophysical (Kim et al. 2012; Nobles et al. 2007; Zhuang et al. 2013) suggest that all arrestins undergo extensive conformational changes upon binding to the phosphorylated GPCRs. In their basal free cytosolic state, arrestins are elongated molecules, which consist of two (N- and C-) domains and the C-terminus anchored in a polar core between them, unavailable for interaction with

Y. Kang · K. Melcher · H. Eric Xu (✉)
Laboratory of Structural Sciences, Center for Structural Biology and Drug Discovery, Van Andel Research Institute, Grand Rapids, MI 49503, USA
e-mail: Eric.Xu@vai.org

V.V. Gurevich (✉)
Department of Pharmacology, Vanderbilt University, Nashville, TN 37232, USA
e-mail: vsevolod.gurevich@vanderbilt.edu

V.V. Gurevich (ed.), *The Structural Basis of Arrestin Functions*,
DOI 10.1007/978-3-319-57553-7_13

partner proteins (Granzin et al. 1998; Han et al. 2001; Hirsch et al. 1999; Milano et al. 2002; Sutton et al. 2005; Zhan et al. 2011).

Crystal structure of truncated arrestin-2 (a.k.a. β-arrestin1)[1] in complex with a phosphorylated vasopressin receptor-2 (V2Rpp) carboxy-terminus revealed structural basis for arrestin activation. Activated arrestin-2 demonstrated extensive conformational changes in the C-terminus, which is released, which apparently makes it available for the interactions with clathrin (Goodman et al. 1996) and clathrin adaptor AP2 (Laporte et al. 1999). The V_2Rpp-arrestin-2 crystal structure also revealed a 20° twist between the N- and C-domains. A similar ~20° rotation has been observed in the crystal structure of pre-activated short splice variant of arrestin-1 (Kim et al. 2013), as well as the mouse visual arrestin-1 bound to a constitutively active form of human rhodopsin (Kang et al. 2015; Kim et al. 2013). It has been suggested that the twisting movement of the two domains is part of the general mechanism by which arrestins, upon activation, may expose an additional interface for interacting with their numerous binding partners. Electron microscopy analysis of β2-adrenergic receptor (β2AR) in complex with arrestin-2 (β-arrestin-1) reveals a dynamic interaction reflected in several conformations of the complex (Shukla et al. 2014).

Rhodopsin is a prototypical GPCR that serves as the photon receptor in the visual system. Along with β_2AR, rhodopsin has been a model system for studying GPCR signaling, including its coupling to downstream effectors, i.e., G proteins and arrestins. The structure of the rhodopsin–arrestin complex is the key to understanding both the receptor's conformational changes induced by arrestin binding and how different signaling pathways are activated by binding of G proteins versus arrestins (Kang et al. 2015, 2016). Here, we discuss these recent findings relating to the structural mechanism of arrestin-GPCR interaction.

Arrestin Is Asymmetrically Bound to Rhodopsin

The overall rhodopsin–arrestin structure shows that both rhodopsin and arrestin are in the active conformation (Fig. 13.1a, b). Rhodopsin and arrestin have a similar height from an intracellular view, but the width of arrestin is about three times that of rhodopsin. This arrangement allows compact crystal packing through the soluble parts of the protein complex. Arrestin consists of two β-strand domains, the N-domain and C-domain. The two domains have similar size and form a crescent molecule that interacts with the receptor (Hirsch et al. 1999). The most significant

---

[1]Here we use the systematic names of arrestin arrestin proteins: arrestin-1 (historic names S-antigen, 48 kDa protein, visual or rod arrestin), arrestin-2 (β-arrestin or β-arrestin1), arrestin arrestin-3 (β-arrestin2 or hTHY-ARRX), and arrestin-4 (cone or X-arrestin; for unclear reasons its gene is called "arrestin 3" in the HUGO database).

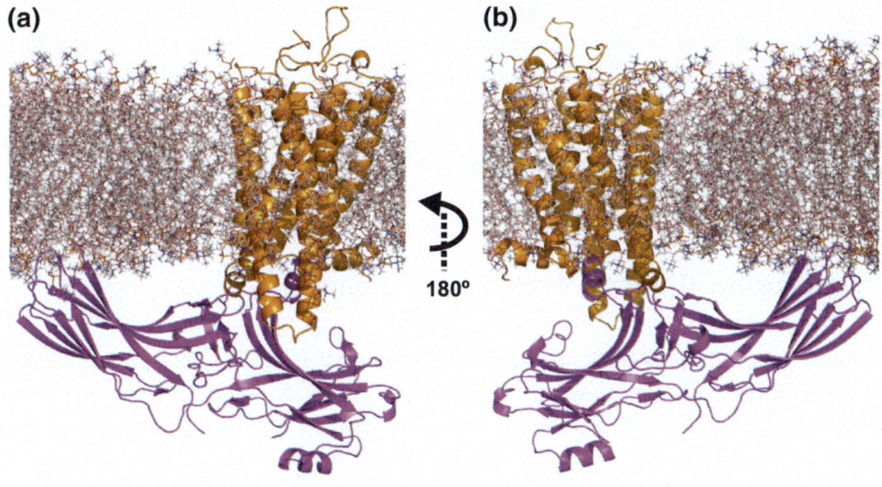

Fig. 13.1 **a**, **b** Asymmetric binding of arrestin to rhodopsin. The C-edge of arrestin likely embeds into the lipid layer through hydrophobic interactions

feature of the rhodopsin–arrestin complex is that arrestin is asymmetrically bound to rhodopsin and also asymmetrically oriented relative to the membrane (Fig. 13.1a, b). Distal tip of the C-domain contacts the cell membrane (Fig. 13.1a, b). There are many hydrophobic residues at this C-tip region, which can embed into the membrane (see Chap. 7). It is well known that other GPCR-binding proteins, the G protein heterotrimers and GPCR kinases (GRKs) need to be anchored into the membrane bilayer to perform their functions. Palmitoylation of the Gα subunit [and of GRKs 4, 5, and 6 (Gurevich et al. 2012)], and the prenylation of the Gγ subunit are involved in tethering the G proteins (or GRKs 1 and 7) to the inner surface of the plasma membrane to enable them to interact with the receptor. In contrast to GRKs and G protein subunits, there is no evidence of any lipid modification of arrestin. This asymmetric positioning of arrestin might play important role(s) in GPCR desensitization, internalization, and arrestin-mediated signaling.

The asymmetric assembly of the rhodopsin-arrestin complex was supported by three pairs of intermolecular distances between rhodopsin and arrestin measured by DEER with non-fused individual proteins (Kang et al. 2016). These results indicate that the structure of the fused rhodopsin-arrestin complex closely resembles the complex assembled by non-fused rhodopsin and arrestin. Conceivably, the asymmetric arrangement of the arrestin-rhodopsin complex might allow the distal C-tip to serve as the binding site of a second rhodopsin, which has been proposed to form dimers in the rod outer-segment disc membrane (Fotiadis et al. 2006). Although this model remains controversial (Gurevich and Gurevich 2008a, b), arrestin-1 is expressed in rod photoreceptors at the level of $\sim$8 molecules per 10 rhodopsins

(Song et al. 2011; Strissel et al. 2006). Some experimental evidence suggests that at least at high bleach levels, when there are more light-activated rhodopsins than arrestin-1, one arrestin-1 might block two rhodopsin molecules (Sommer et al. 2011, 2014). However, high-affinity binding of arrestin-1 to monomeric rhodopsin in nanodiscs has been reported (Bayburt et al. 2011; Tsukamoto et al. 2010). Thus, the resolution of this issue awaits additional evidence.

How Arrestin Binds to Activated Rhodopsin

In comparison to the GPCR–G protein complex (Rasmussen et al. 2007), rhodopsin–arrestin complex shows a very different binding model. There are four important interfaces in the rhodopsin–arrestin complex. The first interface includes the finger loop region of arrestin and the intercellular loop 1 (ICL1), the transmembrane helix 7 (TM7), and helix 8 of rhodopsin (Fig. 13.2a). The most surprising observation is that the original finger loop region of arrestin changes its conformation to a short helix, which enables arrestin to lower the finger loop region to fit into the cleft formed by the TM bundle of rhodopsin. However, helical conformation of rhodopsin-bound finger loop of arrestin was observed in co-structure of active rhodopsin and arrestin-1 peptide (Szczepek et al. 2014). The second interface involves the middle loop and C-loop of arrestin and the ICL2 of rhodopsin. In arrestin's basal state, the middle loop and C-loop are in close proximity. When arrestin is activated, the two loops are open and provide enough space to accommodate the ICL2 of rhodopsin, which changes its conformation to a short helix (Fig. 13.2b). In the third interface, the beta strand (residues 79–86) of arrestin interacts with TM5, TM6, and ICL3 of rhodopsin (Fig. 13.2c). Because of the resolution limit of the structure, the density of the C-terminal tail of rhodopsin is not visible (Kang et al. 2015). However, based on the cysteine-cysteine crosslinking data, a computation model shows that the fourth interface involves the N-terminal β-strand of arrestin-1 and the C-terminal tail of rhodopsin (Fig. 13.2d).

Recently a number of arrestin-3 residues were implicated in receptor preference based on mutagenesis and BRET-based in-cell receptor-arrestin interaction assay (Gimenez et al. 2012b, 2014). Interestingly, the homologues of some of the C-domain residues are not even close to rhodopsin in the arrestin-1-rhodopsin structure (Kang et al. 2015). This suggests that either the pose of the non-visual arrestins bound to their cognate GPCRs is significantly different, or that identified receptor-discriminator residues act allosterically, so that their substitutions affect other residues directly engaged by the receptor. Crystal structures of at least some of the complexes of non-visual arrestins with other GPCRs are necessary to address this isuue.

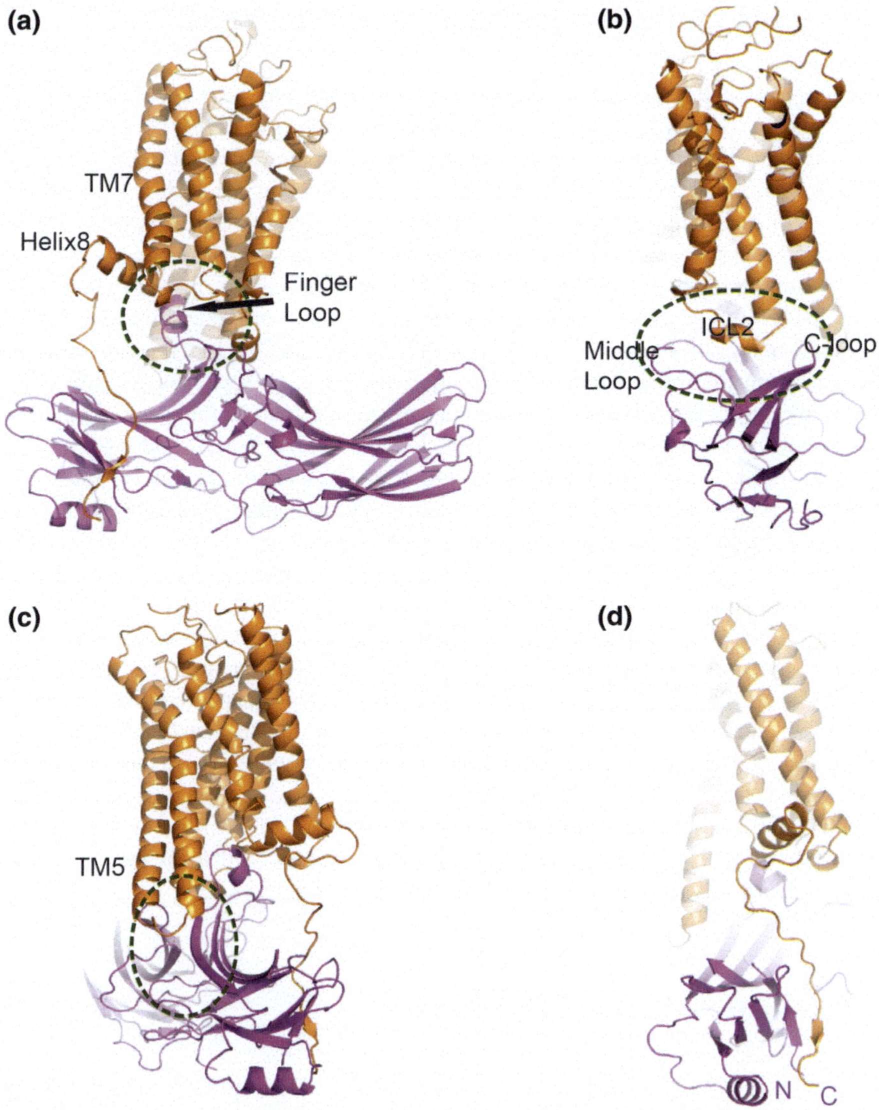

Fig. 13.2 Detailed view of the rhodopsin–arrestin interface. **a** The finger loop of arrestin-1 interacts with ICL1, TM7, and helix 8 of rhodopsin. **b** The middle and lariat loops of arrestin-1 bind to ICL2 of rhodopsin. **c** The β-strand VI of arrestin binds to TM5, TM6, and ICL3. **d** The N-C lock between the N-terminus of arrestin and the C-tail of rhodopsin

Mechanism of Arrestin Recruitment and Activation

Both G protein (Rasmussen et al. 2007) and arrestin (Kang et al. 2015) engage the inter-helical cavity that opens on the cytoplasmic side of GPCRs upon activation (Farrens et al. 1996). This explains why arrestins compete with G proteins for active GPCRs (Krupnick et al. 1997; Wilden 1995; Wilden et al. 1986). The crystal structure of the complex provides a visual model of how the GPCR is desensitized by arrestin. Compared with inactive rhodopsin, the largest difference is a 10 Å outward movement of TM6 when measured at the Cα carbon of Q244 (Standfuss et al. 2011; Zhou et al. 2012). There is also an extension of the cytoplasmic end of the TM5 helix. In addition, TM1, TM4, TM5, and TM7 show a small outward movement on the intracellular side, which creates enough space for arrestin binding. In the basal state, the two domains of arrestin are held in the basal orientation by several inter-domain interactions (Hirsch et al. 1999). When rhodopsin is activated by light, serine and threonine residues of the C-terminal tail are phosphorylated by G protein–coupled receptor kinase. The negatively charged rhodopsin C-tail displaces the arrestin C-terminus, which allows electrostatic interaction between the cationic N-terminus of arrestin and rhodopsin's C-tail, destabilizing arrestin's polar core. Once arrestin loses the polar core constraint, the N- and C-domains rotate against each other by approximately 20°, which opens up the cleft between the middle loop and the C-loop to adopt the short helix formed by the rhodopsin ICL2 (Fig. 13.2c). Because complementarily charged residues between rhodopsin and arrestin initiate the interaction, this binding is sensitive to high salt (Gurevich and Benovic 1993; Vishnivetskiy et al. 1999). There are over 800 GPCRs but only four arrestins in humans. Inspecting the charge distribution of available GPCR structures reveals a conserved pattern of positive charge on the inner side of their transmembrane helix bundle, which could form the molecular basis of promiscuous pairing between GPCRs and arrestins.

Conclusions and Future Directions

The rhodopsin–arrestin crystal structure provides the first atomic resolution view of the assembly of the rhodopsin-arrestin complex. The structure reveals an asymmetric arrangement of arrestin binding to rhodopsin and molecular basis for rhodopsin recruitment of arrestin. The intermolecular interactions and conformational changes in rhodopsin and arrestin can be used as a model for understanding other GPCR-arrestin interactions. The structural availability of both GPCR–G protein and GPCR–arrestin complexes offer an insight into different signaling pathways, which could help improve and accelerate the drug design process aiming at generating signaling-biased GPCR ligands favoring G protein- or arrestin-mediated signaling.

There remain many unanswered questions. First, what is the difference between GPCR-bound visual arrestin-1 and GPCR-bound non-visual-arrestins 2 and 3?

High-resolution crystal structures of complexes of other GPCRs with their cognate arrestins are necessary to address this question. Some GPCRs don't have long C-terminal tail, such as serotonin receptors. The serines located in the intracellular loop region can be phosphorylated (Karaki et al. 2014). The same is true for the M2 muscarinic receptor, where the third cytoplasmic loop has two clusters of phosphorylatable serines and threonines (Lee et al. 2000; Pals-Rylaarsdam et al. 1997). How are receptors phosphorylated in this element recognized by arrestin? Do they share similar binding mechanism with rhodopsin-arrestin binding? Second, what is the binding mechanism for a phosphorylated GPCR and arrestin? The structure of a pre-activated arrestin bound to a phosphorylated C-terminal tail of V2 vasopressin receptor partially addresses this question, but a complete answer still waits for a structure of arrestin in complexes with full-length phosphorylated GPCRs (Shukla et al. 2013). Interestingly, arrestin binding to some GPCRs was shown to be more dependent on receptor activation than on its phosphorylation (Gimenez et al. 2012a). New structures are needed to test an exciting hypothesis that the positions of receptor-attached phosphates create a "barcode" that determines the shape of the receptor-bound arrestin and the direction of arrestin-mediated signaling (Nobles et al. 2011; Tobin et al. 2008). Even in case of rhodopsin binding to visual arrestin-1 the role of distinct phosphorylated residues in the rhodopsin C-terminus appears to be different (Azevedo et al. 2015). Co-structures of arrestins with the same differentially phosphorylated receptor are needed to yield data shedding light on this issue. Third, what is the function of phosphoinositides in rhodopsin-mediated arrestin signaling? Phosphoinositides are the proposed physiological partners for arrestins at the plasma membrane. The crystal structure of arrestin-2 with inositol hexakisphosphate (inositol 1,2,3,4,5,6-hexakisphosphate [IP6]) has been reported (Milano et al. 2006), and IP6 is known to interact with all four arrestins, but the functional aspects are unknown. It was shown that arrestin-1 (Hanson et al. 2007, 2008a; Kim et al. 2011), as well as non-visual arrestins-2 and -3 (Chen et al. 2014; Milano et al. 2006) oligomerize, and that oligomers likely represent a storage form that does not bind receptors (Chen et al. 2014; Hanson et al. 2007). Interestingly, IP6 has opposing role in arrestin oligomerization: it promotes self-association of arrestin-2/3 (Hanson et al. 2008b; Milano et al. 2006), while inhibiting the oligomerization of visual arrestin-1 (Hanson et al. 2008b). Pursuing the answers to these questions promises to uncover important insights into biochemical mechanisms of GPCR signaling and its exploitation for biomedicine and drug discovery.

References

Azevedo AW, Doan T, Moaven H, Sokal I, Baameur F, Vishnivetskiy SA, Homan KT, Tesmer JJ, Gurevich VV, Chen J, Rieke F (2015) C-terminal threonines and serines play distinct roles in the desensitization of rhodopsin, a G protein-coupled receptor. Elife 4:eLife.05981. doi:10.7554/eLife.05981

Bayburt TH, Vishnivetskiy SA, McLean M, Morizumi T, Huang C-C, Tesmer JJ, Ernst OP, Sligar SG, Gurevich VV (2011) Rhodopsin monomer is sufficient for normal rhodopsin kinase (GRK1) phosphorylation and arrestin-1 binding. J Biol Chem 286:1420–1428

Chen Q, Zhuo Y, Kim M, Hanson SM, Francis DJ, Vishnivetskiy SA, Altenbach C, Klug CS, Hubbell WL, Gurevich VV (2014) Self-association of arrestin family members. Handb Exp Pharmacol 219:205–223

DeWire SM, Ahn S, Lefkowitz RJ, Shenoy SK (2007) Beta-arrestins and cell signaling. Annu Rev Physiol 69:483–510

Farrens DL, Altenbach C, Yang K, Hubbell WL, Khorana HG (1996) Requirement of rigid-body motion of transmembrane helices for light activation of rhodopsin. Science 274:768–770

Fotiadis D, Jastrzebska B, Philippsen A, Muller DJ, Palczewski K, Engel A (2006) Structure of the rhodopsin dimer: a working model for G-protein-coupled receptors. Curr Opin Struct Biol 16:252–259

Gimenez LE, Kook S, Vishnivetskiy SA, Ahmed MR, Gurevich EV, Gurevich VV (2012a) Role of receptor-attached phosphates in binding of visual and non-visual arrestins to G protein-coupled receptors. J Biol Chem 287:9028–9040

Gimenez LE, Vishnivetskiy SA, Baameur F, Gurevich VV (2012b) Manipulation of very few receptor discriminator residues greatly enhances receptor specificity of non-visual arrestins. J Biol Chem 287:29495–29505

Gimenez LE, Babilon S, Wanka L, Beck-Sickinger AG, Gurevich VV (2014) Mutations in arrestin-3 differentially affect binding to neuropeptide Y receptor subtypes. Cell Signal 26:1523–1531

Goodman OB Jr, Krupnick JG, Santini F, Gurevich VV, Penn RB, Gagnon AW, Keen JH, Benovic JL (1996) Beta-arrestin acts as a clathrin adaptor in endocytosis of the beta2-adrenergic receptor. Nature 383:447–450

Granzin J, Wilden U, Choe HW, Labahn J, Krafft B, Buldt G (1998) X-ray crystal structure of arrestin from bovine rod outer segments. Nature 391:918–921

Gurevich VV, Benovic JL (1993) Visual arrestin interaction with rhodopsin: Sequential multisite binding ensures strict selectivity towards light-activated phosphorylated rhodopsin. J Biol Chem 268:11628–11638

Gurevich VV, Gurevich EV (2004) The molecular acrobatics of arrestin activation. Trends Pharmacol Sci 25:105–111

Gurevich VV, Gurevich EV (2006) The structural basis of arrestin-mediated regulation of G-protein-coupled receptors. Pharmacol Ther 110:465–502

Gurevich VV, Gurevich EV (2008a) GPCR monomers and oligomers: it takes all kinds. Trends Neurosci 31:74–81

Gurevich VV, Gurevich EV (2008b) How and why do GPCRs dimerize? Trends Pharmacol Sci 29:234–240

Gurevich VV, Gurevich EV (2014) Extensive shape shifting underlies functional versatility of arrestins. Curr Opin Cell Biol 27:1–9

Gurevich EV, Tesmer JJ, Mushegian A, Gurevich VV (2012) G protein-coupled receptor kinases: more than just kinases and not only for GPCRs. Pharmacol Ther 133:40–69

Han M, Gurevich VV, Vishnivetskiy SA, Sigler PB, Schubert C (2001) Crystal structure of beta-arrestin at 1.9 A: possible mechanism of receptor binding and membrane Translocation. Structure 9:869–880

Hanson SM, Van Eps N, Francis DJ, Altenbach C, Vishnivetskiy SA, Arshavsky VY, Klug CS, Hubbell WL, Gurevich VV (2007) Structure and function of the visual arrestin oligomer. EMBO J 26:1726–1736

Hanson SM, Dawson ES, Francis DJ, Van Eps N, Klug CS, Hubbell WL, Meiler J, Gurevich VV (2008a) A model for the solution structure of the rod arrestin tetramer. Structure 16:924–934

Hanson SM, Vishnivetskiy SA, Hubbell WL, Gurevich VV (2008b) Opposing effects of inositol hexakisphosphate on rod arrestin and arrestin2 self-association. Biochemistry 47:1070–1075

Hirsch JA, Schubert C, Gurevich VV, Sigler PB (1999) The 2.8 A crystal structure of visual arrestin: a model for arrestin's regulation. Cell 97:257–269

Kang Y, Zhou XE, Gao X, He Y, Liu W, Ishchenko A, Barty A, White TA, Yefanov O, Han GW et al (2015) Crystal structure of rhodopsin bound to arrestin by femtosecond X-ray laser. Nature 523:561–567

Kang Y, Gao X, Zhou XE, He Y, Melcher K, Xu HE (2016) A structural snapshot of the rhodopsin-arrestin complex. FEBS J 283:816–821

Karaki S, Becamel C, Murat S, Mannoury la Cour C, Millan MJ, Prezeau L, Bockaert J, Marin P, Vandermoere F (2014) Quantitative phosphoproteomics unravels biased phosphorylation of serotonin 2A receptor at Ser280 by hallucinogenic versus nonhallucinogenic agonists. Mol Cell Proteomics 13:1273–1285

Kim M, Hanson SM, Vishnivetskiy SA, Song X, Cleghorn WM, Hubbell WL, Gurevich VV (2011) Robust self-association is a common feature of mammalian visual arrestin-1. Biochemistry 50:2235–2242

Kim M, Vishnivetskiy SA, Van Eps N, Alexander NS, Cleghorn WM, Zhan X, Hanson SM, Morizumi T, Ernst OP, Meiler J et al (2012) Conformation of receptor-bound visual arrestin. Proc Nat Acad Sci USA 109:18407–18412

Kim YJ, Hofmann KP, Ernst OP, Scheerer P, Choe HW, Sommer ME (2013) Crystal structure of pre-activated arrestin p44. Nature 497:142–146

Krupnick JG, Gurevich VV, Benovic JL (1997) Mechanism of quenching of phototransduction. Binding competition between arrestin and transducin for phosphorhodopsin. J Biol Chem 272:18125–18131

Laporte SA, Oakley RH, Zhang J, Holt JA, Ferguson SG, Caron MG, Barak LS (1999) The 2-adrenergic receptor/arrestin complex recruits the clathrin adaptor AP-2 during endocytosis. Proc Nat Acad Sci USA 96:3712–3717

Lee KB, Ptasienski JA, Pals-Rylaarsdam R, Gurevich VV, Hosey MM (2000) Arrestin binding to the M2 muscarinic acetylcholine receptor is precluded by an inhibitory element in the third intracellular loop of the receptor. J Biol Chem 275:9284–9289

Milano SK, Pace HC, Kim YM, Brenner C, Benovic JL (2002) Scaffolding functions of arrestin-2 revealed by crystal structure and mutagenesis. Biochemistry 41:3321–3328

Milano SK, Kim YM, Stefano FP, Benovic JL, Brenner C (2006) Nonvisual arrestin oligomerization and cellular localization are regulated by inositol hexakisphosphate binding. J Biol Chem 281:9812–9823

Nobles KN, Guan Z, Xiao K, Oas TG, Lefkowitz RJ (2007) The active conformation of beta-arrestin1: direct evidence for the phosphate sensor in the N-domain and conformational differences in the active states of beta-arrestins1 and -2. J Biol Chem 282:21370–21381

Nobles KN, Xiao K, Ahn S, Shukla AK, Lam CM, Rajagopal S, Strachan RT, Huang TY, Bressler EA, Hara MR et al (2011) Distinct phosphorylation sites on the β(2)-adrenergic receptor establish a barcode that encodes differential functions of β-arrestin. Sci Signal 4:ra51

Pals-Rylaarsdam R, Gurevich VV, Lee KB, Ptasienski J, Benovic JL, Hosey MM (1997) Internalization of the m2 muscarinic acetylcholine receptor: arrestin-independent and -dependent pathways. J Biol Chem 272:23682–23689

Rasmussen SG, Choi HJ, Rosenbaum DM, Kobilka TS, Thian FS, Edwards PC, Burghammer M, Ratnala VR, Sanishvili R, Fischetti RF et al (2007) Crystal structure of the human beta2 adrenergic G-protein-coupled receptor. Nature 450:383–387

Shukla AK, Manglik A, Kruse AC, Xiao K, Reis RI, Tseng WC, Staus DP, Hilger D, Uysal S, Huang LY et al (2013) Structure of active beta-arrestin-1 bound to a G-protein-coupled receptor phosphopeptide. Nature 497:137–141

Shukla AK, Westfield GH, Xiao K, Reis RI, Huang LY, Tripathi-Shukla P, Qian J, Li S, Blanc A, Oleskie AN et al (2014) Visualization of arrestin recruitment by a G-protein-coupled receptor. Nature 512:218–222

Sommer ME, Hofmann KP, Heck M (2011) Arrestin-rhodopsin binding stoichiometry in isolated rod outer segment membranes depends on the percentage of activated receptors. J Biol Chem 286:7359–7369

Sommer ME, Hofmann KP, Heck M (2014) Not just signal shutoff: the protective role of arrestin-1 in rod cells. Handb Exp Pharmacol 219:101–116

Song X, Vishnivetskiy SA, Seo J, Chen J, Gurevich EV, Gurevich VV (2011) Arrestin-1 expression in rods: balancing functional performance and photoreceptor health. Neuroscience 174:37–49

Standfuss J, Edwards PC, D'Antona A, Fransen M, Xie G, Oprian DD, Schertler GF (2011) The structural basis of agonist-induced activation in constitutively active rhodopsin. Nature 471:656–660

Strissel KJ, Sokolov M, Trieu LH, Arshavsky VY (2006) Arrestin translocation is induced at a critical threshold of visual signaling and is superstoichiometric to bleached rhodopsin. J Neurosci 26:1146–1153

Sutton RB, Vishnivetskiy SA, Robert J, Hanson SM, Raman D, Knox BE, Kono M, Navarro J, Gurevich VV (2005) Crystal structure of cone arrestin at 2.3Å: evolution of receptor specificity. J Mol Biol 354:1069–1080

Szczepek M, Beyriere F, Hofmann KP, Elgeti M, Kazmin R, Rose A, Bartl FJ, von Stetten D, Heck M, Sommer ME et al (2014) Crystal structure of a common GPCR-binding interface for G protein and arrestin. Nat Commun 5:4801

Tobin AB, Butcher AJ, Kong KC (2008) Location, location, location…site-specific GPCR phosphorylation offers a mechanism for cell-type-specific signalling. Trends Pharmacol Sci 29:413–420

Tsukamoto H, Sinha A, Dewitt M, Farrens DL (2010) Monomeric rhodopsin is the minimal functional unit required for arrestin binding. J Mol Biol 399:501–511

Vishnivetskiy SA, Paz CL, Schubert C, Hirsch JA, Sigler PB, Gurevich VV (1999) How does arrestin respond to the phosphorylated state of rhodopsin? J Biol Chem 274:11451–11454

Wilden U (1995) Duration and amplitude of the light-induced cGMP hydrolysis in vertebrate photoreceptors are regulated by multiple phosphorylation of rhodopsin and by arrestin binding. Biochemistry 34:1446–1454

Wilden U, Hall SW, Kühn H (1986) Phosphodiesterase activation by photoexcited rhodopsin is quenched when rhodopsin is phosphorylated and binds the intrinsic 48-kDa protein of rod outer segments. Proc Natl Acad Sci USA 83:1174–1178

Zhan X, Gimenez LE, Gurevich VV, Spiller BW (2011) Crystal structure of arrestin-3 reveals the basis of the difference in receptor binding between two non-visual arrestins. J Mol Biol 406:467–478

Zhou XE, Melcher K, Xu HE (2012) Structure and activation of rhodopsin. Acta Pharmacol Sin 33:291–299

Zhuang T, Chen Q, Cho M-K, Vishnivetskiy SA, Iverson TI, Gurevich VV, Hubbell WL (2013) Involvement of distinct arrestin-1 elements in binding to different functional forms of rhodopsin. Proc Nat Acad Sci USA 110:942–947

Part IV
Arrestin-Mediated Signaling and Biased Mutants

Chapter 14
Scaffolding c-Jun N-Terminal Kinase Cascades: Mechanistic Insights from the Reconstituted Arrestin-JNK Cascades

Xuanzhi Zhan, Vsevolod V. Gurevich and Eugenia V. Gurevich

Abstract Although other subtypes appear to interact at least with some kinases in the ASK1-MKK4/7-JNK1/2/3 cascades, only arrestin-3 facilitates JNK activation in cells. While cell-based assays are suggestive, only the experiments with purified proteins can definitively show which protein-protein interactions are direct. Pure arrestin-3 was shown to facilitate the phosphorylation of JNK3 and other isoforms by MKK4 and MKK7. Biphasic dependence of the effect on arrestin-3 concentration showed that arrestin acts as a simple scaffold, bringing the kinases (MKKs) and their substrates (JNKs) into close proximity, rather than by activating MKKs or their substrate JNKs. Dissections of arrestin-3 elements involved allowed the identification of a surprisingly small 25-residue peptide that binds all relevant kinases and promotes JNK3 activation in cells.

Keywords Arrestin · JNKs · MKKs · MAP kinases · Scaffolding

Mitogen activated protein kinase (MAPK) pathways are a collection of signaling cascades responding to a wide variety of extracellular stimuli, including cytokines, growth factors, and environmental stress. In mammalian cells, there are at least four subfamilies of MAPKs, namely extracellular signal-regulated kinases (ERKs), c-Jun N-terminal Kinases (JNKs), p38, and noncanonical ERK5 (Chang and Karin 2001; Seger and Krebs 1995). Each MAPK is activated through a three-tier MAP3K-MAP2K-MAPK kinase cascade. There are more than 20 MAP3Ks in mammalian cells belonging to several fairly diverse families. A relatively small number of MAP2Ks (also called MKKs) activate downstream MAPKs: MEK1 and MEK2 activate ERK family MAPKs; MKK4 (also called MEK4) and MKK7 (also called MEK7) activate JNK family MAPKs; MKK3 (MEK3), MKK6 (MEK6) and

X. Zhan
Department of Chemistry, Tennessee Technological University, Cookeville, TN 38505, USA

V.V. Gurevich (✉) · E.V. Gurevich
Department of Pharmacology, Vanderbilt University, Nashville, TN 46795, USA
e-mail: vsevolod.gurevich@vanderbilt.edu

V.V. Gurevich (ed.), *The Structural Basis of Arrestin Functions*,
DOI 10.1007/978-3-319-57553-7_14

MKK4 activate p38 family MAPKs; and MEK5 activates ERK5 (Bardwell 2006). These highly conserved MAPK networks consist of distinct as well as overlapping signaling pathways. In many instances, the same signaling proteins are employed to elicit distinctly different functional responses. Therefore, the signal flow within a complex array of signaling networks needs to be precisely regulated with minimal or no crossover with adjacent signaling pathways.

Several mechanisms are likely involved in ensuring the fidelity and efficiency of signal transduction via MAPK cascades (Breitkreutz and Tyers 2002; Takekawa et al. 2005). The kinase-substrate specific recognition and interaction is certainly important. For instance, the activated Raf1, a MAP3K member, can phosphorylate MEK1 and MEK2, which then activate ERKs, but cannot activate other MKKs from JNK and p38 pathways (Good et al. 2009; Macdonald et al. 1993). However, it is clear that the specific interaction between kinase catalytic center and the substrate phosphoacceptor site alone is not sufficient to ensure the signal transduction specificity. Scaffold proteins have been found to play a central role in coordinating the signal flow within MAPK cascades (Bardwell 2006; Good et al. 2011; Whitmarsh 2006). Scaffold proteins provide a simple, flexible strategy for mediating selectivity in signaling networks by physically assembling the appropriate kinase component complexes.

One of the essential signaling events involving MAPK cascades is the G protein coupled receptor (GPCR) signaling (Kolch 2005; Marinissen and Gutkind 2001). GPCRs receive many extracellular signals and transduce them to appropriate downstream effectors (including ERK cascade) through heterotrimeric G proteins. In a two-step desensitization mechanism, activated GPCRs are phosphorylated by GPCR kinases (GRKs) (Gurevich et al. 2011), and the phosphorylation of receptor leads to the recruitment of arrestin, a small family multi-functional adaptor proteins (Drake et al. 2006; Gurevich and Gurevich 2006a). The binding of arrestin turns off the G protein-mediated signaling (Krupnick et al. 1997; Wilden et al. 1986). Interestingly, now it is clear that there is a second wave of G protein-independent signal flow during the desensitization and internalization of GPCR. Both non-visual arrestins (arrestin-2 and -3[1]) initiate this novel GPCR signaling by recruiting kinases, E3 ubiquitin ligases, phosphatases, and phosphodiesterases to into multi-protein signalosomes (Fig. 14.1) (Miller and Lefkowitz 2001). The signaling network mediated by arrestin appears to be as diverse as that regulated by G proteins.

The scaffolding function of arrestins in MAPK cascades has attracted extensive attention in the last couple of decades due to their pivotal physiological functions (Hanson et al. 2006a; Lefkowitz and Shenoy 2005; Miller and Lefkowitz 2001). Non-visual arrestins are among very few scaffold proteins which can scaffold all three canonical MAPK cascades: ERK (Raf-MEK1-ERK1/2), JNK

[1]Here we use the systematic names of arrestin proteins: arrestin-1 (historic names S-antigen, 48 kDa protein, visual or rod arrestin), arrestin-2 (β-arrestin or β-arrestin1), arrestin-3 (β-arrestin2 or hTHY-ARRX), and arrestin-4 (cone or X-arrestin; for unclear reasons its gene is called “arrestin 3” in the HUGO database).

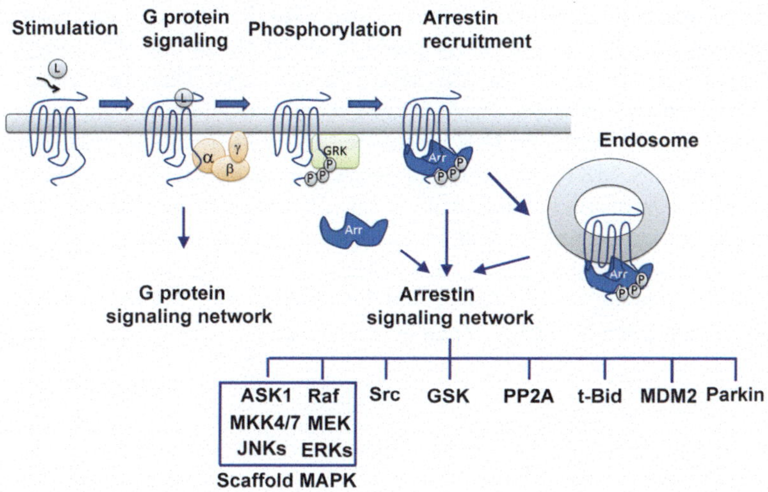

Fig. 14.1 Arrestin-mediated G protein-independent second wave of GPCR signaling

(ASK1-MKK4/7-JNKs) and p38 (ASK1-MKK3-p38) (Whitmarsh and Davis 1998). Among all these MAPK cascades, the scaffolding function of arrestin in the JNK cascade has been best studied. Arrestin-dependent activation of JNK3 was first discovered in 2000 based on several cell-based functional assays and protein-protein interaction analysis (McDonald et al. 2000). Although cell-based assays have been valuable tools to understand the mechanisms of arrestin-mediated signaling, they have their limits. For instance, it is almost impossible to avoid the interference from unknown factors in the cellular milieu, and it is hard to control the expression level of each component, especially when multiple cDNAs are co-transfected. Strictly controlled reconstructed systems from purified proteins overcome many inherent problems of cell-based assays and often provide more direct and quantitative answers (Good et al. 2009; Macdonald et al. 1993). In recent years, arrestin-JNK cascades reconstructed with purified proteins have proved a very powerful approach to understand the molecular mechanism of these signal regulation processes (Zhan et al. 2011b, 2013, 2014a, b).

Assembly of Arrestin-JNK Cascades

Arrestin Directly Associates with All Kinase Components of JNK Cascades

Arrestin-mediated scaffolding of JNK3 activation requires simultaneous recruitment of all kinase components. The interactions between one of the non-visual arrestins, arrestin-3 (also called β-arrestin2), with ASK1 (MAP3K) and JNK3

(MAPK) were first detected by yeast two-hybrid screen, then verified by co-immunoprecipitation (co-IP) (McDonald et al. 2000). In that study the interaction between arrestin-3 and MKK4 (MAP2K) were only observed when ASK1 and/or JNK3 are co-expressed. These binding data led to an indirect model in which both ASK1, the MAP3K in JNK3 cascade, and JNK3 directly associate with arrestin-3, whereas MKK4 is recruited by ASK1 and/or JNK3 without directly interacting with arrestin-3 (McDonald et al. 2000). However, direct interaction between arrestin-3 and MKK4 was first suggested in follow-up studies by several cell-based assays, including co-IP (Song et al. 2006, 2007, 2009). More recently, direct pull-down assay using purified proteins confirmed that both arrestin-2 and-3 can directly interact with JNK3 (Zhan et al. 2011b).

Determining the MAP Kinases Docking Sites on Arrestin

Arrestins are elongated two-domain (N- and C-domains) molecules with an overall fold that is highly conserved among different isoforms (Fig. 14.2) (Granzin et al. 1998; Han et al. 2001; Hirsch et al. 1999; Milano et al. 2002; Sutton et al. 2005; Zhan et al. 2011a). The interface between arrestin and GPCR has been elucidated fairly well. Receptor binding elements are located on the concave sides of both domains including the elements determining receptor specificity and "phosphate-binding sensors" (Hanson et al. 2006b; Hanson and Gurevich 2006; Vishnivetskiy et al. 2000, 2004, 2011, 2013). It has been known that arrestin undergoes a significant conformational change upon binding receptors (Kim et al. 2012; Schleicher et al. 1989; Zhuo et al. 2014), including the release of C-tail from the N-domain (Hanson et al. 2006a; Palczewski et al. 1991; Zhuo et al. 2014). We now know more about the receptor-binding induced conformational movements of arrestins thanks to several structural studies that revealed the twist of the two

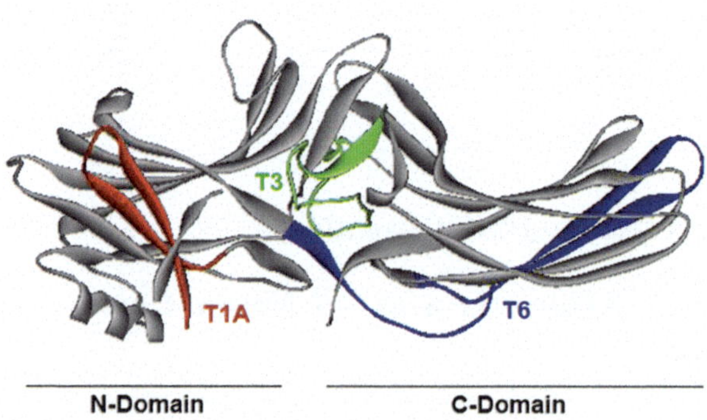

Fig. 14.2 JNK3 interaction sites on arrestin-3. Arrestin-3 structure is from (PDB: 3P2D) (Zhan et al. 2011a). Three regions (T1A from the N-domain; T3 and T6 from the C-domain) have been identified as JNK3 primary binding sites (Zhan et al. 2014b)

domains relative to each other and the movement of several loops on the receptor-binding surface (Kang et al. 2015; Kim et al. 2012, 2013; Shukla et al. 2013). One of the most striking features of the receptor-bound arrestin is an $\sim 20°$ twist between N- and C-domains. These binding-induced movements have very profound effects on the recruitment of non-receptor partners.

In addition to hundreds of different GPCRs, arrestins also interact with at least dozens of non-receptor partners (Gurevich and Gurevich 2006b; Hanson et al. 2006a; Xiao et al. 2007). The most important question from a structural viewpoint is which of these signaling components preferentially bind free arrestin, which bind more tightly to the receptor-bound arrestins and which bind to both conformations with comparable affinity. These questions remain unclear for JNK cascade kinases. Arrestin-dependent scaffolding of JNK3 activation cascade was originally reported as a receptor-dependent process (McDonald et al. 2000); however, robust interaction between arrestin and JNK3 was demonstrated in the follow-up studies by different assays including nuclear exclusion (Song et al. 2006, 2007), co-IP (Seo et al. 2011; Song et al. 2009), direct pull-down (Zhan et al. 2013, 2014b, 2016), and BRET (Breitman et al. 2012). Moreover, free arrestin-3 can interact with upstream ASK1, MKK4 and MKK7 as well (Song et al. 2009; Zhan et al. 2013). The most conclusive evidence that receptor-bound arrestin-3 can bind JNK3 was provided in a reconstituted rhodopsin-arrestin3-JNK3 system (Kook et al. 2013; Zhan et al. 2014b). Strong interaction between arrestin-3 and phosphorylated rhodopsin was employed to bind arrestin-3 to rhodopsin in native membranes, then the binding of JNK3 to the rhodopsin-bound arrestin-3 was detected by a spin-down assay, where un-phosphorylated rhodopsin was used as a negative control. The interaction between JNK3 and rhodopsin-bound arrestin-3 indicates that JNK3 binds arrestin-3 on the non-receptor binding side (Fig. 14.2).

With the exception of clathrin/AP2 (Kim and Benovic 2002), the precise localization of binding sites of non-receptor partners on arrestins remains largely unknown (Hanson et al. 2006a). Separated arrestin-3 N- and C-domains were shown to bind each kinase in the ASK1-MKK4-JNK3 and Raf-MEK1-ERK2 cascades (Song et al. 2009). This appears to be a universal mode of assembly of the kinases in MAPK cascades on arrestin scaffolds. To map the precise JNK3 binding sites on arrestin-3, purified JNK3α2 and maltose binding protein (MBP) fusions containing arrestin-3 peptides exposed on the non-receptor binding surface were employed in the pull-down assays (Zhan et al. 2014b). The results identified one element on the N-domain and two on the C-domain that directly interact with JNK3α2. The first 25 residues from N-domain (named T1A peptide) have demonstrated the strongest binding affinity for JNK3α2, suggesting that this element is the primary binding site. Moreover, this T1A peptide also binds MKK4, MKK7 and ASK1, the upstream kinases of JNK3 (Zhan et al. 2016). This was a surprising findings considering the sizes of all three kinases. This small peptide is sufficient to facilitate JNK activation as a mini-scaffold in cell (Zhan et al. 2016) (see Chap. 18).

Arrestin Conformation and JNK Activation

Many of the binding partners have to compete with each other for arrestin. How arrestin makes the "decision" to recruit the appropriate components remains a very interesting and challenging question to answer. At least three well-defined arrestin conformations are known: free arrestins, receptor-bound (active) arrestins, and microtubule-bound arrestins (Hanson et al. 2006a). Binding of receptor, microtubules or other partners causes significant and often global conformational rearrangements, which change the exposure of arrestin elements. These rearrangements of surface of arrestin have profound effects on the recruitments of arrestin partners.

JNK3, as well as its upstream activating kinases MKK4, MKK7 and ASK1 can bind both free and receptor-bound arrestin-3. Furthermore, arrestin-dependent JNK activation was observed in both receptor-dependent and independent assays (Zhan et al. 2014a). Therefore, receptor stimulation is not obligatory for this arrestin function. Although JNK3 and other kinases can bind both free and receptor-bound arrestin, it is not clear whether JNK3 binds both conformations with comparable affinities or prefers one to the other. To answer this question, careful biochemical experiments with purified proteins need to be performed to evaluate the binding affinities of JNK3, MKK4, MKK7 and ASK1 for arrestin-3 in its free conformation and bound to receptor.

One often overlooked question is the binding-induced conformational changes of arrestins caused by their non-receptor partners. This is an extremely important question in the scaffolding function of arrestin in MAPK cascades. In order to facilitate the signal flow in each cascade, arrestins are required to assemble appropriate combinations of MAPK, MAP2K and MAP3K. It is not clear how simultaneous recruitment of mismatched components is prevented. One possibility is that binding of one kinase to arrestin can significantly alter the recruitment of another (Fig. 14.3). Just like receptor, binding of MAP kinases and other non-receptor partners can cause conformational changes in arrestins, which may favor the recruitment of partner from same pathway to outcompete others.

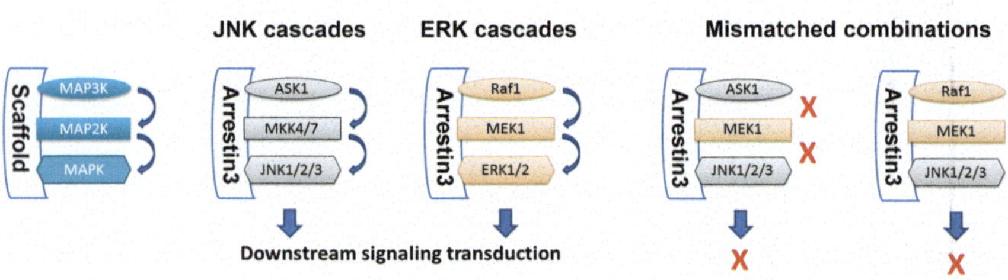

Fig. 14.3 Arrestin assembles multiple signal components simultaneously to facilitate signal transduction. As multi-functional adaptors, arrestins can bind dozens of non-receptor partners. The function of arrestin is regulated in space and time by virtue of its dynamic association with specific signal effectors. Mismatched combinations would terminate signal transduction. The mechanisms to ensure the assembly of appropriate combinations of signaling components by arrestins remain unclear

This type of allosteric regulation was observed in the Arrestin-MKK4/7-JNK3 reconstructed systems. JNKs are activated by concomitant phosphorylation of a threonine and a tyrosine residue within a highly conserved Thr-Pro-Tyr (TPY) motif in the activation loop (Minden et al. 1994). Only two MAP2K can phosphorylate JNKs: MKK4 and MKK7. Interestingly, unlike other MAPK subfamilies, the full activation of JNKs requires both MKKs: MKK4 phosphorylates tyrosine, whereas MKK7 phosphorylates threonine (Haeusgen et al. 2011; Zou et al. 2007). In the cell-based assay, the phosphorylation of both threonine and tyrosine residues are enhanced when arrestin-3 was overexpressed (Zhan et al. 2013). These observations were further confirmed with purified proteins. In the reconstituted protein systems, the concentrations of each protein component were strictly controlled. The arrestin-mediated JNK3 phosphorylation demonstrated a biphasic dependence on the amount of arrestin-3 in both arrestin3-MKK4-JNK3 and arrestin3-MKK7-JNK3 systems (Zhan et al. 2014a), which suggested that arrestin-3 functions as a true scaffold in both cascades to regulate the activation of JNK3 by each upstream MKK. Furthermore, there was a striking difference in the optimal arrestin concentrations, at which the highest JNK3 phosphorylation was detected. The optimal arrestin concentration in MKK4-JNK3 is ~8–10 times lower than that for MKK7-JNK3 module. This could be explained by the different binding affinities between arrestin-3 with each MKK. However, the pull-down assay showed that MKK4 and MKK7 bind to arrestin-3 with comparable affinities. This question was answered by the differential effects of JNK3 binding on the recruitment of MKKs to arrestin-3: the binding of JNK3 enhanced the interaction between MKK4 with arrestin-3, whereas it reduced the binding between MKK7 with arrestin-3 (Zhan et al. 2013). This is consistent with the mathematical models that predict that positive binding cooperativity between the two components lowers the optimal scaffold concentration, whereas negative cooperativity increases the optimal scaffold concentration (Zhan et al. 2011b).

Previous studies show that the biological functions of MKK4 and MKK7 are not completely redundant (Wang et al. 2007). We believe that arrestin-3 can facilitate the JNK3 phosphorylation by both MKK4 and MKK7, but favors the tyrosine phosphorylation catalyzed by MKK4. The level of MKK7 is significantly higher than that of MKK4 in COS-7 cell (Zhan et al. 2011b). This could also be a mechanism to balance the phosphorylation of both residues in the catalytic loop. These studies also suggest the possibility that the prevention of assembling mismatched signal components could be dictated by these signaling components themselves via allosteric regulation. So far, these studies were limited to MAP2K-MAPK modules. The addition of MAP3K and other regulators, such as GPCRs, E3 ubiquitin ligases, and phosphatases, will likely provide novel valuable mechanistic insights.

Arrestin and JNK Isoforms

It is clear now that all four arrestin isoforms can bind JNK3, while only arrestin-3 can facilitate JNK activation in cells (Song et al. 2009). The binding of all these kinases simultaneously does not ensure the facilitation of JNK activation. Two non-visual arrestins have 78% sequence identity and 88% similarity (Attramadal et al. 1992; Sterne-Marr et al. 1993) and their three-dimensional structures are also remarkably similar (Hanson et al. 2006a). It remains a challenging question to answer: why arrestin-3 is the only isoform that is capable of facilitating JNK activation? Again, the strictly controlled protein systems provided some evidence to explain this phenomenon. The binding affinities of both JNK3 and MKK4 were carefully compared in pull-down and FRET (for JNK3) assays (Zhan et al. 2011b). The results showed that JNK3 binds both non-visual arrestins with comparable affinities, while MKK4 significantly prefers arrestin-3 to arrestin-2. MKK4 is known as the weakest arrestin binding partner in the arrestin mediated ASK1-MKK4-JNK3, which might function as a "trigger" to turn on/off the arrestin-JNK modules. The stronger ability of arrestin-3 to recruit MKK4 might be the reason why arrestin-3 is the only isoform that can facilitate JNK activation, although other isoforms can interact with the corresponding kinases, as well (Seo et al. 2011; Song et al. 2006). There is additional evidence showing that both bringing all components into close proximity and holding them in optimal orientation is essential for the scaffolding function of arrestin-3 in JNK cascades.

For more than a decade, it was widely believed that the arrestin-mediated JNK activation is only specific for JNK3. The interaction of arrestin-3 and JNK3 was originally detected in yeast two-hybrid screen and confirmed using co-IP (McDonald et al. 2000). There are ten different JNK isoforms in mammalian cells including three subfamilies: JNK1, JNK2 and JNK3. Two JNK3 isoforms, the original reported to be activated in arrestin-dependent manner, are expressed predominantly in neurons, heart, and testes. In contrast, all JNK1 and JNK2 isoforms are expressed ubiquitously. Furthermore, the unique extended N-terminus of 38 residues in JNK3 was reported as the primary arrestin-3 binding site (Song et al. 2009).

We directly evaluated the interactions between arrestin-3 and JNK1/2 using purified proteins. Robust interactions between arrestin-3 and JNK1/2 were detected in His-tag pull-down assay (Kook et al. 2013). Moreover, the JNK2α2 demonstrated very similar affinity to JNK3α2, whereas JNK1α1 shows a relatively weaker binding to arrestin-3. The interactions between arrestins and these JNK1/2 isoforms were further confirmed by co-IP (Kook et al. 2013). The MKK4/7-JNK1/2 modules were reconstructed to evaluate the scaffolding function of arrestin-3. JNK1/2 activation demonstrated the biphasic dependence on arrestin-3 concentration, which verified that arrestin-3 functions as scaffold for these JNK isoforms (Kook et al. 2013). This biphasic effect of scaffolding proteins was also observed in the intact cells (Kook et al. 2013). These findings demonstrate that arrestin-3 mediates the activation of several isoforms of all three subfamilies of JNKs. The arrestin-dependent scaffolding of JNK signaling modules is not limited to neurons or certain specific tissues (Kook et al. 2013). Therefore, the biological functions of these signaling pathways are more widespread than originally believed. Furthermore, since neither JNK1 nor JNK2 contains the JNK3-specific 38 amino acids, previously identified arrestin-3 docking site, there must be at least one other arrestin-3 interaction site located within JNK kinase domain.

Conclusions

Although the arrestin-dependent JNK activation is the most extensively studied arrestin signaling pathway, many mechanistic questions remain, especially from a structural point of view. We believe the strictly controlled systems are necessary to probe these molecular mechanisms. We strived to reconstruct these cascades with functional purified proteins. These reconstitution systems have offered the great opportunities to elucidate the fine mechanisms of the assembly, allosteric regulation and dynamics of these signaling cascades. From originally envisioned simple ASK1-MKK4-JNK3 cascade, the arrestin mediated JNK cascade has expanded to more kinases (i.e. MKK7) and isoforms (i.e. JNK1/2). Current reconstituted protein systems are mainly limited to MAP2K-MAPK modules. Therefore, reconstructing the whole cascades including MAP3K (i.e., ASK1), GPCRs, and other types of signaling components will be the focus of the future studies.

References

Attramadal H, Arriza JL, Aoki C, Dawson TM, Codina J, Kwatra MM, Snyder SH, Caron MG, Lefkowitz RJ (1992) Beta-arrestin2, a novel member of the arrestin/beta-arrestin gene family. J Biol Chem 267:17882–17890

Bardwell L (2006) Mechanisms of MAPK signalling specificity. Biochem Soc Trans 34:837–841

Breitkreutz A, Tyers M (2002) MAPK signaling specificity: it takes two to tango. Trends Cell Biol 12:254–257

Breitman M, Kook S, Gimenez LE, Lizama BN, Palazzo MC, Gurevich EV, Gurevich VV (2012) Silent scaffolds: inhibition of c-Jun N-terminal kinase 3 activity in the cell by a dominant-negative arrestin-3 mutant. J Biol Chem 287:19653–19664

Chang L, Karin M (2001) Mammalian MAP kinase signalling cascades. Nature 410:37–40

Drake MT, Shenoy SK, Lefkowitz RJ (2006) Trafficking of G protein-coupled receptors. Circ Res 99:570–582

Good M, Tang G, Singleton J, Reményi A, Lim WA (2009) The Ste5 scaffold directs mating signaling by catalytically unlocking the Fus3 MAP kinase for activation. Cell 136:1085–1097

Good MC, Zalatan JG, Lim WA (2011) Scaffold proteins: hubs for controlling the flow of cellular information. Science 332:680–686

Granzin J, Wilden U, Choe HW, Labahn J, Krafft B, Buldt G (1998) X-ray crystal structure of arrestin from bovine rod outer segments. Nature 391:918–921

Gurevich EV, Gurevich VV (2006a) Arrestins are ubiquitous regulators of cellular signaling pathways. Genome Biol 7:236

Gurevich VV, Gurevich EV (2006b) The structural basis of arrestin-mediated regulation of G-protein-coupled receptors. Pharmacol Ther 110:465–502

Gurevich EV, Tesmer JJ, Mushegian A, Gurevich VV (2011) G protein-coupled receptor kinases: more than just kinases and not only for GPCRs. Pharmacol Ther 133:40–69

Haeusgen W, Herdegen T, Waetzig V (2011) The bottleneck of JNK signaling: Molecular and functional characteristics of MKK4 and MKK7. Eur J Cell Biol 90:536–544

Han M, Gurevich VV, Vishnivetskiy SA, Sigler PB, Schubert C (2001) Crystal structure of beta-arrestin at 1.9 A: possible mechanism of receptor binding and membrane translocation. Structure 9:869–880

Hanson SM, Gurevich VV (2006) The differential engagement of arrestin surface charges by the various functional forms of the receptor. J Biol Chem 281:3458–3462

Hanson SM, Francis DJ, Vishnivetskiy SA, Klug CS, Gurevich VV (2006a) Visual arrestin binding to microtubules involves a distinct conformational change. J Biol Chem 281: 9765–9772

Hanson SM, Francis DJ, Vishnivetskiy SA, Kolobova EA, Hubbell WL, Klug CS, Gurevich VV (2006b) Differential interaction of spin-labeled arrestin with inactive and active phosphorhodopsin. Proc Natl Acad Sci USA 103:4900–4905

Hirsch JA, Schubert C, Gurevich VV, Sigler PB (1999) The 2.8 A crystal structure of visual arrestin: a model for arrestin's regulation. Cell 97:257–269

Kang Y, Zhou XE, Gao X, He Y, Liu W, Ishchenko A, Barty A, White TA, Yefanov O, Han GW et al (2015) Crystal structure of rhodopsin bound to arrestin by femtosecond X-ray laser. Nature 523:561–567

Kim YM, Benovic JL (2002) Differential roles of arrestin-2 interaction with clathrin and adaptor protein 2 in G protein-coupled receptor trafficking. J Biol Chem 277:30760–30768

Kim M, Vishnivetskiy SA, Van Eps N, Alexander NS, Cleghorn WM, Zhan X, Hanson SM, Morizumi T, Ernst OP, Meiler J et al (2012) Conformation of receptor-bound visual arrestin. Proc Natl Acad Sci 109:18407–18412

Kim YJ, Hofmann KP, Ernst OP, Scheerer P, Choe HW, Sommer ME (2013) Crystal structure of pre-activated arrestin p44. Nature 497:142–146

Kolch W (2005) Coordinating ERK/MAPK signalling through scaffolds and inhibitors. Nat Rev Mol Cell Biol 6:827–837

Kook S, Zhan X, Kaoud TS, Dalby KN, Gurevich VV, Gurevich EV (2013) Arrestin-3 binds c-Jun N-terminal kinase 1 (JNK1) and JNK2 and facilitates the activation of these ubiquitous JNK isoforms in cells via scaffolding. J Biol Chem 288:37332–37342

Krupnick JG, Gurevich VV, Benovic JL (1997) Mechanism of quenching of phototransduction. Binding competition between arrestin and transducin for phosphorhodopsin. J Biol Chem 272:18125–18131

Lefkowitz RJ, Shenoy SK (2005) Transduction of receptor signals by β-arrestins. Science 308:512–517

Macdonald SG, Crews CM, Wu L, Driller J, Clark R, Erikson RL, McCormick F (1993) Reconstitution of the Raf-1-MEK-ERK signal transduction pathway in vitro. Mol Cell Biol 13:6615–6620

Marinissen MJ, Gutkind JS (2001) G-protein-coupled receptors and signaling networks: emerging paradigms. Trends Pharmacol Sci 22:368–376

McDonald PH, Chow C-W, Miller WE, Laporte SA, Field ME, Lin F-T, Davis RJ, Lefkowitz RJ (2000) β-Arrestin 2: a receptor-regulated MAPK scaffold for the activation of JNK3. Science 290:1574–1577

Milano SK, Pace HC, Kim YM, Brenner C, Benovic JL (2002) Scaffolding functions of arrestin-2 revealed by crystal structure and mutagenesis. Biochemistry 41:3321–3328

Miller WE, Lefkowitz RJ (2001) Expanding roles for β-arrestins as scaffolds and adapters in GPCR signaling and trafficking. Curr Opin Cell Biol 13:139–145

Minden A, Lin A, McMahon M, Lange-Carter C, Derijard B, Davis R, Johnson G, Karin M (1994) Differential activation of ERK and JNK mitogen-activated protein kinases by Raf-1 and MEKK. Science 266:1719–1723

Palczewski K, Buczyłko J, Imami NR, McDowell JH, Hargrave PA (1991) Role of the carboxyl-terminal region of arrestin in binding to phosphorylated rhodopsin. J Biol Chem 266:15334–15339

Schleicher A, Kuhn H, Hofmann KP (1989) Kinetics, binding constant, and activation energy of the 48-kDa protein-rhodopsin complex by extra-metarhodopsin II. Biochemistry 28: 1770–1775

Seger R, Krebs EG (1995) The MAPK signaling cascade. FASEB J 9:726–735

Seo J, Tsakem EL, Breitman M, Gurevich VV (2011) Identification of arrestin-3-specific residues necessary for JNK3 activation. J Biol Chem 286:27894–27901

Shukla AK, Manglik A, Kruse AC, Xiao K, Reis RI, Tseng W-C, Staus DP, Hilger D, Uysal S, Huang L-Y et al (2013) Structure of active [bgr]-arrestin-1 bound to a G-protein-coupled receptor phosphopeptide. Nature 497:137–141

Song X, Raman D, Gurevich EV, Vishnivetskiy SA, Gurevich VV (2006) Visual and Z "inactive" conformation bind JNK3 and Mdm2 and relocalize them from the nucleus to the cytoplasm. J Biol Chem 281:21491–21499

Song X, Gurevich EV, Gurevich VV (2007) Cone arrestin binding to JNK3 and Mdm2: conformational preference and localization of interaction sites. J Neurochem 103:1053–1062

Song X, Coffa S, Fu H, Gurevich VV (2009) How does arrestin assemble MAPKs into a signaling complex? J Biol Chem 284:685–695

Sterne-Marr R, Gurevich VV, Goldsmith P, Bodine RC, Sanders C, Donoso LA, Benovic JL (1993) Polypeptide variants of beta-arrestin and arrestin3. J Biol Chem 268:15640–15648

Sutton RB, Vishnivetskiy SA, Robert J, Hanson SM, Raman D, Knox BE, Kono M, Navarro J, Gurevich VV (2005) Crystal structure of cone arrestin at 2.3 Å: evolution of receptor specificity. J Mol Biol 354:1069–1080

Takekawa M, Tatebayashi K, Saito H (2005) Conserved docking site is essential for activation of mammalian MAP kinase kinases by specific MAP kinase kinase kinases. Mol Cell 18:295–306

Vishnivetskiy SA, Schubert C, Climaco GC, Gurevich YV, Velez M-G, Gurevich VV (2000) An additional phosphate-binding element in arrestin molecule: implications for the mechanism of arrestin activation. J Biol Chem 275:41049–41057

Vishnivetskiy SA, Hosey MM, Benovic JL, Gurevich VV (2004) Mapping the arrestin-receptor interface: structural elements responsible for receptor specificity of arrestin proteins. J Biol Chem 279:1262–1268

Vishnivetskiy SA, Gimenez LE, Francis DJ, Hanson SM, Hubbell WL, Klug CS, Gurevich VV (2011) Few residues within an extensive binding interface drive receptor interaction and determine the specificity of arrestin proteins. J Biol Chem 286:24288–24299

Vishnivetskiy SA, Baameur F, Findley KR, Gurevich VV (2013) Critical role of the central 139-loop in stability and binding selectivity of arrestin-1. J Biol Chem 288:11741–11750

Wang X, Destrument A, Tournier C (2007) Physiological roles of MKK4 and MKK7: insights from animal models. Biochim Biophys Acta (BBA) Mol Cell Res 1773:1349–1357

Whitmarsh AJ (2006) The JIP family of MAPK scaffold proteins. Biochem Soc Trans 34:828–832

Whitmarsh AJ, Davis RJ (1998) Structural organization of MAP-kinase signaling modules by scaffold proteins in yeast and mammals. Trends Biochem Sci 23:481–485

Wilden U, Hall SW, Kühn H (1986) Phosphodiesterase activation by photoexcited rhodopsin is quenched when rhodopsin is phosphorylated and binds the intrinsic 48-kDa protein of rod outer segments. Proc Natl Acad Sci USA 83:1174–1178

Xiao K, McClatchy DB, Shukla AK, Zhao Y, Chen M, Shenoy SK, Yates JR, Lefkowitz RJ (2007) Functional specialization of beta-arrestin interactions revealed by proteomic analysis. Proc Natl Acad Sci USA 104:12011–12016

Zhan X, Gimenez LE, Gurevich VV, Spiller BW (2011a) Crystal structure of arrestin-3 reveals the basis of the difference in receptor binding between two non-visual arrestins. J Mol Biol 406:467–478

Zhan X, Kaoud TS, Dalby KN, Gurevich VV (2011b) Nonvisual arrestins function as simple scaffolds assembling the MKK4–JNK3a2 signaling complex. Biochemistry 50:10520–10529

Zhan X, Kaoud TS, Kook S, Dalby KN, Gurevich VV (2013) JNK3 enzyme binding to arrestin-3 differentially affects the recruitment of upstream mitogen-activated protein (MAP) kinase kinases. J Biol Chem 288:28535–28547

Zhan X, Kook S, Gurevich EV, Gurevich VV (2014a) Arrestin-dependent activation of JNK family kinases. In: Gurevich VV (ed) Arrestins—pharmacology and therapeutic potential. Springer, Berlin, pp 259–280

Zhan X, Perez A, Gimenez LE, Vishnivetskiy SA, Gurevich VV (2014b) Arrestin-3 binds the MAP kinase JNK3a2 via multiple sites on both domains. Cell Signal 26:766–776

Zhan X, Stoy H, Kaoud TS, Perry NA, Chen Q, Perez A, Els-Heindl S, Slagis JV, Iverson TM, Beck-Sickinger AG et al (2016) Peptide mini-scaffold facilitates JNK3 activation in cells. Sci Rep 6:21025

Zhuo Y, Vishnivetskiy SA, Zhan X, Gurevich VV, Klug CS (2014) Identification of receptor binding-induced conformational changes in non-visual arrestins. J Biol Chem 289:20991–21002

Zou H, Li Q, Lin S-C, Wu Z, Han J, Ye Z (2007) Differential requirement of MKK4 and MKK7 in JNK activation by distinct scaffold proteins. FEBS Lett 581:196–202

Chapter 15
Arrestin-Dependent ERK Activation and Its Disruption

Louis M. Luttrell and Benjamin W. Spiller

Abstract Regulation of the ERK1/2 cascade is one of the most studied functions of arrestins and illustrates many of the features that enable them to function as GPCR-regulated scaffolds. While all four arrestins can bind the component kinases of the ERK cascade; c-Raf1, MEK1/2 and ERK1/2, their binding is dependent on arrestin conformation, such that inactive ERK1/2 can be sequestered by a microtubule-bound pool of arrestin, while activated ERK1/2 binds with high affinity only to the GPCR-bound arrestin conformation. The result is both a dampening of basal pathway activity, and the arrestin-dependent activation of a spatially and temporally constrained pool of ERK1/2 that differs in function from ERK1/2 activated by G protein-dependent mechanisms or classical receptor tyrosine kinase growth factor receptors. Arrestin-bound ERK1/2 performs numerous functions in cells, among them contributing to the regulation of GPCR internalization and trafficking, control of cell proliferation and non-proliferative cell growth, and regulation of cytoskeletal dynamics involved in cell migration and chemotaxis. The finding that arrestin binding of c-Raf1 and MEK1/2 can be disrupted by point mutations that eliminate its ability to activate ERK1/2 without disrupting its other functions indicates that the two major functions of arrestins, GPCR desensitization and signaling, are dissociable, and offers tools to probe arrestin's diverse cellular functions.

L.M. Luttrell
Department of Medicine, Medical University of South Carolina,
Charleston, SC 29425, USA

L.M. Luttrell
Ralph H. Johnson Veterans Affairs Medical Center, Charleston, SC 29401, USA

B.W. Spiller
Department of Microbiology and Immunology,
Vanderbilt University, Nashville, TN 37232, USA
e-mail: benjamin.spiller@vanderbilt.edu

L.M. Luttrell (✉)
Division of Endocrinology, Diabetes and Medical Genetics,
Medical University of South Carolina, 96 Jonathan Lucas Street,
MSC 624, Charleston, SC 29426, USA
e-mail: luttrell@musc.edu

V.V. Gurevich (ed.), *The Structural Basis of Arrestin Functions*,
DOI 10.1007/978-3-319-57553-7_15

Keywords Arrestin · Extracellular signal-regulated kinase · G protein-coupled receptor · Mitogen-activated protein kinase · Signal transduction

Introduction

The sheer complexity of cell signaling has necessitated the evolution of scaffold proteins whose roles are to increase the efficiency of information transfer between successive enzymes in a signaling cascade; to enhance fidelity by dampening cross talk between parallel cascades; and to target effectors to specific subcellular locations. The arrestins, which were originally discovered for their role in G protein-coupled receptor (GPCR) desensitization and internalization, are now recognized to also serve as signaling scaffolds, linking GPCRs to numerous intracellular signaling pathways independent of heterotrimeric G protein signaling (Gurevich and Gurevich 2006; Luttrell and Gesty-Palmer 2010). Indeed, arrestins perform roles in GPCR signaling that are analogous to the functions of other scaffolds, such as insulin receptor substrate-1, which organizes proteins into signaling complexes around activated insulin receptors (White 1998).

The visual/β-arrestins[1] are semi-bisymmetric soluble proteins composed of β-strand sandwich structures organized into N- and C-domains, that form two concave baskets connected by a short hinge (Vishnivetskiy et al. 2002; Aubry et al. 2009). Despite their relatively small 48–50 kDa size, arrestins are uniquely suited to function as GPCR-regulated scaffolds. First, they have the ability to recognize activated GPCRs and translocate from the cytosol to the plasma membrane upon receptor activation. Second, they have the flexibility to bind multiple cargo proteins on the convex side of the N- and C-domains away from the sites of GPCR interaction, a property that allows them to recruit signaling proteins into GPCR-based 'signalsome' complexes. Third, they can exist in different intracellular pools wherein they adopt different conformations. This is critical to their scaffolding function in that many arrestin-bound signaling proteins exhibit a preference for the cytosolic, microtubule-bound, or GPCR-bound arrestin conformation. Thus, arrestin binding can constrain signaling proteins to one cellular compartment until an external GPCR-mediated stimulus prompts a conformational change that causes them to release some cargos and associate with others.

Among the first reported and most studied arrestin-mediated signals is activation of the extracellular signal-regulated kinase 1 and 2 (ERK1/2) mitogen-activated protein (MAP) kinase cascade (DeFea et al. 2000; Luttrell et al. 2001). The ubiquitous MAP kinases play critical roles in cell cycle regulation/proliferation and survival/apoptotic signaling by controlling phosphorylation of nuclear transcription

[1]Various names of arrestin proteins are used: Arrestin-1 is also called S-antigen, 48 kDa protein, visual or rod arrestin; β-arrestin or β-arrestin1 is also called arrestin-2; β-arrestin2 is also called arrestin-3 and hTHY-ARRX, and arrestin-4 is also called cone or X-arrestin (its gene is called "arrestin 3" in the HUGO database).

factors, e.g. Elk1 and c-Jun, as well as diverse regulatory functions mediated through phosphorylation of cytosolic substrates (Davis 2000; Pearson et al. 2001; Kyriakis and Avruch 2012). Eukaryotic cells possess several different MAP kinases that fall into three groups; the extracellular signal-regulated kinases (ERK), c-Jun N-terminal kinases (JNK), and p38 MAP kinases. Each MAP kinase module consists of three component kinases; a MAP kinase kinase kinase (MAPKKK), MAP kinase kinase (MAPKK), and MAP kinase (MAPK) that must phosphorylate one another in succession. Not surprisingly, MAP kinase regulation commonly involves scaffold proteins that aid in assembling the correct MAPKKK-MAPKK-MAPK sets and targeting activated MAP kinases to their relevant substrates (Burack and Shaw 2000). In this chapter, we discuss the roles that arrestins play in balancing G protein- and arrestin-dependent ERK1/2 activation, the mechanism of arrestin-dependent scaffolding of the ERK1/2 cascade, and how the assembly of GPCR-arrestin-ERK1/2 'signalsomes' control ERK1/2 function by activating a spatially and temporally distinct ERK1/2 pool.

G Protein- and Arrestin-Dependent ERK1/2 Regulation

Among its many functions, ERK1/2 is involved in phosphorylating nuclear transcription factors necessary for G0–G1 cell cycle transition and the passage of cells through mitosis or meiosis (Pearson et al. 2001). Not surprisingly then, it is subject to extensive convergent regulation, including Ras-dependent proliferative signals originating from GPCRs and receptor tyrosine kinases, protein kinase (PK)A and PKC mediated signals downstream of heterotrimeric G proteins, and signals transmitted via arrestin scaffolds (Luttrell and Miller 2013). While receptor and cell type variability is more the rule than the exception, it is clear that most heterotrimeric G protein families signal to ERK1/2. Stimulation of the Gq/11-phospholipase Cβ-PKC pathway can activate ERK1/2 through direct phosphorylation of the MAPKKK c-Raf1 by PKCα (Kolch et al. 1993; Hawes et al. 1995). The consequences of stimulating of cAMP production by Gs-adenylyl cyclase are complex and determined primarily by which Raf isoform, c-Raf1 or B-Raf, is predominantly expressed. Unlike PKC, phosphorylation of cRaf-1 by PKA inhibits its activation (Wu et al. 1993). As a result, activation of Gs-coupled receptors in some cells inhibits, rather than activates, ERK1/2 (Lefkowitz et al. 2002). On the other hand, B-Raf is activated both by PKA-dependent phosphorylation of the Ras-family GTPase, Rap-1 (Vossler et al. 1997) and by cAMP binding to the Rap-1 guanine nucleotide exchange factor, Epac (DeRooij et al. 1998), permitting direct cAMP mediated activation of a B-Raf-MEK1/2-ERK1/2 module. Pertussis toxin-sensitive G protein-dependent activation of ERK1/2 by Gi/o-coupled GPCRs is mediated primarily by Gβγ subunits and typically involves activation of receptor or non-receptor tyrosine kinases leading to Ras-dependent activation of c-Raf1 (van Biesen et al. 1995; Luttrell et al. 1996, 1997, 1999).

ERK1/2 activated by any of these mechanisms is free to translocate into the cell nucleus and regulate transcription by phosphorylating nuclear transcription factors.

Since they are first and foremost negative regulators of G protein signaling, arrestins are positioned to regulate the balance between G protein-mediated signals that activate ERK1/2 dependent transcription, and arrestin-dependent signals that direct the kinase to perform other functions. Because nuclear ERK1/2 is rapidly inactivated by dephosphorylation, the time course of ERK1/2 activation by G protein-regulated effectors parallels the level of heterotrimeric G protein activity, and arrestin-dependent GPCR desensitization limits the duration of G protein-mediated ERK1/2 activation. In β-arrestin1/2 null murine embryo fibroblasts, for example, ERK1/2 activation by lysophosphatidic acid (LPA) receptors results primarily from Gi/o-dependent transactivation of epidermal growth factor (EGF) receptors (Gesty-Palmer et al. 2005). Because LPA receptor desensitization is impaired in the absence of arrestins, the EGF receptor-dependent ERK1/2 signal is persistent, lasting for several hours in the continued presence of LPA. Reintroducing β-arrestin2, which restores desensitization, makes the transactivation-dependent signal transient, such that it contributes significantly to ERK1/2 activation only during the first few minutes of stimulation. But as G protein-dependent signaling wanes, the arrestin-dependent scaffolding of the c-Raf1-MEK1/2-ERK1/2 module provides an alternative mechanism for sustaining ERK1/2 activity.

Arrestin Scaffolding of the ERK1/2 Cascade

The arrestins have been reported to engage all three major MAP kinase modules, but with different effects in terms of the direction of regulation. This likely reflects the differing functions of MAP kinase modules, and a consistent role of arrestins to favor activation of growth/survival pathways and suppression of apoptotic signals (Gurevich and Gurevich 2013). All four visual/β-arrestins can bind the component kinases of the JNK cascade, ASK1-MKK4/7-JNK1/2/3 (McDonald et al. 2000; Song et al. 2009; Kook et al. 2013), but only β-arrestin2 is able to orient the kinases so as to support efficient phosphorylation (Seo et al. 2011; Zhan et al. 2011). Moreover, the JNK cascade binds tightly to the 'inactive' cytosolic arrestin conformation such that GPCR activation is neither required nor apparently involved (Song et al. 2006). The net result is that arrestin expression tends to suppress JNK signaling, either by sequestering pathway elements bound to β-arrestin1 in non-productive 'silent scaffolds', or in the case of β-arrestin2 keeping active JNK out of the nucleus and away from its transcription factor targets by virtue of its intrinsic Nuclear Export Signal (NES) (Scott et al. 2002; Breitman et al. 2012). In contrast, ERK1/2 binding to arrestins is dependent upon the conformational changes that occur upon receptor binding, and thus dependent upon GPCR stimulation.

Assembly of the Arrestin-Bound ERK Activation Complex

One obvious question is how relatively small globular proteins like arrestins can simultaneously engage the three kinases of the ERK1/2 module, each of which is roughly equal in size to the arrestin itself (Gurevich and Gurevich 2006). The earliest observation, obtained by co-immunoprecipitation of overexpressed pathway components from transfected cells, was that c-Raf1 and ERK2 bound independently to β-arrestin2, while binding of MEK1 was enhanced by the presence of either c-Raf1 or ERK2 in the complex (Luttrell et al. 2001). This suggested that MEK1 binding was at least in part indirect. Subsequent work, however, indicates that MEK1 can bind β-arrestin1 directly, both as immobilized peptides and in cells (Meng et al. 2009). In cells, all three kinases interact directly with either the isolated N- or C-terminal domains of β-arrestin1 and β-arrestin2 (Song et al. 2009), suggesting a broad interaction surface involving the cytoplasmic surface of both domains. The same applies to ASK1-MKK4-JNK3, the three components of the JNK3 cascade. ERK2 binding has even been reconstituted using pure proteins in vitro, where ERK2 directly binds free β-arrestin1 and β-arrestin 2, as well as receptor-associated visual arrestin, β-arrestin1 and β-arrestin2 (Coffa et al. 2011a).

Unlike JNK3, the binding of ERK2 to β-arrestins is enhanced by conformational changes that occur upon GPCR binding. Early reports suggested that GPCR stimulation promoted assembly of the complex (Luttrell et al. 2001). This is supported by comparisons of the binding of ERK pathway components to wild type 'cytosolic' β-arrestin versus either 'active' or 'inactive' β-arrestin mutants. Mutations that destabilize the polar core generate 'pre-activated' arrestins whose ability to recognize GPCRs is independent of receptor phosphorylation (Gurevich and Benovic 1997; Kovoor et al. 1999). Conversely, arrestin mutants with deletions in the hinge region that restrict interdomain flexibility exhibit reduced receptor affinity and enhanced microtubule binding (Vishnivetskiy et al. 2002; Hanson et al. 2007). In cells, ERK2 binds preferentially to the 'active' and 'microtubule-bound' β-arrestin conformations, while exhibiting virtually no binding to the 'inactive' cytosolic conformation (Coffa et al. 2011a). Like ERK2, c-Raf1 prefers the receptor-bound conformation, although the difference is less dramatic, while MEK1 binds equivalently to both free arrestin (Meng et al. 2009; Coffa et al. 2011b) and the mutationally constrained conformations (Coffa et al. 2011a).

Kinetics of Arrestin-Dependent ERK Activation

One property of visual/β-arrestins that contributes to their effectiveness as scaffolds is that they are relatively abundant in relation to the GPCRs they desensitize and catalytically active signaling proteins they carry as cargos. The ability to shuttle between abundant but low-affinity microtubule binding sites and smaller numbers of high affinity binding sites on activated GPCR permits them to sequester ERK1/2 away from its site of activation and dampen basal pathway activity until called to respond to an extracellular stimulus (Fig. 15.1). Thus, cRaf1 and ERK1/2 binding to the microtubule-associated pool of arrestin provides a mechanism for dampening basal ERK1/2 activity in the absence of receptor stimulation. In cells, overexpressing visual arrestin, β-arrestin1 and β-arrestin2, but not arrestin4, recruits

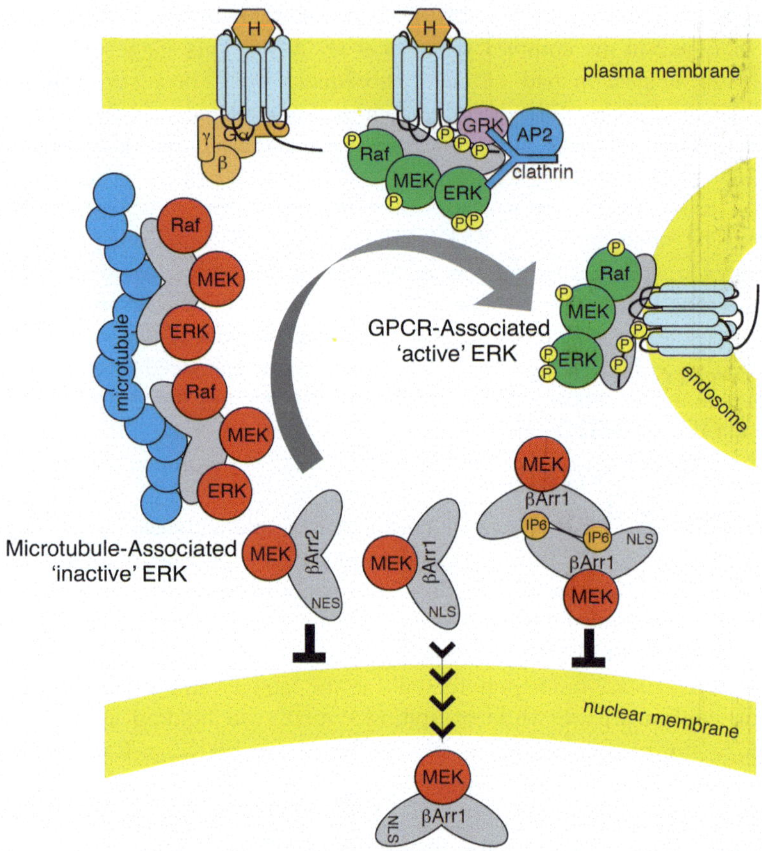

Fig. 15.1 β-Arrestins function as GPCR regulated scaffolds for the cRaf-1–MEK1/2–ERK1/2 MAP kinase module. In quiescent cells, arrestins reside in a either microtubule-bound pool or free in the cytosol, where they are excluded from the cell nucleus either by the intrinsic nuclear export sequence (NES) of β-arrestin2 or through inositol hexakisphosphate (IP6) promoted β-arrestin1 self-association (Hanson et al. 2008). Upon stimulation, the arrestins trade low affinity microtubule and IP6 interactions for high affinity binding sites on GRK phosphorylated receptors. Two of the ERK1/2 cascade components, cRaf-1 and ERK1/2, bind preferentially to the GPCR- and microtubule-bound β-arrestin conformations, while MEK1/2 binds equivalently to all three pools. Thus, inactive ERK1/2 can be sequestered in a microtubule-associated pool until a GPCR stimulus promotes assembly of the ERK1/2 activation complex on the plasma membrane. Since ERK1/2 has very low affinity for the cytosolic β-arrestin conformation, ERK1/2 remains active only while in GPCR-arrestin 'signalsome' complexes

ERK1/2 to microtubules and quenches its activity (Hanson et al. 2007), as does the β-arrestin2 mutant that mimics the microtuble-associated conformation (Coffa et al. 2011a). On the other hand, arrestin-dependent activation of ERK1/2 appears to be dependent upon binding to GPCRs as originally proposed (Coffa et al. 2011a, b). In fact, expressing a neurokinin 1 receptor–β-arrestin1 chimera is sufficient to activate ERK1/2 in the absence of G protein signaling, probably by bringing cRaf-1 into contact with membrane-delimited activators, e.g. Ras, which initiate the ERK1/2 phosphorylation cascade (Jafri et al. 2006).

When a GPCR-activating stimulus is received, the initial burst of heterotrimeric G protein signaling activates ERK1/2 via second messenger-dependent protein kinases, e.g. PKC, a process that is rapidly desensitized by GPCR kinase (GRK)-dependent arrestin recruitment. But the waning of G protein-mediated signaling coincides with the onset of arrestin-dependent ERK1/2 activation. Thus, ERK1/2 activity does not cease as the receptor switches from a G protein-coupled to an arrestin-coupled signaling mode. The contribution of G protein-dependent and arrestin-dependent signals to angiotensin AT_{1A} receptor-stimulated ERK1/2 activation has been dissected using isoform-selective arrestin RNA interference, pharmacologic inhibitors, G protein-uncoupled receptor mutants, and arrestin pathway-selective ligands (Wei et al. 2003; Ahn et al. 2004). When β-arrestin2 expression is silenced in HEK293 cells, AT_{1A} receptor-stimulated ERK1/2 activation becomes transient and sensitive to PKC inhibition, indicating that it occurs via a Gq/11-PLCβ-PKC pathway. In a reciprocal manner, inhibiting PKC in the presence of β-arrestin2 blocks the initial spike in ERK1/2 activity, but does not prevent the persistent late phase response. Exposing the G protein-uncoupled DRY-AAY AT_{1A} receptor mutant to angiotensin II, or the wild type AT_{1A} receptor to the arrestin-selective biased ligand, [Sar^1-Ile^4-Ile^8]-AngII, generates only the sustained signal, which is insensitive to PKC inhibition but abolished by RNA interference targeting β-arrestin2. Studies performed on the β2 adrenergic and type 1 parathyroid hormone (PTH) receptors have produced analogous results (Shenoy et al. 2006; Gesty-Palmer et al. 2006). Thus, arrestin binding ‘switches’ the receptor from G protein-dependent to arrestin-dependent ERK1/2 activation, which proceed with different time courses (Fig. 15.2a).

Subcellular Targeting of ERK

The majority of GPCRs fall into one of two classes based on their selectivity for the two β-arrestin isoforms and the longevity of the receptor-arrestin interaction (Oakley et al. 2000, 2001). One, termed ‘class A’, exhibits higher affinity for β-arrestin2 than β-arrestin1 and forms transient receptor-arrestin complexes that dissociate as the receptor internalizes. These receptors are rapidly resensitized and recycled back to the plasma membrane. The other, ‘class B’ group exhibits equivalent affinities for β-arrestin1 and β-arrestin2 and forms long lasting

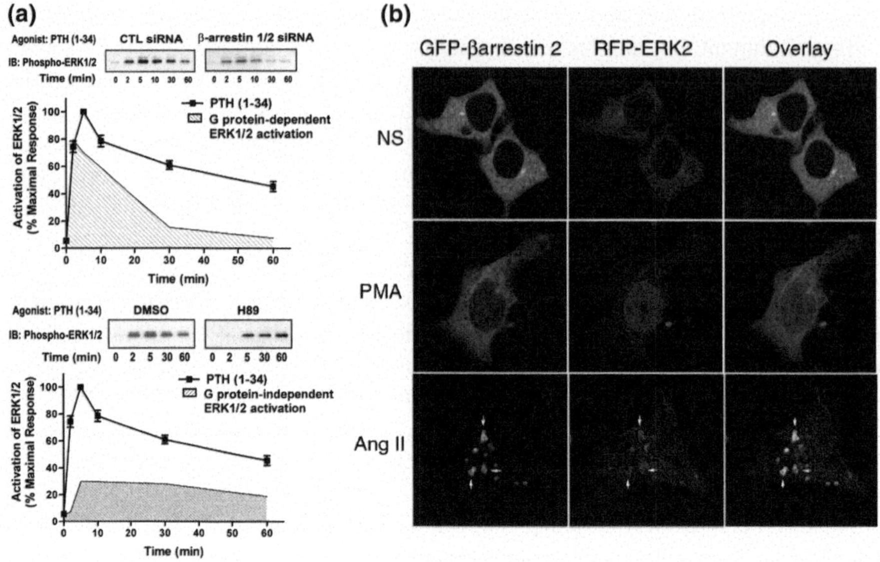

Fig. 15.2 Arrestin scaffolding produces a temporally and spatially discrete pool of active ERK1/2. **a** G protein and arrestin signaling activate ERK1/2 with distinct time courses. In HEK293 cells stimulation of PTH_1 receptors with PTH(1-34) activates ERK1/2 via both PKA-dependent and arrestin-dependent mechanisms. As shown in the *upper panel*, silencing β-arrestin1/2 expression with small interfering RNA (siRNA) inhibits the late phase of ERK1/2 activation, revealing the early component (*shaded area*) to be G protein-mediated. Conversely, as shown in the *lower panel*, pharmacologic inhibition of PKA with H89 blocks the early phase of ERK1/2 activation without affecting the late sustained phase of arrestin-dependent signal (*shaded area*). Data originally published in: Gesty-Palmer et al. (2006). **b** Arrestin scaffolding localizes active ERK1/2 to GPCR-arrestin 'signalsomes'. In non-stimulated (NS) HEK293 cells, both *green* fluorescent protein (GFP)-tagged β-arrestin2 and *red* fluorescent protein (RFP)-tagged ERK2 are diffusely cytosolic. Phorbol ester (PMA) stimulates PKC-dependent activation of ERK and promotes nuclear translocation of RFP-ERK2. In contrast, stimulation of angiotensin AT_{1A} receptors by angiotensin II (Ang II) recruits both GFP-β-arrestin2 and RFP-ERK2 to the receptor where they traffic with it into endosomes (*white arrows*). Data originally published in: Luttrell et al. (2001)

receptor-arrestin complexes that remain intact as the receptor undergoes endosomal sorting. These receptors tend to be sequestered in endosomes and either recycle slowly or are degraded.

Because activated ERK1/2 has little affinity for free cytosolic arrestin and would dissociate soon after the receptor and arrestin disengage, arrestin-bound active ERK1/2 is constrained within receptor-arrestin 'signalsome' complexes (Fig. 15.2b). Thus, ERK1/2 activated by arrestin scaffolds remains part the signalsome complex, at least when activated by 'class B' GPCRs like the angiotensin AT_{1A} and vasopressin V_2 receptors (Tohgo et al. 2003). Because the GPCR-arrestin complex is stable, activated ERK1/2 is targeted to the plasma membrane and early endosomes (DeFea et al. 2000; Luttrell et al. 2001). In fact, an estimated 75–80% of the active ERK1/2 produced in response to short term stimulation of protease activated receptor (PAR)2 is associated with the GPCR-arrestin signalsome (DeFea et al. 2000). As a result, nuclear translocation of active ERK1/2 is retarded and its kinase activity is directed away from nuclear, and toward cytosolic, targets.

Functions of Arrestin-Dependent ERK Activation

While ERK1/2 activated by classical receptor tyrosine kinases and heterotrimeric G protein-mediated pathways is free to translocate to the cell nucleus and gain access to nuclear transcription factors, arrestin-dependent ERK1/2 activation produces a spatially-constrained ERK1/2 pool that is unable to produce a transcriptional response (Tohgo et al. 2003; Wei et al. 2004; Lee et al. 2008). Rather than transcriptional regulation, the arrestin-bound ERK1/2 activity appears to be directed predominantly toward membrane and cytosolic substrates, where it plays a key role in the regulation of vesicle trafficking, cytoskeletal rearrangement, and protein translation.

Regulation of GPCR Internalization and Trafficking

The most conserved arrestin function is the negative regulation of G protein signaling through direct steric hindrance of the GPCR-G protein interaction, and in the case of β-arrestins, mediation of clathrin-dependent GPCR endocytosis by linking GPCRs to clathrin and AP-2 (Goodman et al. 1996; Laporte et al. 1999). Besides these functions, several β-arrestin enzymatic cargos have been reported either to accelerate second messenger degradation or modulate the endocytic process. These include the cAMP phosphodiesterases and diacylglycerol kinase that accelerate second messenger degradation (Perry et al. 2002; Nelson et al. 2007), E3 ubiquitin ligases and de-ubiquitinases that modulate the stability of the GPCR-arrestin interaction (Shenoy and Lefkowitz 2003, 2005; Bhandari et al. 2007; Shenoy et al. 2008, 2009), and c-Src, which regulates GRK2 degradation (Penela et al. 2001), promotes dynamin self assembly (Ahn et al. 1999, 2002), and controls the stability of the β-arrestin-AP2 complex (Fessart et al. 2005, 2007; Zimmerman et al. 2009). Arrestin-bound ERK1/2 also appears to modulate the interaction between β-arrestin1 and the clathrin-dependent endocytic machinery. ERK1/2 phosphorylates Ser^{412} in the C-terminus of β-arrestin1, limiting its ability to bind clathrin (Lin et al. 1997, 1999). β-arrestin1 in the cytosol is almost stoichiometrically phosphorylated on Ser^{412}, and dephosphorylation of Ser^{412} upon receptor binding promotes receptor internalization and ERK1/2 activation. Re-phosphorylation by ERK1/2 in the signalsome complex probably provides either negative feedback regulation of receptor endocytosis or facilitates receptor internalization by promoting dissociation of arrestin and clathrin, allowing the receptor to exit clathrin-coated vesicles. In rodent, but not human, β-arrestin2, ERK1/2 phosphorylation of S^{178} stabilizes the GPCR-arrestin complex in endosomes and delays recycling (Khoury et al. 2014).

Cell Proliferation

Owing in part to their tripartite functions of dampening basal ERK1/2 activity, desensitizing G protein-mediated pathways involved in generating a transcriptionally-active pool of ERK1/2, and activating ERK1/2 in signalsomes, the role of arrestins in the regulation of proliferative ERK1/2 signaling is complex and context-dependent. Primary calvarial preosteoblasts from β-arrestin2 null mice proliferate faster that wild type, and treatment with an arrestin pathway-selective 'biased' agonist of the PTH_1 receptor, [D-Trp^{12},Tyr^{34}]-bPTH(7-34), slows proliferation in a β-arrestin2-dependent manner, suggesting that arrestins restrain pre-osteoblast proliferation (Gesty-Palmer et al. 2013). Such effects are consistent β-arrestin-dependent sequestration of inactive ERK1/2 in a microtubule-bound pool (Hanson et al. 2007; Coffa et al. 2011b) and cytosolic retention of active ERK1/2 bound to stable GPCR-arrestin complexes (DeFea et al. 2000; Tohgo et al. 2002; Ahn et al. 2004). On the other hand, neointimal hyperplasia following carotid endothelial injury is reduced in β-arrestin2 null mice, where its loss is associated with decreased GPCR-stimulated vascular smooth muscle cell proliferation and migration, consistent with a stimulatory role for β-arrestin2 signaling in the proliferative response (Kim et al. 2008). In this system, knockout of β-arrestin1 has the opposite effect, suggesting that β-arrestin1 and 2 play opposing roles.

The proliferative effects of β-arrestin2 probably reflect its role in 'transactivation' of EGF receptors via another arrestin-bound cargo, c-Src, rather than direct scaffolding of the ERK1/2 cascade. Many GPCRs stimulate cell proliferation through crosstalk with the EGF receptor family of tyrosine kinase kinases, which in turn promote Ras-dependent mitogenic ERK1/2 signaling. EGF receptors are activated by matrix metalloprotease-dependent shedding of EGF family growth factors, e.g. heparin-binding EGF, which exist as preformed membrane associated precursors (Carpenter 2000), and GPCRs possess both G protein-dependent and G protein-independent mechanisms of promoting their release. EGF receptor transactivation is the major mechanism underlying ERK1/2 activation by endogenous LPA receptors in β-arrestin1/2 null murine embryo fibroblasts, clearly indicating that arrestin signaling is not essential for GPCR-EGF receptor cross talk (Gesty-Palmer et al. 2005). On the other hand, in HEK293 cells β1-Adrenergic receptor-mediated EGF receptor transactivation and ERK1/2 activation are inhibited by downregulating β-arrestin1/2 or GRK5/6, inhibiting Src or MMP activity, or exposing cells to a heparin-binding EGF neutralizing antibody, suggesting that β1 receptor-mediated EGF receptor transactivation is β-arrestin-dependent (Noma et al. 2007). Consistent with this, a GRK site mutant β1 receptor that cannot undergo arrestin-dependent desensitization fails to transactivate EGF receptors despite exaggerated G protein signaling. In human coronary smooth muscle cells, the arrestin pathway-selective angiotensin AT_{1A} receptor agonist, [Sar^1-Ile^4-Ile^8]-AngII, induces ERK1/2 activation and proliferation by promoting EGF receptor transactivation (Miura et al. 2004), while in rat vascular smooth muscle both angiotensin II and [Sar^1-Ile^4-Ile^8]-AngII stimulate Src-dependent EGF receptor phosphorylation on Tyr^{845}, an effect that is lost when β-arrestin2 is downregulated by RNA interference (Kim et al. 2009). Similarly, the LH receptor activates c-Fyn in an β-arrestin2-dependent manner (Galet and Ascoli 2008). Downregulating arrestin expression reduces the rate of LH receptor internalization and inhibits LH-mediated activation of c-Fyn, phosphorylation of the anti-apoptotic focal adhesion kinase, and the release of EGF-like growth factors.

Non-proliferative Cell Growth

β-Arrestin scaffolding of the ERK1/2 cascade allows it to preferentially target cytosolic substrates involved in the control of protein translation including the ribosomal S6 kinase, p90RSK (Aplin et al. 2007) and the MAP kinase-interacting kinase, MNK1, a regulator of the ribosomal protein translation initiation complex. β-Arrestin2-dependent ERK1/2 activation by the angiotensin AT_{1A} receptor increases phosphorylation of MNK1 and eukaryotic translation initiation factor 4E, increasing rates of mRNA translation (DeWire et al. 2008). In addition, arrestin-dependent assembly of another signaling complex composed of protein phosphatase 2A (PP2A), AKT and glycogen synthase kinase 3β (GSK3β), is involved in regulation of mammalian target of rapamycin (mTOR)-dependent protein translation (Kendall et al. 2014). Increased rates of protein translation in response to angiotensin II [Sar^1-Ile^4-Ile^8]-AngII involves activation of arrestin-bound pools of both ERK1/2 and AKT, AKT-mediated phosphorylation of mTOR and its downstream effector p70/p85 ribosomal S6 kinase, and ERK1/2 phosphorylation of p90RSK.

Cell Migration and Chemotaxis

Several arrestin cargos have been implicated in GPCR-mediated actin cytoskeletal rearrangement leading to enhanced cell migration and chemotaxis, and promoting cancer metastasis. GPCR-stimulated chemotaxis is dependent upon two factors; the ability to sense a chemoattractant gradient and to establish cell polarity through cytoskeletal rearrangement at the leading edge (DeFea 2013). Arrestin-dependent GPCR desensitization and recycling are critical to the former (Tomhave et al. 1994; Aragay et al. 1998), while the coordinated regulation of several processes by β-arrestin scaffolds, including ERK1/2 and p38 MAPK activation, filamin A recruitment, cofilin dephosphorylation, and regulation of small GTPases contribute to the cellular shape changes needed for the latter. During PAR2-induced chemotaxis, PAR2-arrestin-ERK1/2 complexes localize to the leading edge of the cell where ERK1/2 activity is required for actin cytoskeletal reorganization (Ge et al. 2003, 2004). Similarly, assembly of a complex containing the angiotensin AT_{1A} receptor, β-arrestin, ERK1/2, and the actin filament bundling protein, filamin A, coordinates the formation of membrane ruffles in Hep2 cells (Scott et al. 2006). In HeLa and HEK293 cells, overexpression/downregulation of β-arrestin2 reciprocally enhances/attenuates activation of both p38 MAPK and ERK1/2 by the chemokine receptor CXCR4, and inhibition of p38 MAPK blocks β-arrestin2 dependent chemotaxis, (Sun et al. 2002). Similarly, both β-arrestin2 and p38 MAPK have been implicated in AT_{1A} and LPA receptor-mediated chemotaxis in HEK293 cells (Hunton et al. 2005), and arrestin-dependent scaffolding of p38 MAPK is involved in control of cell polarization, actin bundle formation and internalization of platelet-activating factor receptors in polymorphonuclear neutrophils (McLaughlin et al. 2006).

Along with activating localized pools of ERK1/2 and p38 MAPK, β-arrestin scaffolds coordinate the activity of other cargos that drive actin assembly and regulate small GTPase activity. β-Arrestin scaffolding of a cofilin-chronophin-LIM kinase complex is necessary for the localized generation of free barbed ends on actin filaments that supports filament extension and generates membrane protrusions (Zoudilova et al. 2007, 2010). As a result, downregulation of either β-arrestin1 or 2 in MDA-MB-468 cells inhibits PAR2 stimulated cofilin dephosphorylation and chemotaxis. Arrestin-dependent regulation of the small GTPase, RalA, through its interaction with the Ral-guanine nucleotide dissociation stimulator, Ral-GDS, has been implicated in formyl-Met-Leu-Phe receptor stimulated membrane ruffling (Bhattacharya et al. 2002), Ral-dependent activation of PLCδ1 by the angiotensin AT_{1A} receptor (Godin et al. 2010), and LPA receptor mediated proliferation and migration of breast cancer cells (Li et al. 2009a). Regulation of RhoA via the interaction between β-arrestin1 and ARFGAP21 promotes angiotensin AT_{1A} receptor dependent RhoA activation and membrane ruffling (Anthony et al. 2011), while β-arrestin1 and Gq/11 coordinately activate RhoA to promote stress fiber formation in AT_{1A} receptor-expressing HEK293 cells (Barnes et al. 2005).

Disrupting Arrestin-Dependent ERK Regulation

Because arrestins function both as negative regulators of heterotrimeric G proteins and as G protein-independent signaling scaffolds, arrestin knockouts exhibit both prolonged/exaggerated G protein signaling (Kohout et al. 2001) and loss of arrestin-dependent signals (Wei et al. 2003; Ahn et al. 2004). Thus, it is often difficult to attribute phenotypes associated with arrestin loss-of-function to one role or the other (Luttrell and Gesty-Palmer 2010). Dissociating the desensitizing and signaling functions of arrestins, such that desensitization/sequestration are preserved while arrestin signaling is disrupted, is one approach that might help shed light onto the physiological roles of arrestin scaffolds.

Despite the fact that c-Raf1-MEK1/2-ERK1/2 and ASK1-MKK4-JNK3 engage a broad interface involving both the N- and C-domains of β-arrestin1 and β-arrestin2 (Song et al. 2009), binding of ERK1/2 and JNK3 pathway components can be affected by point mutations that disrupt arrestin-dependent scaffolding while leaving the rest of the protein, and its other functions, intact. Peptide array analysis of the binding of JNK3 pathway components to β-arrestin1 and β-arrestin2 has identified conserved ASK1 binding motifs located in the N- and C-domains of both proteins, such that a β-arrestin1 $D^{26,29}A$ mutation in the N-domain eliminates ASK1 binding in vitro and in cells, as does a $A^{391,392}D$ mutation in the C-domain (Li et al. 2009b). Similarly, mutating E^{66}, D^{67}, and D^{69} of β-arrestin1 to alanine ablates its binding to MKK4. Applying a similar approach to β-arrestin-ERK1/2 pathway interactions identified D^{26} and D^{29} as also critical for the β-arrestin1-MEK1 interaction, such that the $D^{26,29}A$ mutant fails to interact with MEK1 in cells and fails to undergo ERK1/2-dependent S^{412} phosphorylation (Meng et al. 2009). Interestingly, a cell permeable 25 amino acid β-arrestin1 peptide encompassing the β-arrestin1-MEK1 binding site prevents MEK1 binding to wild type β-arrestin1 and S^{412} phosphorylation, while promoting β-arrestin1 binding to other cargos, including c-Src and clathrin, and enhancing β2 adrenergic receptor internalization.

Both ERK1/2 and cRaf-1 binding are also sensitive to point mutations. $K^{285}A/R^{286}A$ and $K^{295}A$ mutants of β-arrestin2 exhibit impaired ERK1/2 binding and fail to support β2 adrenergic receptor-mediated ERK1/2 activation in cells (Xu et al. 2008). However, these mutations, which involve the polar core of the protein, also interfere with β-arrestin2 self-association and β2-adrenergic receptor binding. Greater specificity has been obtained by targeting cRaf-1 binding. Alanine-scanning mutagenesis of conserved residues on the non-receptor binding surface of β-arrestin1 and β-arrestin2 identified R^{307} in the C-domain of β-arrestin1 as important for c-Raf1 binding, such that an $R^{307}A$ mutant of β-arrestin1 exhibits significantly reduced c-Raf1 binding with no effect on binding to active phosphorylated rhodopsin or to MEK1 and ERK2 (Coffa et al. 2011b). In β-arrestin1/2 double null cells, the $R^{307}A$ mutant fails to rescue β2 adrenergic receptor-stimulated ERK1/2 activation by the arrestin-biased β2 receptor agonist, ICI118551, indicating that it can function as a dominant negative inhibitor of arrestin-dependent ERK1/2 activation while not impairing receptor desensitization. Interestingly, the analogous $K^{308}A$ mutation in β-arrestin2 does not disrupt c-Raf1 binding or its ability to rescue arrestin-dependent ERK1/2 activation in the null background.

Conclusions

ERK1/2 is a nodal kinase that is critical to many cellular functions, from cell cycle progression and survival to cell growth, migration and metastasis. As such, it is subject to multiple forms of convergent regulation that determine both where it is active and for how long (Stork 2002; Luttrell 2003). By binding components of the

ERK1/2 cascade in a conformation-sensitive manner, arrestins are critical regulators of ERK1/2 pathway activity, both in the presence and absence of GPCR stimulation. Association of cRaf-1-MEK1/2-ERK1/2 with the microtubule-associated storage pool of arrestins keeps ERK1/2 pathway components away from its membrane delimited activators and contributes to dampening basal pathway activity. Stimulus-dependent recruitment of arrestins to high affinity GPCR binding sites has two effects. Arrestin-dependent desensitization limits the duration of G protein-dependent signals that activate a transcriptionally competent pool of ERK1/2 via second messenger-dependent protein kinases and EGF receptor cross talk. At the same time, arrestin scaffolding of the ERK1/2 cascade produces sustained and spatially discrete pool of active ERK1/2 that is preferentially targeted to membrane and cytosolic substrates involved in GPCR trafficking, cytoskeletal rearrangement, and protein translation. Thus, arrestins fulfill every criterion of a GPCR-regulated signaling scaffold.

Acknowledgements Supported by National Institutes of Health Grants R01 DK055524 (LML), R01 GM095497 (LML), Department of Veterans Affairs Merit Review Grant I01 BX003188 (LML), and the Research Service of the Charleston, SC Veterans Affairs Medical Center. The contents of this article do not represent the views of the Department of Veterans Affairs or the United States Government.

References

Ahn S, Maudsley S, Luttrell LM, Lefkowitz RJ, Daaka Y (1999) Src-mediated tyrosine phosphorylation of dynamin is required for beta2-adrenergic receptor internalization and mitogen-activated protein kinase signaling. J Biol Chem 274:1185–1188

Ahn S, Kim J, Lucaveche CL, Reedy MC, Luttrell LM, Lefkowitz RJ, Daaka Y (2002) Src-dependent tyrosine phosphorylation regulates dynamin self-assembly and ligand-induced endocytosis of the epidermal growth factor receptor. J Biol Chem 277:26642–26651

Ahn S, Shenoy SK, Wei H, Lefkowitz RJ (2004) Differential kinetic and spatial patterns of beta-arrestin and G protein-mediated ERK activation by the angiotensin II receptor. J Biol Chem 279:35518–35525

Anthony DF, Sin YY, Vadrevu S, Advant N, Day JP, Byrne AM, Lynch MJ, Milligan G, Houslay MD, Baillie GS (2011) Beta-arrestin 1 inhibits the GTPase-activating protein function of ARHGAP21, promoting activation of RhoA following angiotensin II type 1A receptor stimulation. Mol Cell Biol 31:1066–1075

Aplin M, Christensen GL, Schneider M, Heydorn A, Gammeltoft S, Kjølbye AL, Sheikh SP, Hansen JL (2007) Differential extracellular signal-regulated kinases 1 and 2 activation by the angiotensin type 1 receptor supports distinct phenotypes of cardiac myocytes. Basic Clin Pharmacol Toxicol 100:296–301

Aragay AM, Mellado M, Frade JM, Martin AM, Jimenez-Sainz MC, Martinez AC, Mayor F Jr (1998) Monocyte chemoattractant protein-1-induced CCR2B receptor desensitization mediated by the G protein-coupled receptor kinase 2. Proc Natl Acad Sci U S A 95:2985–2990

Aubry L, Guetta D, Klein G (2009) The arrestin fold: variations on a theme. Curr Genomics 10:133–142

Barnes WG, Reiter E, Violin JD, Ren XR, Milligan G, Lefkowitz RJ (2005) Beta-arrestin 1 and Galphaq/11 coordinately activate RhoA and stress fiber formation following receptor stimulation. J Biol Chem 280:8041–8050

Bhandari D, Trejo J, Benovic JL, Marchese A (2007) Arrestin-2 interacts with the ubiquitin-protein isopeptide ligase atrophin-interacting protein 4 and mediates endosomal sorting of the chemokine receptor CXCR4. J Biol Chem 282:36971–36979

Bhattacharya M, Anborgh PH, Babwah AV, Dale LB, Dobransky T, Benovic JL, Feldman RD, Verdi JM, Rylett RJ, Ferguson SS (2002) Beta-arrestins regulate a Ral-GDS Ral effector pathway that mediates cytoskeletal reorganization. Nat Cell Biol 4:547–555

Breitman M, Kook S, Gimenez LE, Lizama BN, Palazzo MC, Gurevich EV, Gurevich VV (2012) Silent scaffolds: inhibition of c-Jun N-terminal kinase 3 activity in cell by dominant-negative arrestin-3 mutant. J Biol Chem 287:19653–19664

Burack WR, Shaw AS (2000) Signal transduction: hanging on a scaffold. Curr Opin Cell Biol 12:211–216

Carpenter G (2000) EGF receptor transactivation mediated by the proteolytic production of EGF-like agonists. Sci STKE 15:1pe

Coffa S, Breitman M, Hanson SM, Callaway K, Kook S, Dalby KN, Gurevich VV (2011a) The effect of arrestin conformation on the recruitment of c-Raf1, MEK1, and ERK1/2 activation. PLoS ONE 6:e28723

Coffa S, Breitman M, Spiller BW et al (2011b) A single mutation in arrestin-2 prevents ERK1/2 activation by reducing c-Raf1 binding. Biochemistry 50:6951–6958

Davis RJ (2000) Signal transduction by the JNK group of MAP kinases. Cell 103:239–252

DeFea KA (2013) Arrestins in actin reorganization and cell migration. Prog Mol Biol Transl Sci 118:205–222

DeFea KA, Zalevsky J, Thoma MS, Dery O, Mullins RD, Bunnett NW (2000) Beta-arrestin-dependent endocytosis of proteinase-activated receptor 2 is required for intracellular targeting of activated ERK1/2. J Cell Biol 148:1267–1281

DeRooij J, Zwartkruis FL, Verheijen MH, Cool RH, Nijman SM, Wittinghofer A, Bos JL (1998) Epac is a Rap1 guanine nucleotide exchange factor directly activated by cAMP. Nature 396:474–477

DeWire SM, Kim J, Whalen EJ, Ahn S, Chen M, Lefkowitz RJ (2008) Beta-arrestin-mediated signaling regulates protein synthesis. J Biol Chem 283:10611–10620

Fessart D, Simaan M, Laporte SA (2005) c-Src regulates clathrin adapter protein 2 interaction with beta-arrestin and the angiotensin II type 1 receptor during clathrin-mediated internalization. Mol Endocrinol 19:491–503

Fessart D, Simaan M, Zimmerman B, Comeau J, Hamdan FF, Wiseman PW, Bouvier M, Laporte SA (2007) Src-dependent phosphorylation of beta2-adaptin dissociates the beta-arrestin-AP-2 complex. J Cell Sci 120:1723–1732

Galet C, Ascoli M (2008) Arrestin-3 is essential for the activation of Fyn by the luteinizing hormone receptor (LHR) in MA-10 cells. Cell Signal 20:1822–2829

Ge L, Ly Y, Hollenberg M, DeFea K (2003) A beta-arrestin-dependent scaffold is associated with prolonged MAPK activation in pseudopodia during protease-activated receptor-2-induced chemotaxis. J Biol Chem 278:34418–34426

Ge L, Shenoy SK, Lefkowitz RJ, DeFea K (2004) Constitutive protease-activated receptor-2-mediated migration of MDA MB-231 breast cancer cells requires both beta-arrestin-1 and -2. J Biol Chem 279:55419–55424

Gesty-Palmer D, El Shewy H, Kohout TA, Luttrell LM (2005) Beta-arrestin 2 expression determines the transcriptional response to lysophosphatidic acid stimulation in murine embryo fibroblasts. J Biol Chem 280:32157–32167

Gesty-Palmer D, Chen M, Reiter E, Ahn S, Nelson CD, Wang S, Eckhardt AE, Cowan CL, Spurney RF, Luttrell LM, Lefkowitz RJ (2006) Distinct beta-arrestin- and G protein-dependent pathways for parathyroid hormone receptor-stimulated ERK1/2 activation. J Biol Chem 281:10856–10864

Gesty-Palmer D, Yuan L, Martin B, Wood WH 3rd, Lee MH, Janech MG, Tsoi LC, Zheng WJ, Luttrell LM, Maudsley S (2013) β-Arrestin-selective G protein-coupled receptor agonists engender unique biological efficacy in vivo. Mol Endocrinol 27:296–314

Godin CM, Ferreira LT, Dale LB, Gros R, Cregan SP, Ferguson SS (2010) The small GTPase Ral couples the angiotensin II type 1 receptor to the activation of phospholipase C-delta 1. Mol Pharmacol 77:388–395

Goodman OB Jr, Krupnick JG, Santini F, Gurevich VV, Penn RB, Gagnon AW, Keen JH, Benovic JL (1996) Beta-arrestin acts as a clathrin adaptor in endocytosis of the beta2-adrenergic receptor. Nature 383:447–450

Gurevich VV, Benovic JL (1997) Mechanism of phosphorylation-recognition by visual arrestin and the transition of arrestin into a high affinity binding state. Mol Pharmacol 51:161–169

Gurevich VV, Gurevich EV (2006) The structural basis of arrestin-mediated regulation of G-protein-coupled receptors. Pharmacol Ther 110:465–502

Gurevich VV, Gurevich EV (2013) Structural determinants of arrestin functions. Prog Mol Biol Transl Sci 118:57–92

Hanson SM, Cleghorn WM, Francis DJ, Vishnivetskiy SA, Raman D, Song X, Nair KS, Slepak VZ, Klug CS, Gurevich VV (2007) Arrestin mobilizes signaling proteins to the cytoskeleton and redirects their activity. J Mol Biol 368:375–387

Hanson SM, Vishnivetskiy SA, Hubbell WL, Gurevich VV (2008) Opposing effects of inositol hexakisphosphate on rod arrestin and arrestin2 self-association. Biochemistry 47:1070–1075

Hawes BE, van Biesen T, Koch WJ, Luttrell LM, Lefkowitz RJ (1995) Distinct pathways of Gi- and Gq-mediated mitogen activated protein kinase activation. J Biol Chem 270:17148–17153

Hunton DL, Barnes WG, Kim J, Ren XR, Violin JD, Reiter E, Milligan G, Patel DD, Lefkowitz RJ (2005) Beta-arrestin 2-dependent angiotensin II type 1A receptor-mediated pathway of chemotaxis. Mol Pharmacol 67:1229–1236

Jafri F, El-Shewy HM, Lee M-H, Kelly M, Luttrell DK, Luttrell LM (2006) Constitutive ERK1/2 activation by a chimeric neurokinin 1 receptor-beta-arrestin1 fusion protein. Probing the composition and function of the G protein-coupled receptor "signalsome". J Biol Chem 281:19346–19357

Kendall RT, Lee MH, Pleasant DL, Robinson K, Kuppuswamy D, McDermott PJ, Luttrell LM (2014) Arrestin-dependent angiotensin AT1 receptor signaling regulates Akt and mTor-mediated protein synthesis. J Biol Chem 289:26155–26166

Khoury E, Nikolajev L, Simaan M, Namkung Y, Laporte SA (2014) Differential regulation of endosomal GPCR/β-arrestin complexes and trafficking by MAPK. J Biol Chem 289: 23302–23317

Kim J, Zhang L, Peppel K, Wu JH, Zidar DA, Brian L, DeWire SM, Exum ST, Lefkowitz RJ, Freedman NJ (2008) Beta-arrestins regulate atherosclerosis and neointimal hyperplasia by controlling smooth muscle cell proliferation and migration. Circ Res 103:70–79

Kim J, Ahn S, Rajagopal K, Lefkowitz RJ (2009) Independent beta-arrestin2 and Gq/protein kinase Czeta pathways for ERK stimulated by angiotensin type 1A receptors in vascular smooth muscle cells converge on transactivation of the epidermal growth factor receptor. J Biol Chem 284:11953–11962

Kohout TA, Lin FS, Perry SJ, Conner DA, Lefkowitz RJ (2001) Beta-arrestin 1 and 2 differentially regulate heptahelical receptor signaling and trafficking. Proc Natl Acad Sci U S A 98: 1601–1606

Kolch W, Heldecker G, Kochs G, Hummel R, Vahidi H, Mischak H, Finkenzeller G, Marmé D, Rapp UR (1993) Protein kinase C alpha activates Raf-1 by direct phosphorylation. Nature 364:249–255

Kook S, Zhan X, Kaoud TS, Dalby KN, Gurevich VV, Gurevich EV (2013) Arrestin-3 binds c-Jun N-terminal kinase 1 (JNK1) and JNK2 and facilitates the activation of these ubiquitous JNK isoforms in cells via scaffolding. J Biol Chem 288:37332–37342

Kovoor A, Celver J, Abdryashitov RI, Chavkin C, Gurevich VV (1999) Targeted construction of phosphorylation-independent beta-arrestin mutants with constitutive activity in cells. J Biol Chem 274:6831–6834

Kyriakis JM, Avruch J (2012) Mammalian MAPK signal transduction pathways activated by stress and inflammation: a 10-year update. Physiol Rev 92:689–737

Laporte SA, Oakley RH, Zhang J, Holt JA, Ferguson SS, Caron MG, Barak LS (1999) The beta2-adrenergic receptor/betaarrestin complex recruits the clathrin adaptor AP-2 during endocytosis. Proc Natl Acad Sci U S A 96:3712–3717

Lee MH, El-Shewy HM, Luttrell DK, Luttrell LM (2008) Role of beta-arrestin-mediated desensitization and signaling in the control of angiotensin AT1a receptor-stimulated transcription. J Biol Chem 283:2088–2097

Lefkowitz RJ, Pierce KL, Luttrell LM (2002) Dancing with different partners: PKA phosphorylation of seven membrane-spanning receptors regulates their G protein coupling specificity. Mol Pharmacol 62:971–974

Li TT, Alemayehu M, Aziziyeh AI, Pape C, Pampillo M, Postovit LM, Mills GB, Babwah AV, Bhattacharya M (2009a) Beta-arrestin/Ral signaling regulates lysophosphatidic acid-mediated migration and invasion of human breast tumor cells. Mol Cancer Res 7:1064–1077

Li X, MacLeod R, Dunlop AJ, Edwards HV, Advant N, Gibson LC, Devine NM, Brown KM, Adams DR, Houslay MD, Baillie GS (2009b) A scanning peptide array approach uncovers association sites within the JNK/beta arrestin signalling complex. FEBS Lett 583:3310–3316

Lin FT, Krueger KM, Kendall HE, Daaka Y, Fredericks ZL, Pitcher JA, Lefkowitz RJ (1997) Clathrin-mediated endocytosis of the beta-adrenergic receptor is regulated by phosphorylation/dephosphorylation of beta-arrestin1. J Biol Chem 272:31051–31057

Lin FT, Miller WE, Luttrell LM, Lefkowitz RJ (1999) Feedback regulation of beta-arrestin1 function by extracellular signal-regulated kinases. J Biol Chem 274:15971–15974

Luttrell LM (2003) Location, location, location. Spatial and temporal regulation of MAP kinases by G protein-coupled receptors. J Mol Endocrinol 30:117–126

Luttrell LM, Gesty-Palmer D (2010) Beyond desensitization: physiological relevance of arrestin-dependent signaling. Pharmacol Rev 62:305–330

Luttrell LM, Miller WE (2013) Arrestins as regulators of kinases and phosphatases. Prog Mol Biol Transl Sci 118:115–147

Luttrell LM, Hawes BE, van Biesen T, Luttrell DK, Lansing TJ, Lefkowitz RJ (1996) Role of c-Src in G protein-coupled receptor- and Gbetagamma subunit-mediated activation of mitogen activated protein kinases. J Biol Chem 271:19443–19450

Luttrell LM, Della Rocca GJ, van Biesen T, Luttrell DK, Lefkowitz RJ (1997) Gbetagamma subunits mediate Src-dependent phosphorylation of the epidermal growth factor receptor. J Biol Chem 272:4637–4644

Luttrell LM, Ferguson SS, Daaka Y, Miller WE, Maudsley S, Della Rocca GJ, Lin F, Kawakatsu H, Owada K, Luttrell DK, Caron MG, Lefkowitz RJ (1999) Beta-arrestin-dependent formation of beta2 adrenergic receptor-Src protein kinase complexes. Science 283:655–661

Luttrell LM, Roudabush FL, Choy EW, Miller WE, Field ME, Pierce KL, Lefkowitz RJ (2001) Activation and targeting of extracellular signal-regulated kinases by beta-arrestin scaffolds. Proc Natl Acad Sci U S A 98:2449–2454

McDonald PH, Chow CW, Miller WE, Laporte SA, Field ME, Lin FT, Davis RJ, Lefkowitz RJ (2000) Beta-arrestin 2: a receptor-regulated MAPK scaffold for the activation of JNK3. Science 290:1574–1577

McLaughlin NJ, Banerjee A, Kelher MR, Gamboni-Robertson F, Hamiel C, Sheppard FR, Moore EE, Silliman CC (2006) Platelet-activating factor-induced clathrin-mediated endocytosis requires beta-arrestin-1 recruitment and activation of the p38 MAPK signalosome at the plasma membrane for actin bundle formation. J Immunol 176:7039–7050

Meng D, Lynch MJ, Huston E, Beyermann M, Eichhorst J, Adams DR, Klusmann E, Houslay MD, Baillie GS (2009) MEK1 binds directly to beta-arrestin1, influencing both its phosphorylation by ERK and the timing of its isoprenaline-stimulated internalization. J Biol Chem 284:11425–11435

Miura S, Zhang J, Matsuo Y, Saku K, Karnik SS (2004) Activation of extracellular signal-activated kinase by angiotensin II-induced Gq-independent epidermal growth factor receptor transactivation. Hypertens Res 27:65–77

Nelson CD, Perry SJ, Regier DS, Prescott SM, Topham MK, Lefkowitz RJ (2007) Targeting of diacylglycerol degradation to M1 muscarinic receptors by beta-arrestins. Science 315:663–666
Noma T, Lemaire A, Prasad SVN, Barki-Harrington L, Tilley DG, Chen J, Le Corvoisier P, Violin JD, Wei H, Lefkowitz RJ, Rockman HA (2007) Beta-arrestin-mediated beta1-adrenergic receptor transactivation of the EGFR confers cardioprotection. J Clin Invest 117:2445–2458
Oakley RH, Laporte SA, Holt JA, Caron MG, Barak LS (2000) Differential affinities of visual arrestin, beta arrestin1, and beta arrestin2 for G protein-coupled receptors delineate two major classes of receptors. J Biol Chem 275:17201–17210
Oakley RH, Laporte SA, Holt JA, Barak LS, Caron MG (2001) Molecular determinants underlying the formation of stable intracellular G protein-coupled receptor-beta-arrestin complexes after receptor endocytosis. J Biol Chem 276:19452–19460
Pearson G, Robinson F, Beers Gibson T, Xu BE, Karandikar M, Berman K, Cobb MH (2001) Mitogen-activated protein (MAP) kinase pathways: regulation and physiological functions. Endocr Rev 22:153–183
Penela P, Elorza A, Sarnago S, Mayor F Jr (2001) Beta-arrestin- and c-Src-dependent degradation of G-protein-coupled receptor kinase 2. EMBO J 20:5129–5138
Perry SJ, Baillie GS, Kohout TA, McPhee I, Magiera MM, Ang KL, Miller WE, McLean AJ, Conti M, Houslay MD, Lefkowitz RJ (2002) Targeting of cyclic AMP degradation to beta 2-adrenergic receptors by beta-arrestins. Science 298:834–836
Scott MG, Le Rouzic E, Perianin A, Pierotti V, Enslen H, Benichou S, Marullo S, Benmerah A (2002) Differential nucleocytoplasmic shuttling of beta-arrestins. Characterization of a leucine-rich nuclear export signal in beta-arrestin2. J Biol Chem 277:37693–37701
Scott MG, Pierotti V, Storez H, Lindberg E, Thuret A, Muntaner O, Labbé-Jullié C, Pitcher JA, Marullo S (2006) Cooperative regulation of extracellular signal-regulated kinase activation and cell shape change by filamin A and beta-arrestins. Mol Cell Biol 26:3432–3445
Seo J, Tsakem EL, Breitman M, Gurevich VV (2011) Identification of arrestin-3-specific residues necessary for JNK3 kinase activation. J Biol Chem 286:27894–27901
Shenoy SK, Lefkowitz RJ (2003) Trafficking pattern of beta-arrestin and G protein-coupled receptors determined by the kinetics of beta-arrestin deubiquitination. J Biol Chem 278: 14498–14506
Shenoy SK, Lefkowitz RJ (2005) Receptor-specific ubiquitination of beta-arrestin directs assembly and targeting of seven-transmembrane receptor signalosomes. J Biol Chem 280:15315–15324
Shenoy SK, Drake MT, Nelson CD, Houtz DA, Xiao K, Madabushi S, Reiter E, Premont RT, Lichtarge O, Lefkowitz RJ (2006) Beta-arrestin-dependent, G protein-independent ERK1/2 activation by the beta2 adrenergic receptor. J Biol Chem 281:1261–1273
Shenoy SK, Xiao K, Venkataramanan V, Snyder PM, Freedman NJ, Weissman AM (2008) Nedd4 mediates agonist-dependent ubiquitination, lysosomal targeting, and degradation of the beta2-adrenergic receptor. J Biol Chem 283:22166–22176
Shenoy SK, Modi AS, Shukla AK, Xiao K, Berthouze M, Ahn S, Wilkinson KD, Miller WE, Lefkowitz RJ (2009) Beta-arrestin-dependent signaling and trafficking of 7-transmembrane receptors is reciprocally regulated by the deubiquitinase USP33 and the E3 ligase Mdm2. Proc Natl Acad Sci U S A 106:6650–6655
Song X, Raman D, Gurevich EV, Vishnivetskiy SA, Gurevich VV (2006) Visual and both non-visual arrestins in their "inactive" conformation bind JNK3 and Mdm2 and relocalize them from the nucleus to the cytoplasm. J Biol Chem 281:21491–21499
Song X, Coffa S, Fu H, Gurevich VV (2009) How does arrestin assemble MAPKs into a signaling complex? J Biol Chem 284:685–695
Stork PJ (2002) ERK signaling: duration, duration, duration. Cell Cycle 1:315–317
Sun Y, Cheng Z, Ma L, Pei G (2002) Beta-arrestin2 is critically involved in CXCR4-mediated chemotaxis, and this is mediated by its enhancement of p38 MAPK activation. J Biol Chem 277:49212–49219

Tohgo A, Pierce KL, Choy EW, Lefkowitz RJ, Luttrell LM (2002) βarrestin scaffolding of the ERK cascade enhances cytosolic ERK activity but inhibits ERK mediated transcription following angiotensin AT1a receptor stimulation. J Biol Chem 277:9429–9436

Tohgo A, Choy EW, Gesty-Palmer D, Pierce KL, Laporte S, Oakley RH, Caron MG, Lefkowitz RJ, Luttrell LM (2003) The stability of the G protein-coupled receptor-beta-arrestin interaction determines the mechanism and functional consequence of ERK activation. J Biol Chem 278:6258–6267

Tomhave ED, Richardson RM, Didsbury JR, Menard L, Snyderman R, Ali H (1994) Cross-desensitization of receptors for peptide chemoattractants. Characterization of a new form of leukocyte regulation. J Immunol 153:3267–3275

van Biesen T, Hawes BE, Luttrell DK, Krueger KM, Touhara K, Porfiri E, Sakaue M, Luttrell LM, Lefkowitz RJ (1995) Receptor-tyrosine-kinase- and Gbetagamma-mediated MAP kinase activation by a common signalling pathway. Nature 376:781–784

Vishnivetskiy SA, Hirsch JA, Velez MG, Gurevich YV, Gurevich VV (2002) Transition of arrestin into the active receptor-binding state requires an extended interdomain hinge. J Biol Chem 277:43961–43967

Vossler MR, Yao H, York RD, Pan MG, Rim CS, Stork PJ (1997) cAMP activates MAP kinase and Elk-1 through a B-Raf- and Rap-1-dependent pathway. Cell 89:73–82

Wei H, Ahn S, Shenoy SK, Karnik SS, Hunyady L, Luttrell LM, Lefkowitz RJ (2003) Independent G protein and beta-arrestin2 mediated activation of ERK by angiotensin. Proc Natl Acad Sci U S A 100:10782–10787

Wei H, Ahn S, Barnes WG, Lefkowitz RJ (2004) Stable interaction between beta-arrestin 2 and angiotensin type 1A receptor is required for beta-arrestin 2-mediated activation of extracellular signal-regulated kinases 1 and 2. J Biol Chem 279:48255–48261

White MF (1998) The IRS-signalling system: a network of docking proteins that mediate insulin action. Mol Cell Biochem 182:3–11

Wu J, Dent P, Jelinek T, Wolfman A, Weber MJ, Sturgill TW (1993) Inhibition of the EGF-activated MAP kinase signaling pathway by adenosine 3′,5′-monophosphate. Science 62:1065–1068

Xu TR, Baillie GS, Bhari N, Houslay TM, Pitt AM, Adams DR, Kolch W, Houslay MD, Milligan G (2008) Mutations of beta-arrestin 2 that limit self-association also interfere with interactions with the beta2-adrenoceptor and the ERK1/2 MAPKs: implications for beta2-adrenoceptor signalling via the ERK1/2 MAPKs. Biochem J 413:51–60

Zhan X, Kaoud TS, Dalby KN, Gurevich VV (2011) Nonvisual arrestins function as simple scaffolds assembling the MKK4–JNK3α2 signaling complex. Biochemistry 50:10520–10529

Zimmerman B, Simaan M, Lee M-H, Luttrell LM, Laporte SA (2009) c-Src-mediated phosphorylation of AP-2 reveals a general mechanism for receptors internalizing through the clathrin pathway. Cell Signal 21:103–110

Zoudilova M, Kumar P, Ge L, Wang P, Bokoch GM, DeFea KA (2007) Beta-arrestin-dependent regulation of the cofilin pathway downstream of protease-activated receptor-2. J Biol Chem 282:20634–20646

Zoudilova M, Min J, Richards HL, Carter D, Huang T, DeFea KA (2010) Beta-arrestins scaffold cofilin with chronophin to direct localized actin filament severing and membrane protrusions downstream of protease-activated receptor-2. J Biol Chem 285:14318–14329

Chapter 16
The Functional Role of the Conformational Changes in Arrestin Upon Activation

Zhao Yang, Fan Yang, Anthony Nguen, Chuan Liu, Amy Lin, Thomas Joseph Cahill III and Jin-Peng Sun

Abstract G protein-coupled receptors (GPCR) signal primarily through G proteins and arrestins. Arrestin not only intervenes with G protein-mediated signaling, but also initiates a variety of G protein-independent functions. The functional diversity of arrestins has been attributed to their multiple active conformations after interaction with ligand-activated receptors, which are regulated by both the ligand-specific conformational states and the phosphorylation patterns of the receptors. Here we reviewed the current knowledge of the structural features of arrestins that underlie selective arrestin-mediated signaling. Recent breakthroughs in the functional correlation of the conformational changes in arrestin to downstream effector molecules were also highlighted.

Keywords GPCR · Arrestin · Structure · Crystallography · Conformational change · Activation

Introduction

G protein-coupled receptors (GPCRs), also known as seven-transmembrane domain receptors (7TMRs), are the largest family of membrane proteins that communicate extracellular stimuli to intracellular signals and play critical roles in mediating most known physiological functions (Pierce et al. 2002). The classical paradigm of

Z. Yang · F. Yang · C. Liu · J.-P. Sun (✉)
Key Laboratory Experimental Teratology of the Ministry of Education and Department of Biochemistry and Molecular Biology, Shandong University School of Medicine, 44 Wenhua Xi Road, Jinan 250012, Shandong, China
e-mail: sunjinpeng@sdu.edu.cn

Z. Yang · F. Yang · C. Liu · J.-P. Sun
Department of Physiology, Shandong University School of Medicine, Jinan 250012, Shandong, China

A. Nguen · A. Lin · T.J. Cahill III
School of Medicine, Duke University, Durham, NC 27705, USA

V.V. Gurevich (ed.), *The Structural Basis of Arrestin Functions*,
DOI 10.1007/978-3-319-57553-7_16

GPCR signaling is based on a ligand-induced conformational change of the receptor (Farrens et al. 1996), which is recognized by heterotrimeric G proteins to regulate levels of second messengers. The activated receptor is also phosphorylated by GRKs, which results in subsequent recruitment of arrestins for endocytosis and/or G protein-independent signaling (Walther and Ferguson 2013; Thomsen et al. 2016).

The arrestin family consists of four members. In mammals, arrestin-1 and arrestin-4 are located almost exclusively in photoreceptor cells in the retina and interact with very few visual GPCRs (Wilden et al. 1986; Craft et al. 1994), whereas β-arrestin-1 and β-arrestin-2 (also called arrestin-2 and arrestin-3, respectively) are ubiquitously expressed in most tissues and interact with low specificity with the majority of non-visual GPCRs (Lohse et al. 1990; Attramadal et al. 1992). By binding to GPCRs, these four arrestins desensitize G protein-mediated signaling in many cases and also function as adapters to promote receptor internalization by recruiting endocytic proteins, such as clathrin (Goodman et al. 1996), adaptor protein 2 (AP2) (Laporte et al. 1999) and *N*-ethylmaleimide-sensitive fusion protein (NSF) (McDonald et al. 1999), linking the receptors to the clathrin-coated pit machinery (Shenoy and Lefkowitz 2003; Lefkowitz and Shenoy 2005). In addition to their roles in GPCR desensitization, internalization and trafficking, mounting evidence has revealed that β-arrestins also act as signaling transducers that interact with a growing list of intracellular effectors that include, but not limited to, signaling proteins such as c-Src, Raf-1, Akt, ERK1/2, JNK3 and IκB (Luttrell et al. 1999; Gao et al. 2004; Xiao et al. 2010; Reiter et al. 2012; Wang et al. 2014; Ning et al. 2015; Smith and Rajagopal 2016). Through interaction with these signaling molecules, β-arrestins are capable of initiating G-protein-independent signaling via MAPK, PI3K/AKT, and NF-κB pathways, connecting the activated receptor to diverse physiological responses. Although some studies indicate that G protein and arrestin signal synergistically (Thomsen et al. 2016), β-arrestin-dependent and G-protein-dependent signaling pathways could be pharmacologically separable. Biased ligands, which selectively activate one of these two signaling pathways have great therapeutic potential and have been extensively studied (Rajagopal et al. 2010; Reiter et al. 2012).

The remarkable functional diversity of arrestins raises the inevitable question of how distinct arrestin functions are precisely regulated. Aiming at the detailed mechanism, numerous biomedical and biophysical studies have been performed over the past decade using multiple novel techniques and approaches, leading to fruitful results. Recent structural studies of GPCR-arrestin complex provided considerable insight into the activation mechanism of arrestin (Kang et al. 2015). In the present chapter, we will review current knowledge of the structural features that underlie selective arrestin signaling.

Arrestin Undergoes Conformational Changes After Interaction with Activated Receptors

Based on the diverse functional capacities of arrestins, especially β-arrestins, it has been hypothesized that arrestins adopt multiple active conformations after activation that dictate their distinct functions (Gurevich and Gurevich 2006). In cells, the conformational change of β-arrestin-2 was detected by BRET-based biosensors after activation of the angiotensin II receptor type 1a (AT1aR), β_2 adrenergic receptor (β_2AR), and parathyroid hormone 1 receptor (PTH1R) (Shukla et al. 2008). In detail, different conformations of β-arrestin-2 induced by full agonists or β-arrestin-biased agonists in stimulated cells were observed with the BRET sensor. Later, the conformational biosensor of arrestin was improved by the insertion of the FlAsH (intramolecular fluorescein arsenical hairpin) in the specific loops of arrestin. Significant conformational features of arrestins that correlated with their downstream functions were revealed by this new version of probes (Lee et al. 2016). Further, using the FlAsH-decorated arrestins, a rapid activation/deactivation cycle of β-arrestins was disclosed by FRET analysis, which reshapes our thinking of the arrestin signaling (Nuber et al. 2016).

Phosphorylation of the GPCRs plays a key role in the recruitment and activation of arrestins. Since the 1970s, the finding of phosphorylation of rhodopsin and β_2AR on their C-termini during desensitization has led to the identification of a seven-member family of G protein-coupled receptor kinases (GRKs) (Weller et al. 1975; Benovic et al. 1986; Pitcher et al. 1998). Although subsequent studies revealed that receptors could be phosphorylated by multiple intracellular kinases other than GRKs, such as PKA (Tran et al. 2004) and PKC (Garcia et al. 1998), it has been suggested that the phosphorylation mediated by GRKs initiates arrestin recruitment and receptor internalization (Tobin et al. 2008; Walther and Ferguson 2013). Whereas GRK1 and 7 are confined to the retina and GRK4 is distributed primarily in the reproductive system, GRKs 2, 3, 5, and 6 are ubiquitously expressed and thus regulate the phosphorylation patterns of most GPCRs (Krupnick and Benovic 1998; Pitcher et al. 1998). Studies of different GPCRs revealed that siRNA-mediated knockdown of specific GRK or combinations of GRKs leads to clearly distinguishable effects on receptor functions, suggesting that different sets of GRKs might phosphorylate distinct sites on the receptors. Many of the regulatory roles of GRKs on receptors are through arrestin-mediated functions, including receptor desensitization, internalization and non-receptor-binding partners activation (Kim et al. 2005; Ren et al. 2005). Studies regarding specific GRK-regulated GPCR functions through arrestin brought up the barcode concept of the phosphorylation-regulated receptor function, of which the phosphorylation pattern of the receptor resembles the barcode of the beer can, transducing specific information to downstream effectors (Tobin et al. 2008; Nobles et al. 2011). Using the prototypical GPCR β_2AR as a model, mass spectrometry-based quantitative proteomics and phosphorylation site-specific antibodies were exploited to map the phosphorylation sites of GRK2 and GRK6 at C-terminus and the third intracellular loop of the β_2AR expressed in HEK293 cells, whereas RNAi-induced GRK silencing and BRET assay were employed to delineate conformation-specific β-arrestin capabilities imparted by specific phosphorylation patterns induced by GRK2 and GRK6. Results indicated that whereas both GRK2 and GRK6 contribute to desensitization of the β_2AR through β-arrestin-2, GRK2 phosphorylation sites (T360, S364, S396, S401, S407, and S411) are primarily responsible for receptor internalization, whereas GRK6 sites (S355 and S356) are required for β-arrestin2-mediated ERK1/2 activation. Notably, phosphorylation of GRK2 sites appears to inhibit phosphorylation of GRK6 sites as well as β_2AR signaling to ERK1/2. It was also demonstrated that isoproterenol, the full agonist of β_2AR, initiates phosphorylation on both GRK2 and GRK6 sites, whereas carvedilol, a β-arrestin-biased ligand of β_2AR (Wisler et al. 2007), induces phosphorylation only at the GRK6 sites. Moreover, the BRET data suggested that distinct phosphorylation patterns on the β_2AR induced by either GRK2 or GRK6 result in different β-arrestin-2 conformations. Collectively, these results provide evidence that the receptor phosphorylation patterns induced by different GRKs establish a 'barcode' that is related to its functional capabilities. This barcode hypothesis has been strengthened by growing evidence from the studies of many other GPCRs, such as M3 muscarinic receptor (Butcher et al. 2011), cannabinoid 1 receptor (Delgado-Peraza et al. 2016), ghrelin receptor (Bouzo-Lorenzo et al. 2016), and fatty acid receptor GPR120 (Prihandoko et al. 2016).

In vitro, binding of the fully phosphorylated vasopressin 2 receptor (V2R) C-tail induces significant conformational changes of both β-arrestin-1 (Nobles et al. 2007) and β-arrestin-2 (Xiao et al. 2004). Recently, a flute model of how phosphorylation pattern of GPCRs was recognized by arrestin and thereafter transduced to downstream effectors through specific conformational changes were revealed by ^{19}F-NMR and unnatural amino acid incorporation approaches (Yang et al. 2015). Not only the phosphorylation states of the receptor, but the ligand-induced specific receptor conformations also contribute to specific arrestin conformations and thus activation. Studies comparing the binding of both visual and non-visual arrestins to four functional forms (inactive unphosphorylated receptor, inactive phosphoreceptor, active unphosphorylated receptor, and active phosphoreceptor) of rhodopsin, β_2AR and M2 muscarinic receptor suggested the existence of an 'activation sensor' in arrestin, which is proposed to mediate the interaction of arrestin with the ligand-induced activated conformation of the receptor (Gurevich et al. 1995; Gurevich and Gurevich 2004). Correspondingly, electron paramagnetic resonance (EPR) studies on rhodopsin-arrestin interactions revealed strong intermolecular contacts between the finger loop, a highly flexible loop connecting β-strands V and VI of arrestin, and the transmembrane core of light-activated phosphorylated rhodopsin, but not that of inactive phosphorylated rhodopsin (Hanson et al. 2006). Moreover, the light-activated unphosphorylated rhodopsin has been further demonstrated to independently induce a unique conformational change in arrestin, which is different from that induced by phosphorylated rhodopsin (Zhuang et al. 2013). Similarly, A NAPol-reconstituted V2R was demonstrated to interact with purified β-arrestin-2 after stimulation with full agonist AVP or β-arrestin-biased

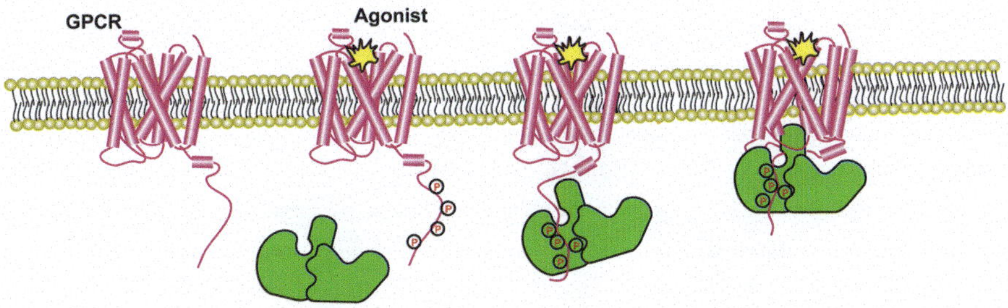

Fig. 16.1 A model of arrestin recruitment and activation by phosphorylated GPCRs. GPCRs are phosphorylated after activation, leading to the recruitment of arrestins. The phosphorylation states of the receptor and ligand-induced receptor conformations collectively determine the active conformations of the arrestins

ligand SR121463, but not with the Gs-biased ligand MCF14 (Rahmeh et al. 2012). Therefore, ligand-specific conformational states of GPCR might not only determine the GRK-mediated phosphorylation pattern but also be directly correlated with the conformation changes of arrestin (Fig. 16.1).

Taken together, these findings revealed how arrestin is activated and then directs distinct functions downstream of the receptor. Ligand-dependent stabilization of distinct receptor conformations leads to specific phosphorylation patterns at the C-terminus and/or intracellular loops, which, together with ligand-specific conformational states of the receptor, define distinct arrestin conformations that might dictate selective signaling pathways (Fig. 16.1).

Structural Insight into Arrestin Conformational Change by Crystallography

Crystal structures of all four arrestins at inactive or basal state have been determined and they demonstrate high sequence and structural homology (Hirsch et al. 1999; Han et al. 2001; Sutton et al. 2005; Zhan et al. 2011). The arrestins essentially consist of N- and C-domains that are built almost entirely from antiparallel β sheets and linked by a relatively flexible 'hinge' region (Fig. 16.2a). The inactive conformation of the arrestin has been suggested to be constrained by two distinct subsets of intramolecular interactions: polar core that consists of five interacting charged residues in the center of arrestin (D26, R169, D290, D297, and R393 in β-arrestin-1) (Vishnivetskiy et al. 1999) and three-element interactions that comprises the β-strand I, α-helix I, and C-terminus of arrestin (Fig. 16.2b, d) (Vishnivetskiy et al. 2000).

Multiple mutagenesis and biophysical studies suggested that visual arrestin undergoes a conformational change upon binding to light-activated phosphorylated

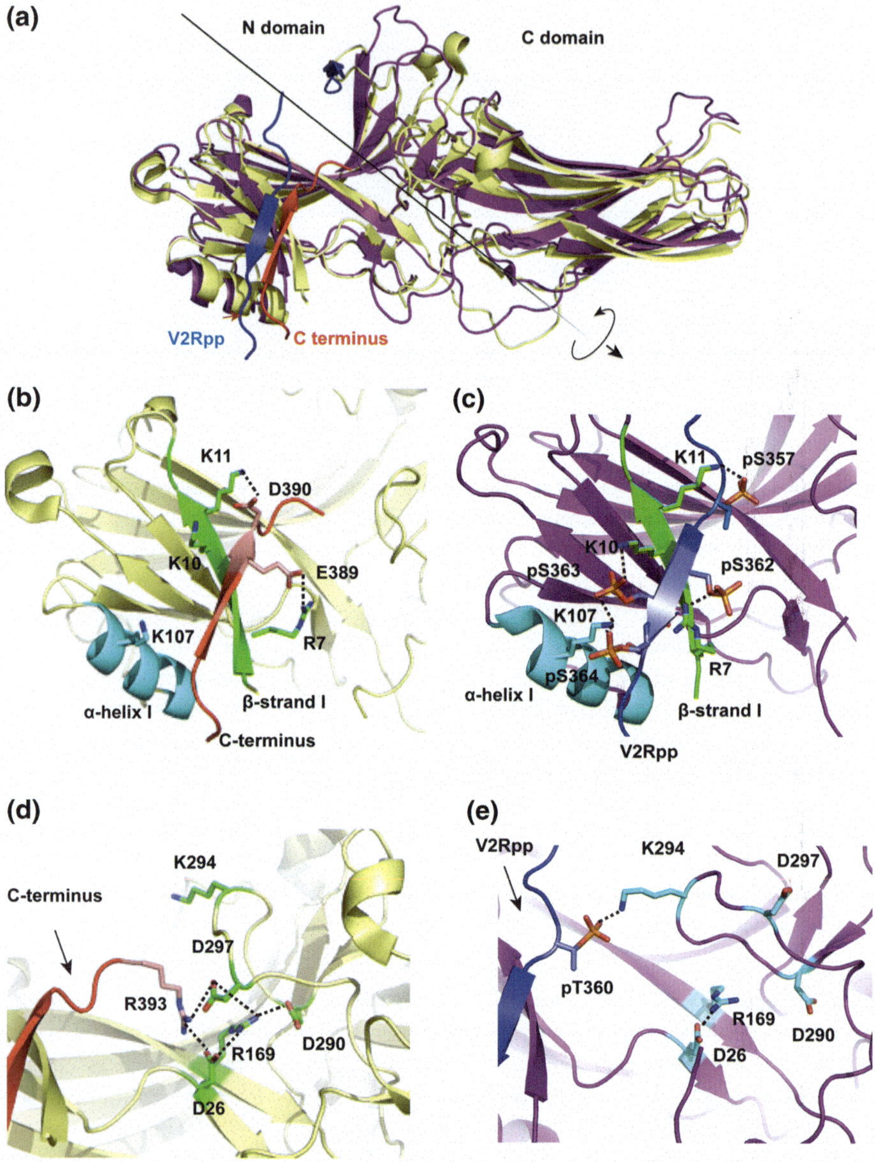

rhodopsin (Gurevich and Gurevich 2003, 2004). Subsequently, direct evidence of a substantial conformational change of both β-arrestins upon activation was obtained using limited tryptic proteolysis and matrix-assisted laser desorption/ionization-time of flight mass spectrometry analysis in the presence of a phosphopeptide derived from the C-terminus of V2R (Xiao et al. 2004; Nobles et al. 2007). Collectively, results from these studies support a model in which both polar core

◂**Fig. 16.2** Comparison of inactive β-arrestin-1 structure and active β-arrestin-1 structure. **a** Overall comparison of the inactive β-arrestin-1 structure (PDB accession code 1G4M; *gold*) and the active β-arrestin-1 structure (PDB accession code 4JQI; *purple*). The β-arrestin-1 C-terminus and the V2Rpp are highlighted in *red* and *blue*, respectively. The three-element interaction in the inactive β-arrestin-1 structure (**b**) and active β-arrestin-1 structure (**c**). The β-arrestin-1 C-terminal b-strand lying along the three-element interaction in the inactive conformation is displaced by the C-terminus of V2Rpp upon activation. The polar core in the inactive β-arrestin-1 structure (**d**) and active β-arrestin-1 structure (**e**). Upon V2Rpp binding, C-terminal residue R393 of β-arrestin-1is displaced, contributing to the disruption of the polar core

and three-element interactions might be disrupted during the activation of the arrestin (Gurevich and Gurevich 2006; Ostermaier et al. 2014).

Two X-ray crystal structures of activated arrestins, including pre-activated arrestin-1 (Kim et al. 2013) and V2R phosphopeptide (V2Rpp)-bound β-arrestin-1 (Shukla et al. 2013), were determined in 2013. In the latter study, the crystal structure of activated β-arrestin-1 revealed the displacement of β-arrestin-1 C-terminus by V2Rpp that binds to the N-domain as an antiparallel β-strand at a similar location to the β-arrestin-1 C-terminus in inactive structure but with a reversed direction (Fig. 16.2a, c). The replacement of the C-terminus of arrestin with V2Rpp disrupts both 3-element interaction and the polar core that restrain the arrestin in the inactive state (Fig. 16.2c, e). It was observed that two highly conserved residues on the β-stand I, the K10 and K11, which interact with F388 and F391 and are proposed to be a key component of three-element interaction in the inactive structure (Gurevich and Gurevich 2006), make instead charge-charge interactions with phosphates of V2Rpp, which therefore disrupt the 3-element interaction (Fig. 16.2c). Displacement of C-terminus of β-arrestin-1 also removes the R393 of the C-terminus, therefore disables the charge-charge interaction of the R393 with the D26 and D297, and distorts the lariat loop to form the new interaction between the pT360 and K294 (Fig. 16.2e). These changes disrupt the polar core. In the active structure of the V2Rpp/arrestin complex, the disruption of the 3-element interaction and the polar core in arrestin release the N-/C-domain associations and enable 20° twisting of the domains relative to each other. The twisting of these two domains of arrestin not only repositions three loops, including the finger loop, the middle loop and the lariat loop, which may facilitate receptor core interactions; but also exposes substantial regions that might enable downstream effectors recruitment (Sommer et al. 2012; Vishnivetskiy et al. 2013; Zhuo et al. 2014).

A similar twisting between the N-/C-domains of the visual arrestin (arrestin-1 or v-arrestin) was also observed in the recently solved crystal structure of rhodopsin/v-arrestin-phosphorylation-independent-active-mutant fusion complex obtained by serial femtosecond X-ray laser crystallography (Fig. 16.3) (Kang et al. 2015). Conformational changes of the finger loop, middle loop and the region between the I157-K168 enable the interaction of the rhodopsin with the v-arrestin. In detail, whereas the interaction between the rhodopsin C-terminus and v-arrestin N-domain was not observed, this structure revealed three patches of

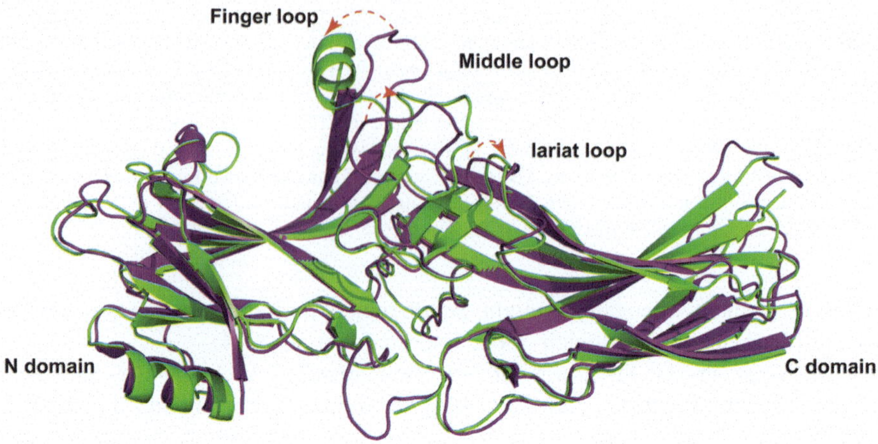

Fig. 16.3 Structural comparison of the active β-arrestin-1 and visual arrestin-1. The phospho-tail (V2Rpp)-activated β-arrestin-1 strucutre (PDB accession code 4JQI; *purple*) and the activated visual arrestin structure (PDB accession code 4ZWJ; *green*) in the complex with rhodopsin are superimposed

rhodopsin-v-arrestin interfaces, which include the interaction of the finger loop of v-arrestin with the intracellular loop-1 (ICL1), TM7 and H8 of the rhodopsin, the accommodation of rhodopsin ICL2 by the middle and C-loops of v-arrestin, and the engagement of β-strand following finger loop and the N-terminal β-strand 6 with the TM5, TM6, and ICL3 of rhodopsin. Interestingly, the structure of the v-arrestin bound to rhodopsin and β-arrestin-1 bound to V2Rpp are very similar other than the receptor interaction regions, suggesting activated arrestins assume similar conformations (Fig. 16.3). However, this result does not support the idea that distinct arrestin functions result from different receptor activation and phosphorylation by different GRKs. Therefore, the low-resolution crystal structures lack important detailed information of structural changes in arrestin to explain how it mediates differential signaling.

Conformational Changes in Arrestin and the Correlation with Downstream Functions

Although mounting evidence has indicated that different arrestin conformations are coupled to distinct functional outcomes, the precise correlation between them is not well established. Since the identification of clathrin as the first non-receptor binding partner of arrestins, a variety of trafficking and signaling proteins, such as c-Src, MAPK, and ASK1, have been reported to interact with arrestin (Goodman et al. 1996; Lefkowitz and Shenoy 2005). The binding sites of clathrin and AP2 on arrestin have been well characterized: both are localized in the C-terminus of β-arrestins (Kim and Benovic 2002; Schmid et al. 2006; Kang et al. 2009). In contrast, few binding sites of other non-receptor binding partners have been precisely mapped, although particular residues in β-arrestins are suggested to participate in the binding of MEK1 or Raf1 (Meng et al. 2009; Coffa et al. 2011a, b). It has been demonstrated that most of these non-receptor binding partners except for clathrin and AP2 interact with both N-/C-domains of arrestin (Song et al. 2007, 2009) and therefore, the relative rotation between these two domains and rearrangement of the arrestin conformation, which have been confirmed in the structural studies, might explain the structural preference of certain partners that serve as effector molecules leading to distinct functional outcomes (Gurevich and Gurevich 2003, 2013). However, despite the continuous improvement and strengthening of barcode hypothesis, the detailed mechanism by which arrestin transmits signals encoded in the receptor to downstream effector molecules still remains largely unknown.

To better detect the conformational changes in arrestin and further explore how arrestin recognizes specific phosphorylation patterns in the receptor and translates them into distinct functional outcomes, we developed a series of structural sensors in β-arrestin-1 by incorporating the unnatural amino acid 3, 5-difluorotyrosin (F2Y) at specific locations and detected phospho-interaction patterns and residue-specific structural information in β-arrestin-1 using ^{19}F-NMR spectroscopy (Yang et al. 2015). In addition to V2Rpp, we also synthesized specific GRK2-, GRK6-, or PKA-phosphorylated β_2AR C-terminal fragments (GRK2pp, GRK6pp, and PKApp, respectively) and examined their effects on the biochemical properties of β-arrestin-1. We demonstrated that β-arrestin-1 reads phospho-messages in receptor C-tail with its concave surface, which harbors at least ten potential phosphate-binding sites (numbered 1–7 according to the binding mode of V2Rpp to β-arrestin-1 in the V2Rpp/β-arrestin-1 complex (Shukla et al. 2014), with A1–A3 indicating potential additional phosphate-binding sites at the N-terminus). Although all the GRK-phosphopeptides (GRKpps) but not PKApp interact with phosphate-binding site 1 of β-arrestin-1, our results revealed distinct phospho-interaction patterns between different GRKpps and β-arrestin-1 at the other binding sites, which are coupled to selective functional outcomes (Fig. 16.4a). Whereas GRK2pp interacts with β-arrestin-1 through binding sites 1–4–6–7 and promotes clathrin recruitment and receptor endocytosis, GRK6pp interacts with β-arrestin-1 in a different 1–5 pattern and elicits Src signaling (Fig. 16.4a). Mutations of key residues in specific phosphate-binding sites selectively eliminate corresponding β-arresin-1-mediated functions downstream of several GPCRs, including β2AR, cholecystokinin type-A receptor (CCKAR), and somatostatin receptor type 2 (SSTR2). These data suggested that specific phospho-patterns were recognized by arrestins and then translated to different cellular signaling pathways through distinct downstream effectors.

Using 7 ^{19}F-NMR probes incorporated into the specific arrestin sites other than phospho-interaction regions, we were able to detect different conformational states that were induced by binding of different GRKpps. Binding of the two distinct GRK2pp phospho-peptides induced conformational change at the Y249 in the loop

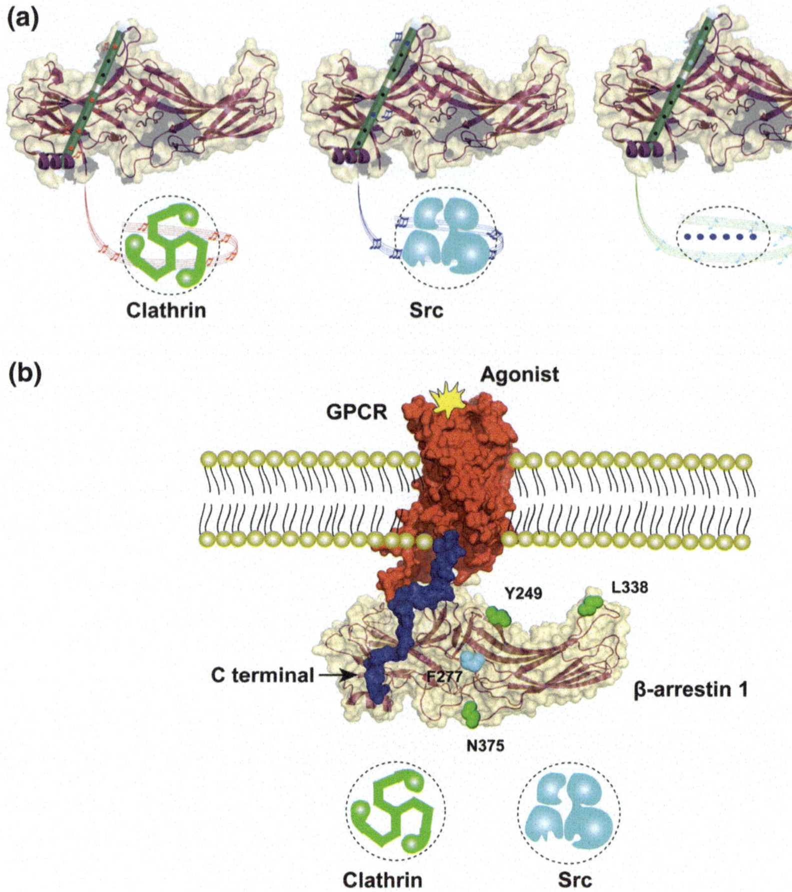

Fig. 16.4 Conformational changes of arrestin induced by different receptor phosphorylation patterns and their correlation with distinct cellular functions. **a** The flute model for the phospho-decision mechanism of the arrestin signaling. A phospho-barcode of 1–4–6–7 directs clathrin recruitment and endocytic function of arrestin, whereas a phospho-barcode of 1–5 directs Src recruitment and signaling. There are potentially more than 1000 phospho-patterns that a single arrestin could recognize. **b** Conformationl changes of arrestin correlate to its distinct functions. The conformational states of the Y249, L338 and N375 are recognized by clathrin and correlated to the endocytic function of β-arrestin-1, whereas the conformational state of F277 is correlated to arrestin-mediated Src signaling

between β-strands XV and XVI, the L338 in the splice loop, and the N375 in the C-terminal region of β-arrestin-1 (Fig. 16.4b). These structural states could be specifically recognized by the subsequent binding of clathrin. In contrast, binding of the GRK6pp to β-arrestin-1 caused a chemical shift at F277 in the end of the lariat loop, whereas binding of the GRK2pps caused no significant conformational change at this site (Fig. 16.4b). Deletion of the partial lariat loop encompassing

F277 abolished the GRK6pp-induced recruitment of Src to arrestin. Collectively, these data have clarified the receptor phospho-coding mechanism, by which changes in a GPCR phosphorylation pattern are translated to distinct conformations of arrestin that could be further recognized by different downstream effector molecules.

In addition to β-arrestin-1, the conformational change of β-arrestin-2 have also been investigated downstream of several receptors in cellular systems. A very recent study monitored the conformational changes of β-arrestin-2 using a panel of intramolecular fluorescein arsenical hairpin (FlAsH) BRET reporters in cells (Lee et al. 2016). Studies of six different types of GPCRs demonstrated the existence of β-arresin-2 "conformational signature" that was indicated by the changes in BRET efficacy from multiple vantage points. This β-arresin-2-FlAsH signature is conserved between GPCRs with similar arrestin binding/signaling characteristics and the changes in BRET efficacy at selected positions correlate with distinct arrestin functions, for example, trafficking pattern of GPCR-arrestin complex and arrestin-dependent ERK1/2 activation. Therefore, these data, together with the results of our study, provide considerable insight into the correlation between specific arrestin conformations and selective arrestin functions and pave the way to the future studies of the detailed functional roles of the conformational changes of arrestin.

Although receptor binding and arrestin conformational change are necessary for the engagement of arrestin with multiple downstream effectors, such as ERK1/2 (Luttrell et al. 2001; Coffa et al. 2011a, b), it has been reported that some binding partners could interact with arrestin in both inactive and active conformations (Song et al. 2009; Ahmed et al. 2011; Gurevich and Gurevich 2013), with some proteins, such as MEK1, even binding to arrestins in both conformations equally well (Coffa et al. 2011a, b; Gurevich and Gurevich 2014). This has added a new layer of complexity in the arrestin-dependent conformational signaling. Moreover, the two isoforms of β-arrestin, which share more than 70% sequence identity and high structural similarity, have been demonstrated to be functionally non-redundant in GPCR regulation by accumulating evidence (Srivastava et al. 2015). Whereas both β-arrestins facilitate the signaling in c-Raf1-MEK1-ERK1/2 cascade, only β-arrestin-2 promotes the signaling in ASK1-MKK4-JNK3 cascade (McDonald et al. 2000). Receptor-specific reciprocal regulation of selective signaling pathways by these two β-arrestins has also been reported. For example, siRNA knockdown of β-arrestin-2 attenuates PTH1R-mediated ERK1/2 activation, whereas knockdown of β-arrestin-1 yields opposite effect (Ahn et al. 2004). Therefore, the interaction between arrestin and effector molecules and the arrestin-dependent signal transduction might be much more complicated than expected. To further decipher the molecular mechanism underlying arrestin conformational signaling, more detailed information from the potential structures, using crystallography, electron microscopy or NMR approaches, of GPCR with arrestin and effectors is required.

References

Ahmed MR, Zhan X, Song X, Kook S, Gurevich VV, Gurevich EV (2011) Ubiquitin ligase parkin promotes Mdm2-arrestin interaction but inhibits arrestin ubiquitination. Biochemistry 50 (18):3749–3763

Ahn S, Wei H, Garrison TR, Lefkowitz RJ (2004) Reciprocal regulation of angiotensin receptor-activated extracellular signal-regulated kinases by beta-arrestins 1 and 2. J Biol Chem 279(9):7807–7811

Attramadal H, Arriza JL, Aoki C, Dawson TM, Codina J, Kwatra MM, Snyder SH, Caron MG, Lefkowitz RJ (1992) Beta-arrestin2, a novel member of the arrestin/beta-arrestin gene family. J Biol Chem 267(25):17882–17890

Benovic JL, Strasser RH, Caron MG, Lefkowitz RJ (1986) Beta-adrenergic receptor kinase: identification of a novel protein kinase that phosphorylates the agonist-occupied form of the receptor. Proc Natl Acad Sci USA 83(9):2797–2801

Bouzo-Lorenzo M, Santo-Zas I, Lodeiro M, Nogueiras R, Casanueva FF, Castro M, Pazos Y, Tobin AB, Butcher AJ, Camina JP (2016) Distinct phosphorylation sites on the ghrelin receptor, GHSR1a, establish a code that determines the functions of ss-arrestins. Sci Rep 6:22495

Butcher AJ, Prihandoko R, Kong KC, McWilliams P, Edwards JM, Bottrill A, Mistry S, Tobin AB (2011) Differential G-protein-coupled receptor phosphorylation provides evidence for a signaling bar code. J Biol Chem 286(13):11506–11518

Coffa S, Breitman M, Hanson SM, Callaway K, Kook S, Dalby KN, Gurevich VV (2011a) The effect of arrestin conformation on the recruitment of c-Raf1, MEK1, and ERK1/2 activation. PLoS One 6(12):e28723

Coffa S, Breitman M, Spiller BW, Gurevich VV (2011b) A single mutation in arrestin-2 prevents ERK1/2 activation by reducing c-Raf1 binding. Biochemistry 50(32):6951–6958

Craft CM, Whitmore DH, Wiechmann AF (1994) Cone arrestin identified by targeting expression of a functional family. J Biol Chem 269(6):4613–4619

Delgado-Peraza F, Ahn KH, Nogueras-Ortiz C, Mungrue IN, Mackie K, Kendall DA, Yudowski GA (2016) Mechanisms of biased beta-arrestin-mediated signaling downstream from the cannabinoid 1 receptor. Mol Pharmacol 89(6):618–629

Farrens DL, Altenbach C, Yang K, Hubbell WL, Khorana HG (1996) Requirement of rigid-body motion of transmembrane helices for light activation of rhodopsin. Science 274:768–770

Gao H, Sun Y, Wu Y, Luan B, Wang Y, Qu B, Pei G (2004) Identification of beta-arrestin2 as a G protein-coupled receptor-stimulated regulator of NF-kappaB pathways. Mol Cell 14(3): 303–317

Garcia DE, Brown S, Hille B, Mackie K (1998) Protein kinase C disrupts cannabinoid actions by phosphorylation of the CB1 cannabinoid receptor. J Neurosci 18(8):2834–2841

Goodman OB Jr, Krupnick JG, Santini F, Gurevich VV, Penn RB, Gagnon AW, Keen JH, Benovic JL (1996) Beta-arrestin acts as a clathrin adaptor in endocytosis of the beta2-adrenergic receptor. Nature 383(6599):447–450

Gurevich VV, Dion SB, Onorato JJ, Ptasienski J, Kim CM, Sterne-Marr R, Hosey MM, Benovic JL (1995) Arrestin interactions with G protein-coupled receptors. Direct binding studies of wild type and mutant arrestins with rhodopsin, beta 2-adrenergic, and m2 muscarinic cholinergic receptors. J Biol Chem 270(2):720–731

Gurevich VV, Gurevich EV (2003) The new face of active receptor bound arrestin attracts new partners. Structure 11(9):1037–1042

Gurevich VV, Gurevich EV (2004) The molecular acrobatics of arrestin activation. Trends Pharmacol Sci 25(2):105–111

Gurevich VV, Gurevich EV (2006) The structural basis of arrestin-mediated regulation of G-protein-coupled receptors. Pharmacol Ther 110(3):465–502

Gurevich VV, Gurevich EV (2013) Structural determinants of arrestin functions. Prog Mol Biol Transl Sci 118:57–92

Gurevich VV, Gurevich EV (2014) Extensive shape shifting underlies functional versatility of arrestins. Curr Opin Cell Biol 27:1–9

Han M, Gurevich VV, Vishnivetskiy SA, Sigler PB, Schubert C (2001) Crystal structure of beta-arrestin at 1.9 A: possible mechanism of receptor binding and membrane translocation. Structure 9(9):869–880

Hanson SM, Francis DJ, Vishnivetskiy SA, Kolobova EA, Hubbell WL, Klug CS, Gurevich VV (2006) Differential interaction of spin-labeled arrestin with inactive and active phospho-rhodopsin. Proc Natl Acad Sci U S A 103(13):4900–4905

Hirsch JA, Schubert C, Gurevich VV, Sigler PB (1999) The 2.8 A crystal structure of visual arrestin: a model for arrestin's regulation. Cell 97(2):257–269

Kang DS, Kern RC, Puthenveedu MA, von Zastrow M, Williams JC, Benovic JL (2009) Structure of an arrestin2-clathrin complex reveals a novel clathrin binding domain that modulates receptor trafficking. J Biol Chem 284(43):29860–29872

Kang Y, Zhou XE, Gao X, He Y, Liu W, Ishchenko A, Barty A, White TA, Yefanov O, Han GW, Xu Q, de Waal PW, Ke J, Tan MH, Zhang C, Moeller A, West GM, Pascal BD, Van Eps N, Caro LN, Vishnivetskiy SA, Lee RJ, Suino-Powell KM, Gu X, Pal K, Ma J, Zhi X, Boutet S, Williams GJ, Messerschmidt M, Gati C, Zatsepin NA, Wang D, James D, Basu S, Roy-Chowdhury S, Conrad CE, Coe J, Liu H, Lisova S, Kupitz C, Grotjohann I, Fromme R, Jiang Y, Tan M, Yang H, Li J, Wang M, Zheng Z, Li D, Howe N, Zhao Y, Standfuss J, Diederichs K, Dong Y, Potter CS, Carragher B, Caffrey M, Jiang H, Chapman HN, Spence JC, Fromme P, Weierstall U, Ernst OP, Katritch V, Gurevich VV, Griffin PR, Hubbell WL, Stevens RC, Cherezov V, Melcher K, Xu HE (2015) Crystal structure of rhodopsin bound to arrestin by femtosecond X-ray laser. Nature 523(7562):561–567

Kim J, Ahn S, Ren XR, Whalen EJ, Reiter E, Wei H, Lefkowitz RJ (2005) Functional antagonism of different G protein-coupled receptor kinases for beta-arrestin-mediated angiotensin II receptor signaling. Proc Natl Acad Sci U S A 102(5):1442–1447

Kim YJ, Hofmann KP, Ernst OP, Scheerer P, Choe HW, Sommer ME (2013) Crystal structure of pre-activated arrestin p44. Nature 497(7447):142–146

Kim YM, Benovic JL (2002) Differential roles of arrestin-2 interaction with clathrin and adaptor protein 2 in G protein-coupled receptor trafficking. J Biol Chem 277(34):30760–30768

Krupnick JG, Benovic JL (1998) The role of receptor kinases and arrestins in G protein-coupled receptor regulation. Annu Rev Pharmacol Toxicol 38:289–319

Laporte SA, Oakley RH, Zhang J, Holt JA, Ferguson SG, Caron MG, Barak LS (1999) The 2-adrenergic receptor/arrestin complex recruits the clathrin adaptor AP-2 during endocytosis. Proc Nat Acad Sci U S A 96:3712–3717

Lee MH, Appleton KM, Strungs EG, Kwon JY, Morinelli TA, Peterson YK, Laporte SA, Luttrell LM (2016) The conformational signature of beta-arrestin2 predicts its trafficking and signalling functions. Nature 531(7596):665–668

Lefkowitz RJ, Shenoy SK (2005) Transduction of receptor signals by beta-arrestins. Science 308 (5721):512–517

Lohse MJ, Benovic JL, Codina J, Caron MG, Lefkowitz RJ (1990) beta-Arrestin: a protein that regulates beta-adrenergic receptor function. Science 248(4962):1547–1550

Luttrell LM, Ferguson SS, Daaka Y, Miller WE, Maudsley S, Della Rocca GJ, Lin F, Kawakatsu H, Owada K, Luttrell DK, Caron MG, Lefkowitz RJ (1999) Beta-arrestin-dependent formation of beta2 adrenergic receptor-Src protein kinase complexes. Science 283(5402):655–661

Luttrell LM, Roudabush FL, Choy EW, Miller WE, Field ME, Pierce KL, Lefkowitz RJ (2001) Activation and targeting of extracellular signal-regulated kinases by beta-arrestin scaffolds. Proc Natl Acad Sci U S A 98(5):2449–2454

McDonald PH, Chow CW, Miller WE, Laporte SA, Field ME, Lin FT, Davis RJ, Lefkowitz RJ (2000) Beta-arrestin 2: a receptor-regulated MAPK scaffold for the activation of JNK3. Science 290(5496):1574–1577

McDonald PH, Cote NL, Lin FT, Premont RT, Pitcher JA, Lefkowitz RJ (1999) Identification of NSF as a beta-arrestin1-binding protein. Implications for beta2-adrenergic receptor regulation. J Biol Chem 274:10677–10680

Meng D, Lynch MJ, Huston E, Beyermann M, Eichhorst J, Adams DR, Klussmann E, Houslay MD, Baillie GS (2009) MEK1 binds directly to betaarrestin1, influencing both its phosphorylation by ERK and the timing of its isoprenaline-stimulated internalization. J Biol Chem 284(17):11425–11435

Ning SL, Zheng WS, Su J, Liang N, Li H, Zhang DL, Liu CH, Dong JH, Zhang ZK, Cui M, Hu QX, Chen CC, Liu CH, Wang C, Pang Q, Chen YX, Yu X, Sun JP (2015) Different downstream signalling of CCK1 receptors regulates distinct functions of CCK in pancreatic beta cells. Br J Pharmacol 172(21):5050–5067

Nobles KN, Guan Z, Xiao K, Oas TG, Lefkowitz RJ (2007) The active conformation of beta-arrestin1: direct evidence for the phosphate sensor in the N-domain and conformational differences in the active states of beta-arrestins1 and -2. J Biol Chem 282(29):21370–21381

Nobles KN, Xiao K, Ahn S, Shukla AK, Lam CM, Rajagopal S, Strachan RT, Huang TY, Bressler EA, Hara MR, Shenoy SK, Gygi SP, Lefkowitz RJ (2011) Distinct phosphorylation sites on the beta(2)-adrenergic receptor establish a barcode that encodes differential functions of beta-arrestin. Sci Signal 4(185):ra51

Nuber S, Zabel U, Lorenz K, Nuber A, Milligan G, Tobin AB, Lohse MJ, Hoffmann C (2016) beta-Arrestin biosensors reveal a rapid, receptor-dependent activation/deactivation cycle. Nature 531(7596):661–664

Ostermaier MK, Schertler GF, Standfuss J (2014) Molecular mechanism of phosphorylation-dependent arrestin activation. Curr Opin Struct Biol 29:143–151

Pierce KL, Premont RT, Lefkowitz RJ (2002) Seven-transmembrane receptors. Nat Rev Mol Cell Biol 3(9):639–650

Pitcher JA, Freedman NJ, Lefkowitz RJ (1998) G protein-coupled receptor kinases. Annu Rev Biochem 67:653–692

Prihandoko R, Alvarez-Curto E, Hudson BD, Butcher AJ, Ulven T, Miller AM, Tobin AB, Milligan G (2016) Distinct phosphorylation clusters determine the signaling outcome of free fatty acid receptor 4/G protein-coupled receptor 120. Mol Pharmacol 89(5):505–520

Rahmeh R, Damian M, Cottet M, Orcel H, Mendre C, Durroux T, Sharma KS, Durand G, Pucci B, Trinquet E, Zwier JM, Deupi X, Bron P, Baneres JL, Mouillac B, Granier S (2012) Structural insights into biased G protein-coupled receptor signaling revealed by fluorescence spectroscopy. Proc Natl Acad Sci U S A 109(17):6733–6738

Rajagopal S, Rajagopal K, Lefkowitz RJ (2010) Teaching old receptors new tricks: biasing seven-transmembrane receptors. Nat Rev Drug Discov 9(5):373–386

Reiter E, Ahn S, Shukla AK, Lefkowitz RJ (2012) Molecular mechanism of beta-arrestin-biased agonism at seven-transmembrane receptors. Annu Rev Pharmacol Toxicol 52:179–197

Ren XR, Reiter E, Ahn S, Kim J, Chen W, Lefkowitz RJ (2005) Different G protein-coupled receptor kinases govern G protein and beta-arrestin-mediated signaling of V2 vasopressin receptor. Proc Natl Acad Sci U S A 102(5):1448–1453

Schmid EM, Ford MG, Burtey A, Praefcke GJ, Peak-Chew SY, Mills IG, Benmerah A, McMahon HT (2006) Role of the AP2 beta-appendage hub in recruiting partners for clathrin-coated vesicle assembly. PLoS Biol 4(9):e262

Shenoy SK, Lefkowitz RJ (2003) Multifaceted roles of beta-arrestins in the regulation of seven-membrane-spanning receptor trafficking and signalling. Biochem J 375(Pt 3):503–515

Shukla AK, Manglik A, Kruse AC, Xiao K, Reis RI, Tseng WC, Staus DP, Hilger D, Uysal S, Huang LY, Paduch M, Tripathi-Shukla P, Koide A, Koide S, Weis WI, Kossiakoff AA, Kobilka BK, Lefkowitz RJ (2013) Structure of active beta-arrestin-1 bound to a G-protein-coupled receptor phosphopeptide. Nature 497(7447):137–141

Shukla AK, Violin JD, Whalen EJ, Gesty-Palmer D, Shenoy SK, Lefkowitz RJ (2008) Distinct conformational changes in beta-arrestin report biased agonism at seven-transmembrane receptors. Proc Natl Acad Sci U S A 105(29):9988–9993

Shukla AK, Westfield GH, Xiao K, Reis RI, Huang LY, Tripathi-Shukla P, Qian J, Li S, Blanc A, Oleskie AN, Dosey AM, Su M, Liang CR, Gu LL, Shan JM, Chen X, Hanna R, Choi M, Yao XJ, Klink BU, Kahsai AW, Sidhu SS, Koide S, Penczek PA, Kossiakoff AA, Woods VL Jr, Kobilka BK, Skiniotis G, Lefkowitz RJ (2014) Visualization of arrestin recruitment by a G-protein-coupled receptor. Nature 512(7513):218–222

Smith JS, Rajagopal S (2016) The beta-arrestins: multifunctional regulators of G protein-coupled receptors. J Biol Chem 291(17):8969–8977

Sommer ME, Hofmann KP, Heck M (2012) Distinct loops in arrestin differentially regulate ligand binding within the GPCR opsin. Nat Commun 3:995

Song X, Coffa S, Fu H, Gurevich VV (2009) How does arrestin assemble MAPKs into a signaling complex? J Biol Chem 284(1):685–695

Song X, Gurevich EV, Gurevich VV (2007) Cone arrestin binding to JNK3 and Mdm2: conformational preference and localization of interaction sites. J Neurochem 103(3):1053–1062

Srivastava A, Gupta B, Gupta C, Shukla AK (2015) Emerging functional divergence of beta-arrestin isoforms in GPCR function. Trends Endocrinol Metab 26(11):628–642

Sutton RB, Vishnivetskiy SA, Robert J, Hanson SM, Raman D, Knox BE, Kono M, Navarro J, Gurevich VV (2005) Crystal structure of cone arrestin at 2.3 A: evolution of receptor specificity. J Mol Biol 354(5):1069–1080

Thomsen AR, Plouffe B, Cahill TJ 3rd, Shukla AK, Tarrasch JT, Dosey AM, Kahsai AW, Strachan RT, Pani B, Mahoney JP, Huang L, Breton B, Heydenreich FM, Sunahara RK, Skiniotis G, Bouvier M, Lefkowitz RJ (2016) GPCR-G protein-beta-arrestin super-complex mediates sustained G protein signaling. Cell

Tobin AB, Butcher AJ, Kong KC (2008) Location, location, location…site-specific GPCR phosphorylation offers a mechanism for cell-type-specific signalling. Trends Pharmacol Sci 29 (8):413–420

Tran TM, Friedman J, Qunaibi E, Baameur F, Moore RH, Clark RB (2004) Characterization of agonist stimulation of cAMP-dependent protein kinase and G protein-coupled receptor kinase phosphorylation of the beta2-adrenergic receptor using phosphoserine-specific antibodies. Mol Pharmacol 65(1):196–206

Vishnivetskiy SA, Baameur F, Findley KR, Gurevich VV (2013) Critical role of the central 139-loop in stability and binding selectivity of arrestin-1. J Biol Chem 288(17):11741–11750

Vishnivetskiy SA, Paz CL, Schubert C, Hirsch JA, Sigler PB, Gurevich VV (1999) How does arrestin respond to the phosphorylated state of rhodopsin? J Biol Chem 274(17):11451–11454

Vishnivetskiy SA, Schubert C, Climaco GC, Gurevich YV, Velez MG, Gurevich VV (2000) An additional phosphate-binding element in arrestin molecule. Implications for the mechanism of arrestin activation. J Biol Chem 275(52):41049–41057

Walther C, Ferguson SS (2013) Arrestins: role in the desensitization, sequestration, and vesicular trafficking of G protein-coupled receptors. Prog Mol Biol Transl Sci 118:93–113

Wang HM, Dong JH, Li Q, Hu Q, Ning SL, Zheng W, Cui M, Chen TS, Xie X, Sun JP, Yu X (2014) A stress response pathway in mice upregulates somatostatin level and transcription in pancreatic delta cells through Gs and beta-arrestin 1. Diabetologia 57(9):1899–1910

Weller M, Virmaux N, Mandel P (1975) Light-stimulated phosphorylation of rhodopsin in the retina: the presence of a protein kinase that is specific for photobleached rhodopsin. Proc Natl Acad Sci U S A 72(1):381–385

Wilden U, Wust E, Weyand I, Kuhn H (1986) Rapid affinity purification of retinal arrestin (48 kDa protein) via its light-dependent binding to phosphorylated rhodopsin. FEBS Lett 207(2): 292–295

Wisler JW, DeWire SM, Whalen EJ, Violin JD, Drake MT, Ahn S, Shenoy SK, Lefkowitz RJ (2007) A unique mechanism of beta-blocker action: carvedilol stimulates beta-arrestin signaling. Proc Natl Acad Sci U S A 104(42):16657–16662

Xiao K, Shenoy SK, Nobles K, Lefkowitz RJ (2004) Activation-dependent conformational changes in {beta}-arrestin 2. J Biol Chem 279(53):55744–55753

Xiao K, Sun J, Kim J, Rajagopal S, Zhai B, Villen J, Haas W, Kovacs JJ, Shukla AK, Hara MR, Hernandez M, Lachmann A, Zhao S, Lin Y, Cheng Y, Mizuno K, Ma'ayan A, Gygi SP, Lefkowitz RJ (2010) Global phosphorylation analysis of beta-arrestin-mediated signaling downstream of a seven transmembrane receptor (7TMR). Proc Natl Acad Sci U S A 107 (34):15299–15304

Yang F, Yu X, Liu C, Qu CX, Gong Z, Liu HD, Li FH, Wang HM, He DF, Yi F, Song C, Tian CL, Xiao KH, Wang JY, Sun JP (2015) Phospho-selective mechanisms of arrestin conformations and functions revealed by unnatural amino acid incorporation and (19)F-NMR. Nat Commun 6:8202

Zhan X, Gimenez LE, Gurevich VV, Spiller BW (2011) Crystal structure of arrestin-3 reveals the basis of the difference in receptor binding between two non-visual subtypes. J Mol Biol 406 (3):467–478

Zhuang T, Chen Q, Cho MK, Vishnivetskiy SA, Iverson TM, Gurevich VV, Sanders CR (2013) Involvement of distinct arrestin-1 elements in binding to different functional forms of rhodopsin. Proc Natl Acad Sci U S A 110(3):942–947

Zhuo Y, Vishnivetskiy SA, Zhan X, Gurevich VV, Klug CS (2014) Identification of receptor binding-induced conformational changes in non-visual arrestins. J Biol Chem 289(30):20991–21002

Chapter 17
Is Signaling Specificity Encoded in Arrestin Conformation?

Carsten Hoffmann, Martin J. Lohse and Louis M. Luttrell

Abstract The visual/β-arrestins perform remarkably diverse roles in cells. Possessing the capacity to detect and bind hundreds of different activated GPCRs, they are integral to the control of GPCR desensitization, internalization and intracellular trafficking. At the same time, they are able to bind and localize dozens of cargo proteins, including signaling pathway intermediates that affect the tonic level of pathway activity and in some cases confer GPCR-dependent regulation. A fundamental question in biology is how proteins like arrestins can achieve the conformational flexibility necessary to interact with so many different partners while maintaining the specificity necessary for fidelity in signal transduction.

C. Hoffmann
Institute of Pharmacology and Toxicology, University of Würzburg,
Versbacher Str. 9, 97078 Würzburg, Germany
e-mail: c.hoffmann@toxi.uni-wuerzburg.de

C. Hoffmann
Rudolf Virchow Center, University of Würzburg, Versbacher Str. 9,
97078 Würzburg, Germany

M.J. Lohse
Max Delbrück Center for Molecular Medicine, Robert-Rössle-Str. 10,
13092 Berlin, Germany
e-mail: Martin.Lohse@mdc-berlin.de

L.M. Luttrell
Department of Medicine, Medical University of South Carolina,
Charleston, SC 29425, USA

L.M. Luttrell
Research Service of the Ralph H. Johnson Veterans Affairs Medical Center,
Charleston, SC 29401, USA

L.M. Luttrell (✉)
Division of Endocrinology, Diabetes and Medical Genetics,
Medical University of South Carolina, 96 Jonathan Lucas Street,
MSC 624, Charleston, SC 29426, USA
e-mail: luttrell@musc.edu

V.V. Gurevich (ed.), *The Structural Basis of Arrestin Functions*,
DOI 10.1007/978-3-319-57553-7_17

Recent data, both from static structures of arrestins in their basal and 'activated' conformations, and dynamic resonance energy transfer measurements, have begun to provide answers. Crystallographic and mutagenesis data indicate that the conformational shifts occurring upon receptor engagement are determined by contact between the arrestin globular domains and a relatively few residues within the GPCR intracellular loops. These contacts, in turn, impose receptor-specific arrestin conformations that influence the avidity with which it binds the receptor and its ability to engage certain downstream partners. Differences in ligand structure are likewise encoded in receptor conformation and transmitted to the arrestin, providing the physical basis for ligand 'bias'. Resonance energy transfer data also indicate that arrestins maintain their 'active' conformation for a period of time after letting go of the receptor, raising the possibility that arrestins, like heterotrimeric G proteins, might be activated catalytically.

Keywords Arrestin · GPCR · Ligand bias · Signal transduction · Conformational change

Introduction

G protein-coupled receptors (GPCRs) are viewed broadly as components of an allosterically regulated signal transduction machinery whose function is to bind molecules at one location, i.e. extracellular ligands, and change shape so as to affect the conformation of other molecules at another location, i.e. intracellular effectors (Kenakin 2012). The plasma membrane of a cell may display receptors for dozens of different ligands, and each receptor may couple to multiple G protein and non-G protein effectors. Information about the external environment is contained both in ligand structure, which determines which GPCRs are activated, and concentration, which determines the proportion of receptors in the active state. Further subtlety arises in that most GPCRs can adopt multiple 'active' conformations that vary in the efficiency with which they couple to downstream effectors, such that differences in the structure of ligands binding the same GPCR can also change, or 'bias', the cellular response (Kenakin and Miller 2010; Luttrell and Kenakin 2011).

Despite their relatively simple architecture and lack of intrinsic catalytic activity, the visual/β-arrestins play remarkably diverse roles in the modulation of GPCR signaling (Ferguson 2001; Gurevich and Gurevich 2013). They possess three properties essential to these roles: the flexibility to bind multiple cargo proteins; the capacity to exist in different intracellular pools wherein they adopt different conformations; and the ability to recognize activated GPCRs. This enables them to move between cellular compartments and deliver different cargos to different locations under the control of environmental stimuli. Ligand-induced changes in GPCR conformation, followed by G protein-coupled receptor kinase (GRK)-mediated phosphorylation, create arrestin docking sites on the plasma membrane.

In response, arrestins relinquish their low affinity interactions with microtubules (Nair et al. 2004; Hanson et al. 2007) and cytosolic Ca^{2+}-calmodulin (Wu et al. 2006) and inositol hexakisphosphate (Milano et al. 2006; Zhuang et al. 2010) in favor of higher affinity interactions with GPCRs. Because the sites of interaction with receptors overlap (Kang et al. 2015), the current view is that arrestin binding displaces heterotrimeric G proteins, promoting receptor desensitization, and in the case of the β-arrestins targets receptors for internalization via clathrin-coated pits (Goodman et al. 1996; Laporte et al. 1999). Arrestins also function as ligand-regulated scaffolds that recruit a multitude of cargos, including numerous kinases, phosphatases, ubiquitin ligases, and other enzymatic effectors into GPCR-based 'signalsomes' (Shenoy and Lefkowitz 2005a; Luttrell and Gesty-Palmer 2010).

While it is clear that the major function(s) performed by arrestins vary between receptors, the mechanistic basis for how different GPCRs specify divergent arrestin functions is not well understood. The developing story, based upon crystallographic structures of free and GPCR-bound arrestins, mutagenesis studies, and dynamic resonance energy measurements, is that arrestin function is encoded by specific conformational shifts that occur when they dock with activated phosphorylated GPCRs. Consistent with a general allosteric model of GPCR function (Kenakin 2009; Kenakin and Miller 2010) several factors, including ligand structure, intracellular domain architecture, and potentially activation-dependent post translational receptor modifications, are represented in the texture of the arrestin binding surface, which in turn stabilizes arrestin conformations that reflect the avidity of the GPCR-arrestin interaction and influence its ability to recruit and activate signaling cargos. In this chapter, we discuss the factors that determine the conformation of activated arrestins and evidence that arrestins can adopt distinct conformations that permit them to perform their functions in a ligand and receptor-specific manner.

The Structural Basis of Arrestin Activation

The four visual/β-arrestins are semi-bisymmetric soluble proteins composed of 20 β strands condensing into two β strand sandwich structures that form N- and C-domains, each with a concave basket, connected by a short hinge (Vishnivetskiy et al. 2002; Aubry et al. 2009; Fig. 17.1). Arrestin activation is a two-step process; an initial interaction with receptor-attached phosphates that destabilizes the inactive conformation, followed by conformational rearrangements that enable the arrestin to interact with receptor intracellular domains. These changes not only enable the arrestin to bind activated receptors, but also change its affinity for various cargo proteins.

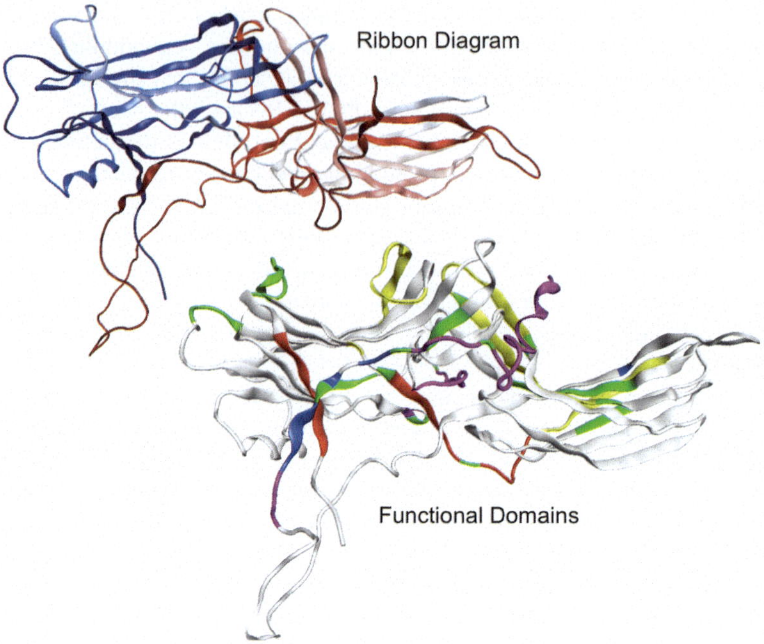

Fig. 17.1 Visual/β-arrestins share a conserved domain architecture. The *upper panel* depicts a ribbon diagram indicating visual arrestin folding from amino (*blue*) to carboxy (*red*). The *lower panel* is a domain diagram showing areas of functional importance from X-ray and mutagenesis studies. Domains are colored such that *red regions* are involved in receptor binding, *green regions* are involved in self-association, *blue* are important in arrestin activation, and *yellow regions* interact with microtubules. Analysis and visualization was performed using MOE 2014.0

Arrestin Activation: A Stepwise Process

In the inactive state, the polar core of arrestin is stabilized by charge interactions involving the gate loop (residues 296–305 of visual arrestin) and C-terminal arm, which collectively maintain the orientation of the N- and C-domain baskets (Han et al. 2001; Kim et al. 2013). Arrestin activation commences upon the interaction of receptor-attached phosphates with 'phosphate sensor' residues located on the positively charged concave surface of the two arrestin β-baskets (Vishnivetskiy et al. 2011). The phosphate sensor acts as a charge-operated trigger that drives arrestin activation though propagating structural rearrangements. The first step, displacement of the arrestin C-terminus, destabilizes the polar core and primes the arrestin for subsequent conformational rearrangements (Schröder et al. 2002; Kirchberg et al. 2011; Gimenez et al. 2012a).

Once the C-terminus moves aside, the open activated conformation forms as the finger loop/motif II (residues 68–79 of visual arrestin) engages a binding crevice on the cytosolic surface of the receptor that stabilizes it as a short α helix and leaves the polar core exposed (Gurevich and Benovic 1993, 1997). Spin-labeling and fluorescence quenching data indicate that in the inactive state motif II makes close contact with the arrestin N-domain, but when bound to phosphorylated light activated rhodopsin it is extended and buried in the rhodopsin-arrestin interface (Hanson et al. 2006; Sommer et al. 2007). Analysis of conformational shifts in β-arrestin1 and 2 binding to rhodopsin in vitro using double electron electron resonance (DEER) spectroscopy, confirms release of the C-terminal tail and movement of motif II towards the receptor (Zhuo et al. 2014). Likewise, the crystal structure of a peptide analog of motif II of visual arrestin in complex with rhodopsin demonstrates that arrestins engage the activated receptor in much the same manner as the Gαt C-terminus, using a common interface (Szczepek et al. 2014).

The movement of motif II in turn leads to rearrangements in motifs I, III and IV of the hinge region that disrupt hinge salt bridges and allow the protein to flex (Granzin et al. 1998). What follows is a rotation of the N- and C-domain baskets relative to one another that exposes receptor-binding residues located on the concave surface of the N- and C-domain baskets. The crystal structure of a 'pre-activated' C-terminally truncated variant of visual arrestin (Arr1-370A) that was 'activated' during crystallization by incubation with retinal-free rhodopsin (PDB code: 4JQ2), shows a dramatic 21° twist between the N- and C-terminal domains and local changes in loop conformation and hydrogen-bonding networks (Kim et al. 2013). Similar rearrangements are seen in the structure of β-arrestin-1 co-crystallized in the presence of an Fab fragment and a phosphopeptide corresponding to the last 29 amino acids of the V2 vasopressin receptor (PDB code: 4JQI), where a 20° rotation between the N- and C-terminal domains is present compared to the structure of inactive β-arrestin1 (Shukla et al. 2013). Finally, the crystal structure of a fusion protein composed of T4 lysozyme, mutationally stabilized rhodopsin, and a 'pre-activated' visual arrestin mutant (PDB code: 4ZWJ) demonstrates that arrestin activation involves a 20° rotation between the N- and C-domains that opens a cleft in the arrestin surface to accommodate part of the rhodopsin second intracellular loop (ICL) (Kang et al. 2015).

This 'two-step' activation model wherein arrestin first engages the phosphorylated GPCR C-terminus, then undergoes a conformational rearrangement that enables it to dock with the heptahelical receptor core has been visualized using single-particle negative-stain electron microscopy. Images of a β_2 adrenergic-V_2 vasopressin receptor chimera in complex with β-arrestin1 indicate two major arrestin docking poses, an initial state wherein arrestin has only engaged the receptor C-tail and a fully docked state where the N- and C-domain baskets have enveloped the intracellular receptor domains (Shukla et al. 2014). Thus, arrestin activation involves large conformational shifts in the protein that simultaneously expose receptor-interacting motifs on the N- and C-domain baskets and release the arrestin C-terminus, which in β-arrestins can then bind clathrin and AP-2 leading to receptor clustering and internalization.

Selectivity Versus Promiscuity

The GPCR selectivity of visual/β-arrestins varies. Visual arrestin, which is expressed exclusively in photoreceptors and desensitizes rhodopsin and cone opsins, is highly dependent upon receptor phosphorylation and exhibits the largest differences in affinity between rhodopsin and other GPCRs (Gurevich and Benovic 1995). The ubiquitously expressed β-arrestin1 and 2, on the other hand, must regulate hundreds of different GPCRs with widely divergent intracellular domains, a task that requires greater conformational flexibility and opens the potential for receptor-specific arrestin conformations. Accordingly, the interdomain hydrogen bonds that stabilize the polar core are weaker in β-arrestins, such that the hydrogen bond networks in β-arrestin1 and 2 resemble that seen in the 'pre-activated' p44 visual arrestin structure (Granzin et al. 2012; Kim et al. 2013). This suggests β-arrestins exist in a partially pre-activated state, accounting for their lower dependence upon receptor phosphorylation (Gurevich and Benovic 1995). β-Arrestin2 is the most promiscuous member of the family in terms of GPCR interaction and this correlates with increased flexibility within the N-terminal basket (Zhan et al. 2011). This area has less defined secondary structure in β-arrestin2, probably reflecting greater flexibility to accommodate differences in GPCR structure.

Information transfer between GPCRs and arrestins must occur through direct contact, and currently available data suggest that binding involves relatively few such contacts. The clearest example to date, the T4 lysozyme-rhodopsin-visual arrestin chimera structure (PDB code: 4ZWJ) identified four principal rhodopsin-arrestin interface patches involving mostly the N-terminal arrestin basket (Kang et al. 2015). When fully engaged, motif II of arrestin interacts with the C-terminus of transmembrane (TM) domain 7, the N-terminus of helix 8, and ICL1 of rhodopsin, while the adjacent arrestin β-strand (residues 79–86) makes contact with residues from TM5, TM6 and ICL3. Another interface patch is formed by the middle loop of the N-domain (around residue 140) and the C-loop of the C-domain (around residue 251) that interact with ICL2 of rhodopsin, and the arrestin back loop (residues 319–320) that contacts the C-terminus of TM5. Although not visualized in the crystal structure, additional contacts between the arrestin N-terminal β-strand (residues 11–19) and the C-terminal tail of rhodopsin can be modeled and detected by nuclear magnetic resonance spectroscopy (Kisselev et al. 2004; Kang et al. 2015).

While the rhodopsin-visual arrestin crystal structure suggests a generally conserved mode of GPCR-arrestin docking, mutagenesis studies support the concept that differences do exist in the relative contribution of residues on the arrestin surface to GPCR affinity and selectivity. Mutagenesis studies have identified a handful of such 'receptor discriminator' residues. Predictably, given the phosphorylation dependence of visual arrestin-rhodopsin binding, alanine substitution of the two conserved phosphate-sensing lysine residues in the visual arrestin N-terminus ($K^{14,15}A$) dramatically reduces binding to active phosphorylated rhodopsin (Vishnivetskiy et al. 2000; Gimenez et al. 2012a). Interestingly, the analogous mutation in β-arrestin2 ($K^{11,12}A$) produces receptor-specific effects. For example, [$K^{11,12}A$]-β-arrestin2 is impaired in binding to the neuropeptide Y_2 receptor, but not the closely related Y_1 receptor (Gimenez et al. 2014). Point mutations in the C-terminal central loop of β-arrestin2 (residues 230–260) also dramatically affect GPCR selectivity. For example, $Y^{239}T$ substitution enhances binding to the β_2 adrenergic receptor at the cost of affinity for M_2 muscarinic, D_1 dopamine, and D_2 dopamine receptors. Introducing a $D^{260}K/Q^{262}P$ substitution has the opposite effect, virtually eliminating β_2 adrenergic receptor binding while preserving M_2, D_1 and D_2 receptor binding, $Q^{256}Y$ substitution selectively reduces affinity for the D_2 receptor, and $Y^{239}T/Q^{256}Y$ substitution preserves D_1 receptor binding while reducing affinity for the β_2 adrenergic and M_2 receptors and eliminating D_2 receptor binding (Gimenez et al. 2012b). For the Y_1 and Y_2 receptors, a $Y^{238}T$ point mutation introduces several-fold selectivity for the Y_1 over Y_2 receptor (Gimenez et al. 2014). To the extent that these results reflect differences between GPCRs in their in the mode of arrestin recognition, it seems possible that an arrestin may be able to 'tell' which GPCR it has bound and that differences in binding affinity across the GPCR-arrestin interface might support receptor-specific arrestin conformations.

The Effect of Arrestin Conformation on Cargo Binding

While GPCR recognition involves primarily the concave surface of the N- and C-domain baskets and clathrin/AP-2 bind to the β-arrestin C-terminal tail, most arrestin signaling cargos appear to interact with the cytosolic face of the protein that is left exposed when docked to a receptor. An astounding array of proteins has been reported to bind arrestins (Xiao et al. 2007; Luttrell and Gesty-Palmer 2010). Besides clathrin and AP-2, the list of arrestin cargos includes Src family tyrosine kinases (Luttrell et al. 1999; DeFea et al. 2000a), components of the extracellular signal regulated kinase 1 and 2 (ERK1/2) and c-Jun N-terminal Kinase 3 (JNK3) mitogen-activated protein (MAP) kinase cascades (McDonald et al. 2000; DeFea et al. 2000b; Luttrell et al. 2001), the Ser/Thr protein phosphatase 2A (Beaulieu et al. 2005), E3 ubiquitin ligases and de-ubiquitinases (Shenoy et al. 2001, 2008, 2009), second messenger degrading cAMP phosphodiesterases (Perry et al. 2002) and diacylglycerol kinase (Nelson et al. 2007), elements of the nuclear factor κB signaling pathway (Witherow et al. 2004), and regulators of small GTPase activity (Claing et al. 2001; Bhattacharya et al. 2002). This multiplicity of binding partners, combined with the fact that the surface of arrestins is simply not big enough to bind everything all at once, inevitably raises the question of how can signaling specificity possibly be achieved?

Part of the answer lies in the conformation-dependence of arrestin-cargo interactions. In the absence of activated GPCRs, most of the visual/β-arrestin pool resides either in the cytosol or bound to microtubules, and while some arrestin cargos appear to be constitutively associated, others exhibit a distinct preference for cytosolic, microtubule-bound, or GPCR-bound arrestin. This capacity to bind different cargos in different locations is central to arrestin function. Under basal conditions they are able to function as 'silent scaffolds' that constrain signaling proteins to specific cellular compartments and dampen basal pathway activity (Breitman et al. 2012; Lin and Defea 2013). But in response to an external stimulus, they can redistribute to high affinity binding sites on activated GPCRs, whereupon they undergo conformational changes that cause them to release some cargos and associate with others, bringing arrestin-bound effectors under the control of environmental cues.

The importance of arrestin conformation in controlling cargo interactions is illustrated in studies examining the effect of arrestin mutations on cargo affinity and pathway activity in cells. 'Constitutively inactive' β-arrestin mutants with deletions in the hinge region that restrict interdomain flexibility exhibit enhanced microtubule binding and reduced receptor affinity compared to the wild type 'cytosolic' protein (Vishnivetskiy et al. 2002; Hanson et al. 2007). Conversely, mutations that destabilize the polar core by disrupting the phosphate sensor or removing the auto-inhibitory C-terminus, produce 'constitutively active' arrestins that lose the ability to recognize receptor phosphorylation while retaining selectivity for the ligand activated receptor conformation (Gurevich and Benovic 1997; Kovoor et al. 1999).

Using these mutants to probe arrestin-cargo interactions indicates that some cargos exhibit distinct preferences for the 'cytosolic', 'microtubule-bound', or 'active' arrestin conformation. JNK3, for example, has highest affinity for the cytosolic form of β-arrestin2 (Song et al. 2006; Breitman et al. 2012). Since β-arrestin2 possesses a nuclear export sequence that excludes it from the nucleus, arrestin-bound JNK3 is sequestered in the cytosol away from its nuclear transcription factor targets. In contrast, inactive ERK2 has significant affinity only for the active and microtubule-bound arrestin conformations, while active ERK2 binds tightly only to active arrestin (Hanson et al. 2007; Coffa et al. 2011). Of the other ERK1/2 pathway components, c-Raf1 also shows a preference for the active conformation, while MEK1 binds equally to active and cytosolic arrestins. Since the microtubule-bound conformation binds ERK1/2 and c-Raf1 better than the cytosolic form, overexpressing the 'constitutively inactive' microtubule-bound mutant redirects inactive ERK1/2 to microtubules where it is sequestered away from membrane-generated activating signals. Thus, the sensitivity of ERK1/2 binding and activation to arrestin conformation, enables arrestin scaffolds both to dampen basal ERK1/2 pathway activity and to support the assembly of GPCR-associated ERK1/2 activation complexes. In yet another example, the E3 ubiquitin ligase Mdm2 has highest affinity for microtubule-bound arrestin, such that overexpressing 'inactive' arrestin markedly increases ubiquitination of microtubule-associated substrates (Song et al. 2006; Hanson et al. 2007). In this case, the differential affinity probably permits dynamic regulation, in that Mdm2 preferentially ubiquitinates GPCR-bound arrestin, a step that stabilizes the GPCR-arrestin complex (Shenoy and Lefkowitz 2005b). Once the arrestin is in the ubiquitinated GPCR-bound conformation, the drop in affinity may allow Mdm2 to dissociate, allowing other cargos to take its place in the receptor-associated complex.

Dynamic Aspects of Arrestin Activation

While structural studies reveal that arrestins undergo conformational rearrangement upon binding receptors, they do not offer insights into the kinetics of the process. Nor do we currently possess sufficient high-resolution structures of arrestins bound to different GPCRs to directly visualize the subtler effects receptor structure may have on arrestin conformation. Nonetheless, live cell studies performed using resonance energy transfer techniques provide evidence that arrestins, like the receptors themselves, have more than one 'active' conformation and that arrestin conformation reflects the avidity with which it binds the receptor and its ability to engage certain downstream partners.

Are Arrestins Activated Catalytically?

The dynamics of arrestin recruitment and the corresponding conformational shifts in the arrestin molecule have been probed using resonance energy transfer. Intermolecular bioluminescence resonance energy transfer (BRET) between *Renilla* luciferase (rLuc)-tagged β-arrestin2 and yellow fluorosecent protein (YFP)-tagged-GPCRs indicates a $t_{1/2}$ of 1–2.5 min (at 25 °C) for arrestin recruitment to the receptor upon agonist stimulation (Charest and Bouvier 2003; Charest et al. 2005). Measured by intermolecular fluorescence resonance energy transfer (FRET) the process appears even faster, taking only seconds to occur (Nuber et al. 2016). As shown in Fig. 17.2a, the τ for cyan fluorescent protein (CFP) tagged β-arrestin2-binding to YFP-tagged β_2 adrenergic receptor is only 1.3 s, indicating the process of GPCR desensitization begins very shortly after receptor activation. Recruitment is followed closely in time by a conformational shift in the arrestin that can be measured using intramolecular BRET or FRET (Charest et al. 2005; Nuber et al. 2016). Consistent with the two-step model that arrestins undergo structural rearrangement after the phosphate sensor is triggered by phosphorylated receptor domains, the intramolecular shifts in arrestin conformation that follow its recruitment to the receptor are slower than the initial binding. When measured by intramolecular fluorescent arsenical hairpin (FlAsH) FRET, the τ for β-arrestin2 conformational rearrangement upon binding the β_2-adrenergic receptor is 2.2 s, nearly twice as long as the initial recruitment (Nuber et al. 2016).

Once docked, the lifetime of the GPCR-arrestin complex is determined by several factors, including receptor structure (Oakley et al. 2001), ligand off-rate (Krasel et al. 2005) and posttranslational modifications such as arrestin

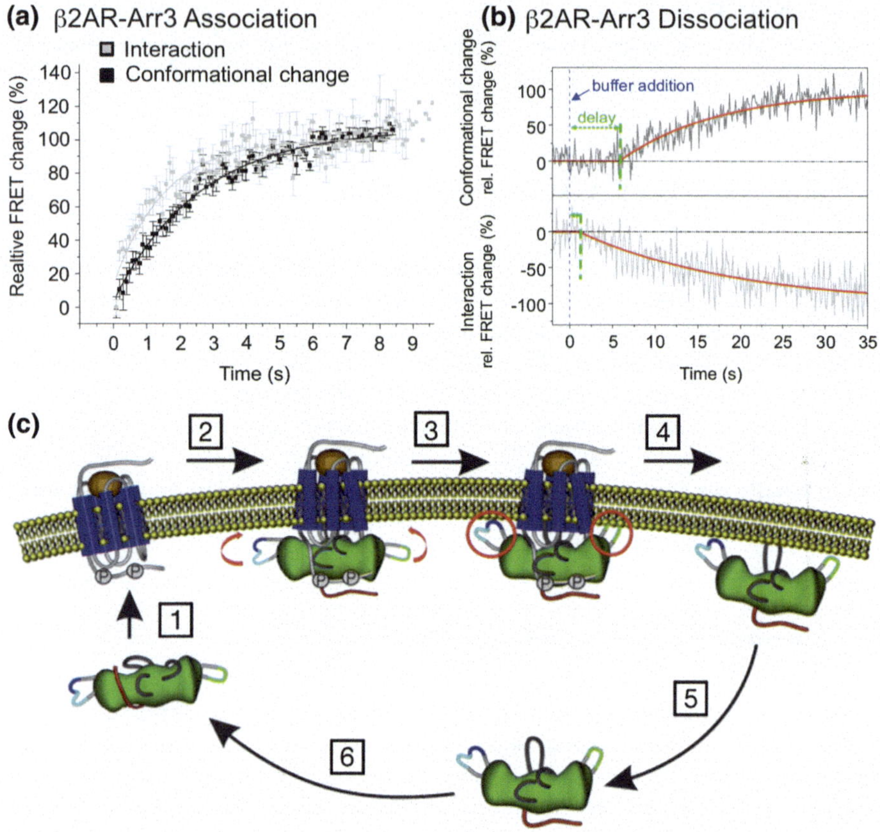

Fig. 17.2 β-Arrestin2 undergoes an activation/deactivation cycle that may permit it to be activated catalytically. **a** Kinetics of the agonist-induced interaction of β-arrestin2 with the β_2 adrenergic receptor (*light gray*) determined by intermolecular FRET between a β-arrestin2-FlAsH-CFP biosensor and YFP-tagged β_2 adrenergic receptor, compared with the kinetics of β-arrestin2 conformational rearrangement (*black*) measured by intramolecular FRET within a β-arrestin2-FlAsH-CFP biosensor interacting with untagged β_2 adrenergic receptor. The lag between initial receptor binding and conformational rearrangement of β-arrestin2 is consistent with a two-step model of arrestin activation. **b** Kinetics of β-arrestin2 dissociation from β_2 adrenergic receptors upon agonist removal. Dissociation from the receptor was recorded as the decay in intermolecular FRET between β-arrestin2 and β_2 adrenergic receptor (*lower trace*), while return of β-arrestin2 to its inactive conformation (*upper trace*) was recorded as change in intramolecular FRET within a β-arrestin2-FlAsH-CFP biosensor. The data suggest that β-arrestin2 retains it 'active' conformation for several seconds after it dissociates from the receptor. **c** Model of the β-arrestin2 activation/deactivation cycle upon interaction with a 'class A' GPCR. Cytosolic arrestin is recruited to an active GRK-phosphorylated receptor, whereupon its initial docking (*1*) promotes a conformational change (*2*) that allows it to adopt a fully activated receptor-bound conformation (*3*). Upon removal of agonist, arrestin dissociates from the receptor but retains it activated conformation (*4*) for some time before reverting to the inactive state (*5*). Data originally published in: Nuber et al. (2016)

ubiquitination (Shenoy et al. 2001) and phosphorylation (Lin et al. 1997; Khoury et al. 2014). Most GPCRs fall into one of two classes based on their selectivity for the two β-arrestin isoforms and the longevity of the receptor-arrestin interaction (Oakley et al. 2000). One, termed 'class A', exhibits higher affinity for β-arrestin2 than β-arrestin1 and forms transient receptor-arrestin complexes that dissociate soon after the receptor internalizes. These receptors are rapidly resensitized and recycled back to the plasma membrane. The other, 'class B', group exhibits equivalent affinities for β-arrestin1 and β-arrestin2 and forms long lasting receptor-arrestin complexes that remain intact as the receptor undergoes endosomal sorting. These receptors tend to be sequestered in endosomes and either recycle slowly or are degraded. Intriguingly, the conformational shift in β-arrestin2 induced by binding to the class A β_2-adrenergic receptor persists for as much as 5 s after the receptor-arrestin complex itself dissociates (Fig. 17.2b). During this period, β-arrestin2 appears to linger in clathrin-coated pits, where it may continue to 'signal' after letting go of the receptor (Nuber et al. 2016). Such findings suggest that while class B GPCRs may tie up arrestins in stoichiometric 'signalsome' complexes that distribute to endosomes, class A receptors might be able to act upon arrestins catalytically, much in the same manner as heterotrimeric G proteins. Figure 17.2c depicts a model of such an activation/deactivation and translocation cycle.

The 'Conformational Signature' of Activated Arrestin

By inserting a six amino acid FlAsH motif at different positions on the surface of the arrestin protein and measuring FRET or BRET efficiency from multiple vantage points, it is possible observe stimulus-induced changes in the population average conformation of the cellular pool of arrestin. Figure 17.3a schematically depicts such a series of six intramolecular rLuc–β-arrestin2–FlAsH BRET probes. Recruitment of these biosensors to an activated GPCR changes the efficiency of resonance energy transfer between an N-terminal fluorescence donor and fluorescent arsenical acceptor, such that increases or decreases in BRET along the protein yield a 'conformational signature' that represents the net effect of conformational shifts and/or steric interference introduced by the binding or release of receptor and cargos.

When β-arrestin2 FlAsH FRET or BRET probes are used to compare the 'conformational signatures' imposed by different GPCRs it is apparent that β-arrestin2 adopts receptor-specific poses (Nuber et al. 2016; Lee et al. 2016). Class B receptors, such as the angiotensin AT_{1A}, V_2 vasopressin, and type 1 parathyroid hormone receptors, all of which form long-lived β-arrestin complexes, produce similar β-arrestin2 FlAsH BRET signatures, with a decrease in net BRET between sensors located on the N- and C-domain baskets, and an increase in BRET within the C-terminal tail. Other receptors, such as the Class A α_{1B} adrenergic and

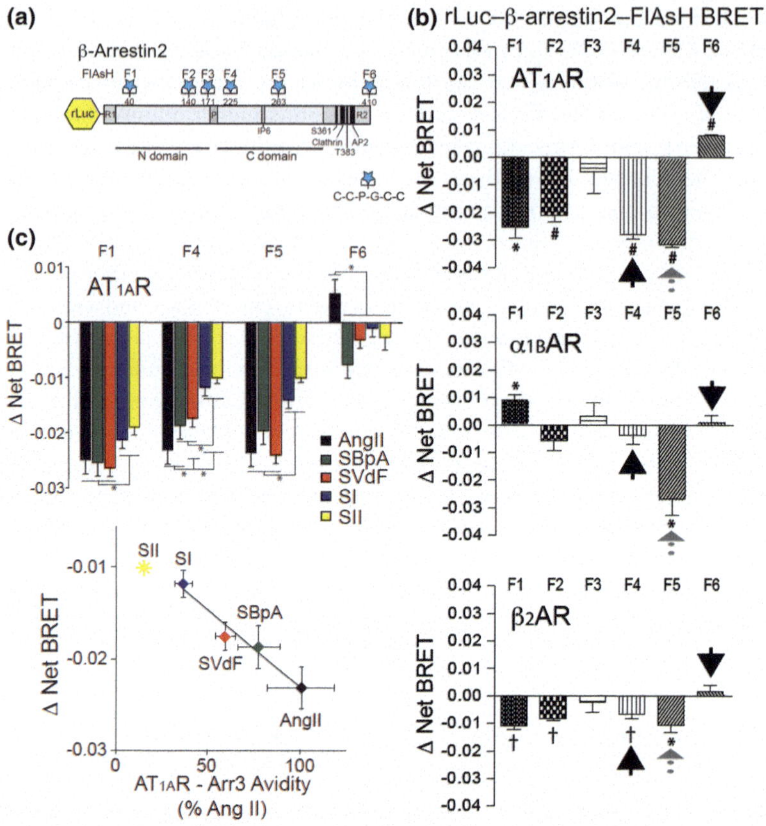

Fig. 17.3 Receptor and ligand structure impose distinct β-arrestin2 'conformational signatures'. **a** Linear diagram showing the position of each of six introduced FlAsH motifs in relation to the N- and C-domain baskets of β-arrestin2 and its C-terminal binding/regulatory regions. **b** Receptor-specific β-arrestin2 intramolecular FlAsH BRET signatures. The panel of six rLuc–β-arrestin2–FlAsH1-6 biosensors (F1–F6) was used to interrogate changes in β-arrestin2 conformation upon binding to the angiotensin AT_{1A} receptor ($AT_{1A}R$), α_{1B} adrenergic receptor ($\alpha_{1B}AR$) and β_2 adrenergic receptor (β_2AR). A decrease in BRET at the F4 position and increase at F6 discriminated between 'class B' and 'class A' GPCRs (*black arrows*), while the magnitude of change at the F5 position correlated with β-arrestin-dependent activation of ERK1/2 (*gray arrows*). **c** Ligand-specific effects on the 'conformation signature' of β-arrestin2. The *upper panel* depicts the F1–4–5–6 profiles generated upon binding of a panel of 'biased' angiotensin analogs (AngII, SBpA, SVdF, SI, and SII; Zimmerman et al. 2012) to the AT_{1A} receptor. The *lower panel* depicts the relationship between the amplitude of the F4 signal and the independently determined avidity of $AT_{1A}R$ and arrestin3 measured by fluorescence recovery after photobleaching. The linear fit demonstrates that the intramolecular BRET signal at the F4 position is reporting on the avidity of the receptor-arrestin interaction. Data originally published in: Lee et al. (2016)

β_2 adrenergic receptors that form transient β-arrestin complexes, yield different patterns (Fig. 17.3b). Interestingly, the magnitude of the β-arrestin2 FlAsH BRET signal at some positions, e.g. FlAsH4, correlates with the avidity of GPCR-arrestin binding and at others correlates with the extent to which the arrestin is engaged in ERK1/2 activation, e.g. FlAsH5 (Lee et al. 2016). Consistent with the structural and mutagenesis data that suggest that receptors and arrestins interface via distinct contact patches and that binding of certain cargos is sensitive to arrestin conformation, resonance energy transfer data support the hypothesis that information about the identity of the receptor is encoded within the activated arrestin structure and influences its subsequent function.

Encoding Ligand-Specific Information

Embedded within the general allosteric model of GPCR function are the concepts of 'pluridimensional' efficacy, i.e. that GPCRs signal by engaging multiple G protein and non-G protein effectors (Galandrin et al. 2007), and 'functional selectivity', i.e. that structurally distinct ligands can 'bias' GPCR signaling by preferentially stabilizing different 'active' receptor conformations (Kenakin 2012). While the concept of ligand bias accounts for differences in the efficiency with which a receptor couples to different effectors, it does not imply that different ligands can activate the same effector in different ways. If, however, effectors have multiple 'active states', then ligand structure may also introduce 'bias' by influencing how an effector is activated.

When measured by intramolecular BRET between the β-arrestin2 N- and C-terminus (Charest et al. 2005), conventional GPCR agonists produce similar increases in BRET signal, while 'biased' agonists that recruit arrestin without activating heterotrimeric G proteins trigger modest shifts in the opposite direction (Shukla et al. 2008). When comparing the β-arrestin2 FlAsH BRET conformational signatures elicited by a panel of 'biased' angiotensin AT_{1A} receptor peptides, it is evident that ligands can differ in the magnitude of the BRET signal observed at certain positions, e.g. FlAsH4 (Fig. 17.3c). Interestingly, the signal from the FlAsH4 position, but not from other locations, e.g. FlAsH1 and FlAsH6, correlates closely with the avidity of AT_{1A} receptor-β-arrestin2 binding measured independently by fluorescence recovery after photobleaching (Lee et al. 2016; Zimmerman et al. 2012). This concept that ligand 'bias' extends to influencing the manner of effector activation is supported by recent work on G_s activation by the human calcitonin receptor demonstrating that human and salmon calcitonin differentially modulate G_s-adenylyl cyclase-cAMP signaling by inducing ligand-specific G_s conformations that differ in the GTP affinity of the calcitonin receptor-Gs ternary complex (Furness et al. 2016).

While the picture is emerging that the structure of the ligand-GPCR complex plays a role in determining arrestin conformation and downstream function, the contributions of other factors are less understood. It is clear, for example, that in some systems different GRKs specify arrestin signaling versus desensitization (Kim et al. 2005; Ren et al. 2005; Nobles et al. 2011; Zimmerman et al. 2012), leading to the hypothesis that they imprint a 'phosphorylation code' written in different patterns of phosphorylation (Tobin et al. 2008; Liggett 2011). To the extent that receptor phosphorylation might alter the conformation of docked arrestin, it may promote differences in the program of arrestin ubiquitination/de-ubiquitination that stabilizes the GPCR-arrestin complex or changes the avidity of the complex for signalsome cargos (Shenoy and Lefkowitz 2005b; Shenoy et al. 2007).

Conclusions

Arrestin binding marks the dividing line between early GPCR signaling mediated via heterotrimeric G proteins and longer lasting arrestin-dependent signals transmitted by GPCR-arrestin 'signalsome' complexes. While G protein signaling accounts for short-term effects mediated by soluble second messengers, arrestin signaling produces longer-term effects on processes such as cell proliferation/growth, survival/apoptosis, and migration/chemotaxis, many of which arise at the transcriptional level (Gesty-Palmer et al. 2013; Maudsley et al. 2015). Given the diverse functions of arrestins, one might imagine that it is not physiologically adaptive for arrestins to execute all of their desensitizing/trafficking/signaling functions under all circumstances, and that mechanisms must exist to modulate arrestin function that are more nuanced than simple transition between 'inactive' and 'active' conformations induced by receptor docking.

The ligand-GPCR-effector complex is an allosterically regulated signal transduction unit in which energy imparted by ligand binding prompts rearrangement the receptor transmembrane and intracellular domains that permit the receptor to engage heterotrimeric G proteins, GRKs and arrestins, which in turn must change conformation to perform their functions (Manglik and Kobilka 2014). Despite commonalities in the mode of ligand-induced GPCR activation, it is clear that GPCR domain structure determines which intracellular effectors are activated. It is also well established that ligand structure can be as important as receptor structure, in that ligands can 'bias' receptor-effector coupling based on the efficiency with which they stabilize different active receptor conformations (Kenakin 2012). What is less clear is how far these allosteric effects, arising as they do from within the ligand binding pocket of the receptor, extend into the cell. Does ligand-receptor conformation control not just which effectors are activated, but also how they are activated? The answer, for both arrestins and heterotrimeric G proteins, appears to be yes (Nuber et al. 2016; Lee et al. 2016; Furness et al. 2016). Developing a deeper understanding of the nuances of how receptor structure and ligand bias favor different effector activation modes, perhaps even specifying which of its many functions arrestins will perform, may prove important to the development of tailored GPCR-targeted therapeutics.

Acknowledgements Supported by National Institutes of Health Grants R01 DK055524 (LML), R01 GM095497 (LML), Department of Veterans Affairs Merit Review Grant I01 BX003188 (LML), and the Research Service of the Charleston, SC Veterans Affairs Medical Center. The contents of this article do not represent the views of the Department of Veterans Affairs or the United States Government.

References

Aubry L, Guetta D, Klein G (2009) The arrestin fold: variations on a theme. Curr Genomics 10:133–142

Bhattacharya M, Anborgh PH, Babwah AV, Dale LB, Dobransky T, Benovic JL, Feldman RD, Verdi JM, Rylett RJ, Ferguson SS (2002) Beta-arrestins regulate a Ral-GDS Ral effector pathway that mediates cytoskeletal reorganization. Nat Cell Biol 4:547–555

Beaulieu JM, Sotnikova TD, Marion S, Lefkowitz RJ, Gainetdinov RR, Caron MG (2005) An Akt/beta-arrestin 2/PP2A signaling complex mediates dopaminergic neurotransmission and behavior. Cell 122:261–273

Breitman M, Kook S, Gimenez LE, Lizama BN, Palazzo MC, Gurevich EV, Gurevich VV (2012) Silent scaffolds: inhibition of c-Jun N-terminal kinase 3 activity in cell by dominant-negative arrestin-3 mutant. J Biol Chem 287:19653–19664

Charest PG, Bouvier M (2003) Palmitoylation of the V2 vasopressin receptor carboxyl tail enhances beta-arrestin recruitment leading to efficient receptor endocytosis and ERK1/2 activation. J Biol Chem 278:41541–41551

Charest PG, Terrillon S, Bouvier M (2005) Monitoring agonist-promoted conformational changes of beta-arrestin in living cells by intramolecular BRET. EMBO Rep 6:334–340

Claing A, Chen W, Miller WE, Vitale N, Moss J, Premont RT, Lefkowitz RJ (2001) Beta-arrestin-mediated ADP-ribosylation factor 6 activation and beta 2-adrenergic receptor endocytosis. J Biol Chem 276:42509–42513

Coffa S, Breitman M, Hanson SM, Callaway K, Kook S, Dalby KN, Gurevich VV (2011) The effect of arrestin conformation on the recruitment of c-Raf1, MEK1, and ERK1/2 activation. PLoS One 6:e28723

DeFea KA, Vaughn ZD, O'Bryan EM, Nishijima D, Dery O, Bunnett NW (2000a) The proliferative and antiapoptotic effects of substance P are facilitated by formation of a beta-arrestin-dependent scaffolding complex. Proc Natl Acad Sci U S A 97:11086–11091

DeFea KA, Zalevsky J, Thoma MS, Dery O, Mullins RD, Bunnett NW (2000b) Beta-arrestin-dependent endocytosis of proteinase-activated receptor 2 is required for intracellular targeting of activated ERK1/2. J Cell Biol 148:1267–1281

Ferguson SS (2001) Evolving concepts in G protein-coupled receptor endocytosis: the role in receptor desensitization and signaling. Pharmacol Rev 53:1–24

Furness SG, Liang YL, Nowell CJ, Halls ML, Wookey PJ, Dal Maso E, Inoue A, Christopoulos A, Wootten D, Sexton PM (2016) Ligand-dependent modulation of G protein conformation alters drug efficacy. Cell 167:739–749

Galandrin S, Oligny-Longpré G, Bouvier M (2007) The evasive nature of drug efficacy: implications for drug discovery. Trends Pharmacol Sci 28:423–430

Gesty-Palmer D, Yuan L, Martin B, Wood WH 3rd, Lee MH, Janech MG, Tsoi LC, Zheng WJ, Luttrell LM, Maudsley S (2013) β-arrestin-selective G protein-coupled receptor agonists engender unique biological efficacy in vivo. Mol Endocrinol 27:296–314

Gimenez LE, Kook S, Vishnivetskiy SA, Ahmed MR, Gurevich EV, Gurevich VV (2012a) Role of receptor-attached phosphates in binding of visual and non-visual arrestins to G protein-coupled receptors. J Biol Chem 287:9028–9040

Gimenez LE, Vishnivetskiy SA, Baameur F, Gurevich VV (2012b) Manipulation of very few receptor discriminator residues greatly enhances receptor specificity of non-visual arrestins. J Biol Chem 287:29495–29505

Gimenez LE, Babilon S, Wanka L, Beck-Sickinger AG, Gurevich VV (2014) Mutations in arrestin-3 differentially affect binding to neuropeptide Y receptor subtypes. Cell Signal 26:1523–1531

Goodman OB Jr, Krupnick JG, Santini F, Gurevich VV, Penn RB, Gagnon AW, Keen JH, Benovic JL (1996) Beta-arrestin acts as a clathrin adaptor in endocytosis of the beta2-adrenergic receptor. Nature 383:447–450

Granzin J, Cousin A, Weirauch M, Schlesinger R, Buldt G, Batra-Safferling R (2012) Crystal structure of p44, a constitutively active splice variant of visual arrestin. J Mol Biol 416: 611–618

Granzin J, Wilden U, Choe HW, Labahn J, Krafft B, Büldt G (1998) X-ray crystal structure of arrestin from bovine rod outer segments. Nature 391:918–921

Gurevich VV, Benovic JL (1993) Visual arrestin interaction with rhodopsin. Sequential multisite binding ensures strict selectivity toward light-activated phosphorylated rhodopsin. J Biol Chem 268:11628–11638

Gurevich VV, Benovic JL (1995) Visual arrestin binding to rhodopsin. Diverse functional roles of positively charged residues within the phosphorylation-recognition region of arrestin. J Biol Chem 270:6010–6016

Gurevich VV, Benovic JL (1997) Mechanism of phosphorylation-recognition by visual arrestin and the transition of arrestin into a high affinity binding state. Mol Pharmacol 51:161–169

Gurevich VV, Gurevich EV (2013) Structural determinants of arrestin functions. Prog Mol Biol Transl Sci 118:57–92

Han M, Gurevich VV, Vishnivetskiy SA, Sigler PB, Schubert C (2001) Crystal structure of beta-arrestin at 1.9 A: possible mechanism of receptor binding and membrane Translocation. Structure 9:869–880

Hanson SM, Francis DJ, Vishnivetskiy SA, Kolobova EA, Hubbell WL, Klug CS, Gurevich VV (2006) Differential interaction of spin-labeled arrestin with inactive and active phospho-rhodopsin. Proc Natl Acad Sci U S A 103:4900–4905

Hanson SM, Cleghorn WM, Francis DJ, Vishnivetskiy SA, Raman D, Song X, Nair KS, Slepak VZ, Klug CS, Gurevich VV (2007) Arrestin mobilizes signaling proteins to the cytoskeleton and redirects their activity. J Mol Biol 368:375–387

Kang Y, Zhou XE, Gao X, He Y, Liu W, Ishchenko A, Barty A, White TA, Yefanov O, Han GW, Xu Q, de Waal PW, Ke J, Tan MH, Zhang C, Moeller A, West GM, Pascal BD, Van Eps N, Caro LN, Vishnivetskiy SA, Lee RJ, Suino-Powell KM, Gu X, Pal K, Ma J, Zhi X, Boutet S, Williams GJ, Messerschmidt M, Gati C, Zatsepin NA, Wang D, James D, Basu S, Roy-Chowdhury S, Conrad CE, Coe J, Liu H, Lisova S, Kupitz C, Grotjohann I, Fromme R, Jiang Y, Tan M, Yang H, Li J, Wang M, Zheng Z, Li D, Howe N, Zhao Y, Standfuss J, Diederichs K, Dong Y, Potter CS, Carragher B, Caffrey M, Jiang H, Chapman HN, Spence JC, Fromme P, Weierstall U, Ernst OP, Katritch V, Gurevich VV, Griffin PR, Hubbell WL, Stevens RC, Cherezov V, Melcher K, Xu HE (2015) Crystal structure of rhodopsin bound to arrestin by femtosecond X-ray laser. Nature 523:561–567

Kenakin TP (2009) 7TM receptor allostery: putting numbers to shapeshifting proteins. Trends Pharmacol Sci 30:460–469

Kenakin TP (2012) Biased signalling and allosteric machines: new vistas and challenges for drug discovery. Br J Pharmacol 165:1659–1669

Kenakin T, Miller LJ (2010) Seven transmembrane receptors as shapeshifting proteins: the impact of allosteric modulation and functional selectivity on new drug discovery. Pharmacol Rev 62:265–304

Khoury E, Nikolajev L, Simaan M, Namkung Y, Laporte SA (2014) Differential regulation of endosomal GPCR/β-arrestin complexes and trafficking by MAPK. J Biol Chem 289:23302–23317

Kim J, Ahn S, Ren XR, Whalen EJ, Reiter E, Wei H, Lefkowitz RJ (2005) Functional antagonism of different G protein-coupled receptor kinases for beta-arrestin-mediated angiotensin II receptor signaling. Proc Natl Acad Sci U S A 102:1442–1447
Kim YJ, Hofmann KP, Ernst OP, Scheerer P, Choe HW, Sommer ME (2013) Crystal structure of pre-activated arrestin p44. Nature 497:142–146
Kirchberg K, Kim TY, Möller M, Skegro D, Dasara Raju G, Granzin J, Büldt G, Schlesinger R, Alexiev U (2011) Conformational dynamics of helix 8 in the GPCR rhodopsin controls arrestin activation in the desensitization process. Proc Natl Acad Sci U S A 108:18690–18695
Kisselev OG, Downs MA, McDowell JH, Hargrave PA (2004) Conformational changes in the phosphorylated C-terminal domain of rhodopsin during rhodopsin arrestin interactions. J Biol Chem 279:51203–51207
Kovoor A, Celver J, Abdryashitov RI, Chavkin C, Gurevich VV (1999) Targeted construction of phosphorylation-independent beta-arrestin mutants with constitutive activity in cells. J Biol Chem 274:6831–6834
Krasel C, Bünemann M, Lorenz K, Lohse MJ (2005) β-Arrestin binding to the β2-adrenergic receptor requires both receptor phosphorylation and receptor activation. J Biol Chem 280:9528–9535
Laporte SA, Oakley RH, Zhang J, Holt JA, Ferguson SS, Caron MG, Barak LS (1999) The beta2-adrenergic receptor/betaarrestin complex recruits the clathrin adaptor AP-2 during endocytosis. Proc Natl Acad Sci U S A 96:3712–3717
Lee M-H, Appleton KM, Strungs EG, Kwon JY, Morinelli TA, Peterson YK, Laporte SA, Luttrell LM (2016) The conformational signature of activated arrestin3 predicts its trafficking and signaling functions. Nature 531:665–668
Liggett SB (2011) Phosphorylation barcoding as a mechanism of directing GPCR signaling. Sci Signal 4:pe36
Lin A, DeFea KA (2013) β-Arrestin-kinase scaffolds: turn them on or turn them off? Wiley Interdiscip Rev Syst Biol Med 5:231–241
Lin FT, Krueger KM, Kendall HE, Daaka Y, Fredericks ZL, Pitcher JA, Lefkowitz RJ (1997) Clathrin-mediated endocytosis of the beta-adrenergic receptor is regulated by phosphorylation/dephosphorylation of beta-arrestin1. J Biol Chem 272:31051–31057
Luttrell LM, Ferguson SS, Daaka Y, Miller WE, Maudsley S, Della Rocca GJ, Lin F, Kawakatsu H, Owada K, Luttrell DK, Caron MG, Lefkowitz RJ (1999) Beta-arrestin-dependent formation of beta2 adrenergic receptor-Src protein kinase complexes. Science 283:655–661
Luttrell LM, Gesty-Palmer D (2010) Beyond desensitization: physiological relevance of arrestin-dependent signaling. Pharmacol Rev 62:305–330
Luttrell LM, Kenakin TP (2011) Refining efficacy: allosterism and bias in G protein-coupled receptor signaling. Methods Mol Biol 756:3–35
Luttrell LM, Roudabush FL, Choy EW, Miller WE, Field ME, Pierce KL, Lefkowitz RJ (2001) Activation and targeting of extracellular signal-regulated kinases by beta-arrestin scaffolds. Proc Natl Acad Sci U S A 98:2449–2454
Manglik A, Kobilka B (2014) The role of protein dynamics in GPCR function: insights from the β2AR and rhodopsin. Curr Opin Cell Biol 27:136–1143
Maudsley S, Martin B, Gesty-Palmer D, Cheung H, Johnson C, Patel S, Becker KG, Wood WH 3rd, Zhang Y, Lehrmann E, Luttrell LM (2015) Delineation of a conserved arrestin-biased signaling repertoire in vivo. Mol Pharmacol 87:706–717
McDonald PH, Chow CW, Miller WE, Laporte SA, Field ME, Lin FT, Davis RJ, Lefkowitz RJ (2000) Beta-arrestin 2: a receptor-regulated MAPK scaffold for the activation of JNK3. Science 290(5496):1574–1577
Milano SK, Kim YM, Stefano FP, Benovic JL, Brenner C (2006) Nonvisual arrestin oligomerization and cellular localization are regulated by inositol hexakisphosphate binding. J Biol Chem 281:9812–9823

Nair KS, Hanson SM, Kennedy MJ, Hurley JB, Gurevich VV, Slepak VZ (2004) Direct binding of visual arrestin to microtubules determines the differential subcellular localization of its splice variants in rod photoreceptors. J Biol Chem 279:41240–41248

Nelson CD, Perry SJ, Regier DS, Prescott SM, Topham MK, Lefkowitz RJ (2007) Targeting of diacylglycerol degradation to M1 muscarinic receptors by beta-arrestins. Science 315:663–666

Nobles KN, Xiao K, Ahn S, Shukla AK, Lam CM, Rajagopal S, Strachan RT, Huang TY, Bressler EA, Hara MR, Shenoy SK, Gygi SP, Lefkowitz RJ (2011) Distinct phosphorylation sites on the beta2-adrenergic receptor establish a barcode that encodes differential functions of beta-arrestin. Sci Signal 4:ra51

Nuber S, Zabel U, Lorenz K, Nuber A, Milligan G, Tobin AB, Lohse MJ, Hoffmann C (2016) FRET-based β-arrestin biosensors reveal a rapid, receptor-dependent activation/deactivation cycle in living cells. Nature 531:661–664

Oakley RH, Laporte SA, Holt JA, Caron MG, Barak LS (2000) Differential affinities of visual arrestin, beta arrestin1, and beta arrestin2 for G protein-coupled receptors delineate two major classes of receptors. J Biol Chem 275:17201–17210

Oakley RH, Laporte SA, Holt JA, Barak LS, Caron MG (2001) Molecular determinants underlying the formation of stable intracellular G protein-coupled receptor-beta-arrestin complexes after receptor endocytosis. J Biol Chem 276:19452–19460

Perry SJ, Baillie GS, Kohout TA, McPhee I, Magiera MM, Ang KL, Miller WE, McLean AJ, Conti M, Houslay MD, Lefkowitz RJ (2002) Targeting of cyclic AMP degradation to beta 2-adrenergic receptors by beta-arrestins. Science 298:834–836

Ren XR, Reiter E, Ahn S, Kim J, Chen W, Lefkowitz RJ (2005) Different G protein-coupled receptor kinases govern G protein and beta-arrestin-mediated signaling of V2 vasopressin receptor. Proc Natl Acad Sci U S A 102:1448–1453

Schröder K, Pulvermüller A, Hofmann KP (2002) Arrestin and its splice variant Arr1-370A (p44). Mechanism and biological role of their interaction with rhodopsin. J Biol Chem 277:43987–43996

Shenoy SK, Lefkowitz RJ (2005a) Angiotensin II-stimulated signaling through G proteins and beta-arrestin. Sci STKE 2005(308):cm10

Shenoy SK, Lefkowitz RJ (2005b) Receptor-specific ubiquitination of beta-arrestin directs assembly and targeting of seven-transmembrane receptor signalosomes. J Biol Chem 280:15315–15324

Shenoy SK, Barak LS, Xiao K, Ahn S, Berthouze M, Shukla AK, Luttrell LM, Lefkowitz RJ (2007) Ubiquitination of beta-arrestin links seven-transmembrane receptor endocytosis and ERK activation. J Biol Chem 282:29549–29562

Shenoy SK, McDonald PH, Kohout TA, Lefkowitz RJ (2001) Regulation of receptor fate by ubiquitination of activated β2-adrenergic receptor and β-arrestin. Science 294:1307–1313

Shenoy SK, Modi AS, Shukla AK, Xiao K, Berthouze M, Ahn S, Wilkinson KD, Miller WE, Lefkowitz RJ (2009) Beta-arrestin-dependent signaling and trafficking of 7-transmembrane receptors is reciprocally regulated by the deubiquitinase USP33 and the E3 ligase Mdm2. Proc Natl Acad Sci U S A 106:6650–6655

Shenoy SK, Xiao K, Venkataramanan V, Snyder PM, Freedman NJ, Weissman AM (2008) Nedd4 mediates agonist-dependent ubiquitination, lysosomal targeting, and degradation of the beta2-adrenergic receptor. J Biol Chem 283:22166–22176

Shukla AK, Manglik A, Kruse AC, Xiao K, Reis RI, Tseng WC, Staus DP, Hilger D, Uysal S, Huang LY, Paduch M, Tripathi-Shukla P, Koide A, Koide S, Weis WI, Kossiakoff AA, Kobilka BK, Lefkowitz RJ (2013) Structure of active beta-arrestin-1 bound to a G-protein-coupled receptor phosphopeptide. Nature 497:137–141

Shukla AK, Westfield GH, Xiao K, Reis RI, Huang LY, Tripathi-Shukla P, Qian J, Li S, Blanc A, Oleskie AN, Dosey AM, Su M, Liang CR, Gu LL, Shan JM, Chen X, Hanna R, Choi M, Yao XJ, Klink BU, Kahsai AW, Sidhu SS, Koide S, Penczek PA, Kossiakoff AA, Woods VL Jr, Kobilka BK, Skiniotis G, Lefkowitz RJ (2014) Visualization of arrestin recruitment by a G-protein-coupled receptor. Nature 512:218–222

Shukla AK, Violin JD, Whalen EJ, Gesty-Palmer D, Shenoy SK, Lefkowitz RJ (2008) Distinct conformational changes in beta-arrestin report biased agonism at seven-transmembrane receptors. Proc Natl Acad Sci U S A 105:9988–9993

Sommer ME, Farrens DL, McDowell JH, Weber LA, Smith WC (2007) Dynamics of arrestin–rhodopsin interactions: loop movement is involved in arrestin activation and receptor binding. J Biol Chem 282:25560–25568

Song X, Raman D, Gurevich EV, Vishnivetskiy SA, Gurevich VV (2006) Visual and both non-visual arrestins in their "inactive" conformation bind JNK3 and Mdm2 and relocalize them from the nucleus to the cytoplasm. J Biol Chem 281:21491–21499

Szczepek M, Beyriere F, Hofmann KP, Elgeti M, Kazmin R, Rose A, Bartl FJ, von Stetten D, Heck M, Sommer ME, Hildebrand PW, Scheerer P (2014) Crystal structure of a common GPCR-binding interface for G protein and arrestin. Nat Commun 5:4801

Tobin AB, Butcher AJ, Kong KC (2008) Location, location, location…site-specific GPCR phosphorylation offers a mechanism for cell-type-specific signalling. Trends Pharmacol Sci 29:413–420

Vishnivetskiy SA, Schubert C, Climaco GC, Gurevich YV, Velez MG, Gurevich VV (2000) An additional phosphate-binding element in arrestin molecule. Implications for the mechanism of arrestin activation. J Biol Chem 275:41049–41057

Vishnivetskiy SA, Hirsch JA, Velez MG, Gurevich YV, Gurevich VV (2002) Transition of arrestin into the active receptor-binding state requires an extended interdomain hinge. J Biol Chem 277:43961–43967

Vishnivetskiy SA, Gimenez LE, Francis DJ, Hanson SM, Hubbell WL, Klug CS, Gurevich VV (2011) Few residues within an extensive binding interface drive receptor interaction and determine the specificity of arrestin proteins. J Biol Chem 286:24288–24299

Witherow DS, Garrison TR, Miller WE, Lefkowitz RJ (2004) beta-Arrestin inhibits NF-kappaB activity by means of its interaction with the NF-kappaB inhibitor IkappaBalpha. Proc Natl Acad Sci U S A 101:8603–8607

Wu N, Hanson SM, Francis DJ, Vishnivetskiy SA, Thibonnier M, Klug CS, Shoham M, Gurevich VV (2006) Arrestin binding to calmodulin: a direct interaction between two ubiquitous signaling proteins. J Mol Biol 364:955–963

Xiao K, McClatchy DB, Shukla AK, Zhao Y, Chen M, Shenoy SK, Yates JR 3rd, Lefkowitz RJ (2007) Functional specialization of beta-arrestin interactions revealed by proteomic analysis. Proc Natl Acad Sci U S A 104:12011–12016

Zhan X, Gimenez LE, Gurevich VV, Spiller BW (2011) Crystal structure of arrestin-3 reveals the basis of the difference in receptor binding between two non-visual subtypes. J Mol Biol 406:467–478

Zhuang T, Vishnivetskiy SA, Gurevich VV, Sanders CR (2010) Elucidation of inositol hexaphosphate and heparin interaction sites and conformational changes in arrestin-1 by solution nuclear magnetic resonance. Biochemistry 49:10473–10485

Zhuo Y, Vishnivetskiy SA, Zhan X, Gurevich VV, Klug CS (2014) Identification of receptor binding-induced conformational changes in non-visual arrestins. J Biol Chem 289:20991–21002

Zimmerman B, Beautrait A, Aguila B, Charles R, Escher E, Claing A, Bouvier M, Laporte SA (2012) Differential β-arrestin-dependent conformational signaling and cellular responses revealed by angiotensin analogs. Sci Signal 5:ra33

Chapter 18
Monofunctional Elements of Multi-functional Proteins

Nicole A. Perry, Xuanzhi Zhan, T.M. Iverson, Eugenia V. Gurevich and Vsevolod V. Gurevich

Abstract Arrestins can act as activators or molecular scaffolds for various effectors, including MAPK signaling cascades, clathrin and AP2, E3 ubiquitin ligases, and phosphatases. Current investigations have focused on isolating the regions of arrestin that interact with effectors in an attempt to understand the role of arrestin in mediating downstream signaling pathways. The interactions of arrestins with some of their binding partners has recently been tested using monofunctional elements, which only have a particular function of a protein, and range from peptide fragments to separated domains. These elements can be used to recapitulate protein-protein interactions, and have been applied successfully in elucidation of the structural basis of the interactions of arrestin with receptors, clathrin, AP2, kinases, and several other effector proteins.

Keywords Arrestin · Scaffold · Peptides · MAPKs · GPCR

Introduction

The arrestin proteins were initially discovered for their interaction with active, phosphorylated G protein-coupled receptors (GPCRs), which block further G protein activation. In addition to their role in receptor desensitization, the arrestins modulate receptor endocytosis via clathrin and AP2 (Goodman et al. 1996; Laporte et al. 1999), and mediate G protein-independent signaling (Shukla et al. 2011). LC tandem MS experiments using the angiotensin II type 1a receptor identified over 100 arrestin-interacting proteins involved in cellular signaling, organization, and nucleic acid binding (Xiao et al. 2007). Due to the emerging role

N.A. Perry · T.M. Iverson · E.V. Gurevich · V.V. Gurevich (✉)
Department of Pharmacology, Vanderbilt University, Nashville, TN 46795, USA
e-mail: vsevolod.gurevich@vanderbilt.edu

X. Zhan
Department of Chemistry, Tennessee Technological University, Cookeville, TN 38505, USA

V.V. Gurevich (ed.), *The Structural Basis of Arrestin Functions*,
DOI 10.1007/978-3-319-57553-7_18

of the arrestin proteins in signaling, it is of the upmost importance to develop methods that can assess arrestin interactions with their binding partners.

The non-visual arrestins are multi-functional proteins. Recent evidence suggests that they can be divided into mono-functional elements, such as individual domains or peptide segments, which perform only one of the functions of the parental protein. Conventional methods used to study protein function include over-expression, knockout, and knockdown. These methods simultaneously affect all pathways in which the protein is involved, making it difficult to examine a single cellular process (Gurevich and Gurevich 2015). In order to investigate individual functions of an arrestin, it is necessary to engineer mutant arrestins where a particular function is disabled while all other functions remain normal; however, this level of precision has yet to be established in a laboratory setting. One way to circumvent this issue is through the expression of mono-functional elements of proteins, be it separated domains or short peptide segments (Gurevich and Gurevich 2015).

Ideally, a mono-functional element performs only a single function allowing the investigator to elucidate its biological role. In order to engineer a mono-functional element, the molecular mechanisms and folding of the protein must be well understood; for example, mechanistic information regarding protein kinases permits the design of kinase-dead mutants where only the kinase activity is impaired (Gurevich and Gurevich 2015). It is important to note that the mono-functionality of isolated domains or peptides does not limit the function of these elements to one signaling event in the cell. Instead, these elements limit unwanted cellular effects encountered when using more conventional methods to study protein function. Here we examine arrestin interactions that can be localized to specific peptide elements, which include the interaction of arrestins with receptors, clathrin, AP2, and some effectors.

The Discovery of Monofunctional Elements: Ste5-ms

The use of mono-functional elements to study protein-protein interactions between mitogen-activated protein kinase (MAPK) cascades and scaffolding proteins was first implemented in the case of the protein Ste5, which acts as a signaling scaffold for the yeast mating pathway (Ste11, Ste7, and Fus3) (Good et al. 2009). Deletion analysis was performed to locate the minimal region of Ste5 necessary to promote phosphorylation of the kinase Fus3 by Ste7. This analysis revealed an $\sim$200 residue fragment of Ste5 (593–786), termed Ste5-ms (minimal scaffold), formed an independently folding domain sufficient for Fus3 activation. Lack of strong binding between Ste5-ms and Fus3 argued against a classical tethering model to explain the functional role of the Ste5 protein. Kinetic analysis suggested that the Ste5-ms domain functioned as a substrate specific co-catalyst for the phosphorylation reaction, which is similar to the role of cyclin with cyclin-dependent kinases. More importantly, this research demonstrated that protein-protein interactions can be recapitulated using peptide fragments, and laid the foundation for studying various effector interactions in many systems using isolated elements.

Arrestin-Receptor Interactions

G protein-coupled receptors (GPCRs) constitute the largest family of proteins in the human proteome, comprising over 800 members in humans. GPCRs recognize a variety of extracellular stimuli including photons, ions, small molecules, peptides, and proteins (Pin et al. 2003). Upon agonist activation, GPCRs transmit signals across the membrane ($\sim$30 Å) and elicit an intracellular response by coupling to cognate heterotrimeric G proteins, which activate downstream effectors to initiate signaling cascades. Following phosphorylation of the receptor by specific G protein-coupled receptor kinase(s) (GRK) (Gurevich et al. 2012), arrestin complexes with the receptor to block further G protein binding (Wilden 1995; Wilden et al. 1986; Krupnick et al. 1997) and initiate receptor internalization through an interaction with clathrin (Goodman et al. 1996) and clathrin adaptor AP2 (Laporte et al. 1999) to recruit GPCRs to coated pits for endocytosis. GPCR sequestration is necessary both to attenuate the signaling response and to allow for resensitization of the receptor. In addition to these roles, arrestin also functions as an adaptor by recruiting signaling proteins to the agonist-activated GPCRs (Shenoy and Lefkowitz 2011).

The best-characterized arrestin-receptor interaction is between arrestin-1 and rhodopsin (Fig. 18.1). Visual arrestin-1[1] terminates rhodopsin signaling in rod photoreceptors by binding to light-activated, phosphorylated rhodopsin and physically blocking the interaction between heterotrimeric G protein transducin and receptor (Krupnick et al. 1994, 1997). A number of experimental techniques have been used to probe the rhodopsin-G-protein interaction including the use of synthetic peptides of the second, third, and fourth cytoplasmic loop of rhodopsin to assess binding to transducin (König et al. 1989), fluorescent labeling of the $\beta\gamma$ transducin to observe disruptions caused by interaction (Phillips et al. 1992), and mutagenesis (Franke et al. 1990). Similar approaches were used to assess the rhodopsin-arrestin interaction, which ultimately suggested the presence of a multi-site interaction between arrestin and the intracellular loops of rhodopsin (Krupnick et al. 1994). In this series of experiments, synthetic peptides of the four cytoplasmic loops of bovine rhodopsin were used to inhibit the rhodopsin-arrestin interaction. It was determined that the first and third cytoplasmic loops of rhodopsin were involved in arrestin binding, which is similar to the interface for both transducin and rhodopsin kinase (systematic name GRK1) (Krupnick et al. 1994). These

[1]This chapter uses the systematic names of the arrestin proteins: arrestin-1 (historic name *S*-antigen, visual or rod arrestin), arrestin-2 (β-arrestin or β-arrestin1), arrestin-3 (β-arrestin2), and arrestin-4 (cone or X-arrestin).

studies demonstrated that the rhodopsin-arrestin interaction can be successfully recapitulated with receptor peptide elements, although it should be noted that the reconstituted individual interactions had significantly lower affinity than the multi-site interaction of the two native proteins.

Arrestin contains both an activation and phosphorylation sensor that, when engaged by the receptor simultaneously, allow the protein to transition into an active receptor-binding conformation (Gurevich and Gurevich 2004, 2006; Gurevich and Benovic 1993). This explains arrestin selectivity for active, phosphorylated receptor; however, current experimental models are limited to rhodopsin and very few non-visual GPCRs, such as the β_2-adrenergic and M2 muscarinic receptors. The first evidence for the activation and phosphorylation sensor came from experiments testing the binding of visual arrestin-1 to different states of rhodopsin (inactive, phosphorylated, light-activated, or light-activated and phosphorylated) (Gurevich and Benovic 1993). Arrestin binding to light-activated, phosphorylated rhodopsin was significantly greater than even the sum of the other conditions, suggesting a multi-site interaction that allows arrestin to "probe" the state of the receptor. Collectively, experiments assessing the rhodopsin/arrestin interface have implicated a large portion of the arrestin surface, including most of the concave surface for both domains, in the interaction (Krupnick et al. 1994; Gurevich and Benovic 1993; Hanson et al. 2006; Hanson and Gurevich 2006; Vishnivetskiy et al. 2004, 2011; Zhuo et al. 2014).

The crystal structure of constitutively active human rhodopsin with pre-activated mouse visual arrestin-1 was recently determined using serial femtosecond X-ray crystallography (Kang et al. 2015) (Fig. 18.1). In contrast to the contiguous interaction surface of the β_2-adrenergic receptor-G_s complex (Rasmussen et al. 2011) and the adenosine A_{2A}-G_s complex (Carpenter et al. 2016), the rhodopsin-arrestin structure revealed four discrete interface patches that mediate the interaction between the receptor and arrestin. Examination of the interfaces identifies peptide-like regions that are involved in the interaction. These regions include the finger loop of arrestin (residues Q70-L78 of arrestin) (Hirsch et al. 1999), which inserts in the hydrophobic pocket between helices that opens upon GPCR activation (Farrens et al. 1996) through interactions with the C-terminus of transmembrane 7 (TM7), the N terminus of rhodopsin helix 8, which is necessary for high-affinity binding, and intracellular loop 1 (ICL1) of rhodopsin (Kang et al. 2015). The second interface consists of the middle loop (Shukla et al. 2013) [also called 139 loop in arrestin-1 (Kim et al. 2012; Vishnivetskiy et al. 2013)] and C-loop of arrestin, which make contact with intracellular loop 2 (ICL2) of rhodopsin, in addition to the back loop of arrestin that interacts with transmembrane helix 5 (TM5). The third interface visible in the crystal structure shows the β-strand (residues 79–86) interacting with TM5, TM6, and ICL3. The final interface, which consists of the N-terminal β-strand and the C-terminal tail of rhodopsin, is not present in the crystal structure due to the apparent flexibility of the rhodopsin C-tail [which was not phosphorylated in this complex (Kang et al. 2015)].

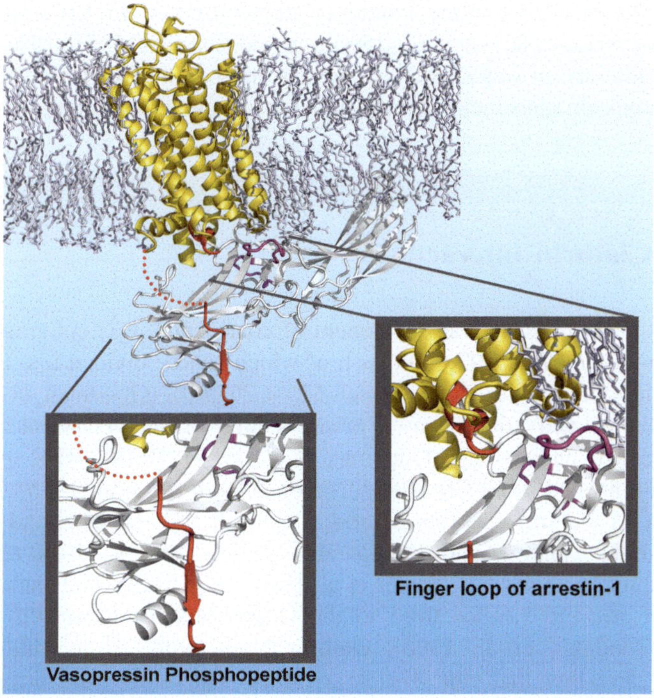

Fig. 18.1 Arrestin-rhodopsin interface. The crystal structure of arrestin in complex with rhodopsin was determined using femtosecond X-ray laser crystallography (PDB 4ZWJ) (Kang et al. 2015). *Left inset* The vasopressin receptor phosphopeptide (*red*) is shown over top of active arrestin-1 (*white*) in complex with rhodopsin (*gold*). This 29-residue phosphopeptide is derived from the carboxy-terminus of the vasopressin receptor (Shukla et al. 2013). *Right inset* The finger loop of arrestin (*red*) is shown inserting into the hydrophobic core of the receptor. The middle loop and C-loop (*magenta*) also form part of the arrestin-rhodopsin interface (Kang et al. 2015). The unstructured region of rhodopsin is depicted using a dashed line (*red*)

Structural insights into the interface between arrestin and receptor have proven extremely valuable in furthering our understanding of receptor-dependent and -independent signaling pathways. Several investigators have probed this interface using receptor-derived phosphopeptides to mimic a phosphorylated GPCR. Shukla et al. used a fully phosphorylated carboxy-terminal peptide derived from human V2 vasopressin receptor (V2Rpp) to induce the active arrestin-2 conformation stabilized by Fab30 specific for this state of arrestin-2 to 2.6 Å resolution (Shukla et al. 2013). The investigators found a number of structural differences compared to the basal state, including a 20° domain rotation and movement of the finger, middle, and lariat loops [the latter provides two out of three negative charges to the inter-domain polar core, which functions as the main phosphate sensor (Hirsch et al. 1999; Vishnivetskiy et al. 1999)]. This "activated" form of arrestin shows similarities with the previously published structure of pre-activated arrestin p44, a C-terminally truncated arrestin-1 (Kim et al. 2013). A second study examined the interaction of arrestin with receptor using a peptide analog of the finger loop of arrestin-1. The arrestin peptide bound to the conserved E(D)RY motif of the cytoplasmic crevice of activated rhodopsin, in addition to TM7/H8 (Szczepek et al. 2014). This interaction was slightly different from the interface between rhodopsin and C terminus of transducin, which suggests a potential mechanism for biased signaling.

Arrestin-Clathrin Interactions

Clathrin is the major structural component of coated pits. It is comprised of three heavy chains (total $M_r \approx 192{,}000$), each of which can be divided into two distinct domains, and three light chains (total $M_r \approx 26{,}000$) (Goodman et al. 1997). Following agonist activation, many G protein-coupled receptors are internalized via clathrin-mediated endocytosis (Goodman et al. 1997). Both GRKs and arrestins have been implicated in GPCR internalization (Goodman et al. 1996). Non-visual arrestins act as clathrin adaptors to bridge clathrin and receptor, and have been shown to bind clathrin with comparable affinity ($K_d \sim$ 10–60 nM) to the clathrin-adaptor protein AP-2 (Peeler et al. 1993) (Fig. 18.2). The clathrin binding interface on arrestin was localized to the C-terminus using arrestin-1/arrestin-2 chimeras (Goodman et al. 1996); clathrin binds with greater affinity to the non-visuals arrestins than the visual arrestins. The non-visual arrestins share approximately 58% sequence identity in the C terminus, and the chimera, which contained residues 1–345 of visual arrestin and 341–418 of arrestin-2, was able to

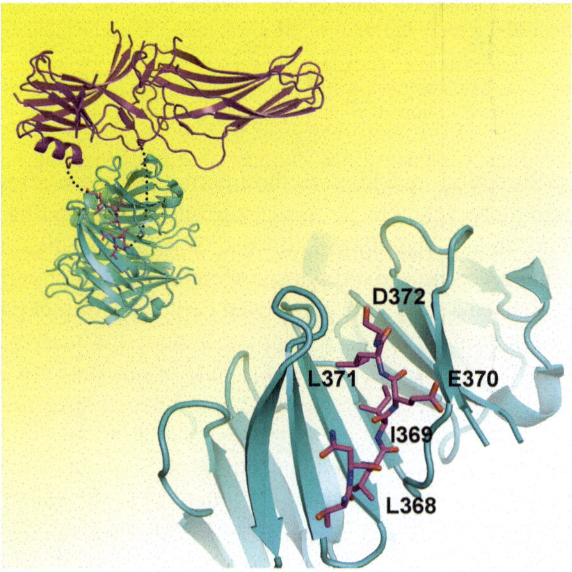

Fig. 18.2 Arrestin-clathrin interaction. The crystal structure of arrestin-2 (*deep purple*) with clathrin (*teal*) was determined using X-ray crystallography to 2.2 Å resolution (PDB 3GC3) (Kang et al. 2009). The LϕXϕ(D/E) motif (*pink*) from arrestin-2 interacts with the clathrin binding box (unstructured regions of arrestin-2 are depicted as a *black dotted line* in the cartoon)

bind clathrin cages with similar affinity to arrestin-2. Together these findings suggested that the C-terminal 77 amino acids comprise the major site for the arrestin/clathrin interaction and could be used as a mono-functional element to recapitulate the arrestin-clathrin interaction.

The terminal domain (TD) of clathrin is an approximately 50 kDa amino-terminal region located on the distal portion of the triskelion leg (Goodman et al. 1997). In addition to containing the binding site for AP-2, this region has a distinct arrestin binding site (heavy chain residues 89–100) that has been tested using deletion and alanine scanning mutagenesis (Goodman et al. 1997). The TD was isolated by subjecting clathrin cages to limited tryptic digestion and assessed for arrestin binding using pull-down assays, which demonstrated that the non-visual arrestins interact with the TD. The arrestin binding site within clathrin was further localized using truncated constructs of the TD, specifically a construct containing heavy chain residues 1–85 that did not bind arrestin, which directed investigators to residues 86–100. It is important to note that, although non-visual arrestins exhibit a strong interaction with clathrin, they do not yield assembled lattices. Arrestin only colocalizes with clathrin in response to agonist activation of GPCRs, which is consistent with the idea that arrestin acts as an adaptor for lattice assembly. Interestingly, expression in cells of the separated arrestin-2 C-terminus, which contains both clathrin and AP2 binding sites, inhibits arrestin-dependent GPCR internalization, apparently by binding to clathrin/AP2 cages and successfully outcompeting the arrestin-receptor complexes (Orsini and Benovic 1998).

Crystallographic studies of the TD of clathrin in complex with an arrestin-3 peptide (ter Haar et al. 2000) and full-length arrestin-2 (Kang et al. 2009) verified the location of the arrestin-clathrin interaction site, which included the N-terminal domain of the clathrin heavy chain and the Lϕxϕ[D/E] motif on the carboxyl terminal region of arrestin (ϕ represents a hydrophobic residue and x represents a polar residue). In addition to these sites, the study using full-length arrestin-2 found that arrestin and clathrin interacted through a previously uncharacterized eight-amino acid loop $\sim$68 Å away from the Lϕxϕ[D/E] motif (Kang et al. 2009). This sequence is not present in arrestin-3, but is present in the two visual arrestins, arrestin-1 and -4, which suggests a mechanism through which the visual arrestins can bind clathrin. However, the interaction of visual arrestins with clathrin has not been tested experimentally. In contrast, the interaction of arrestin-1 peptides with clathrin adaptor AP2 was demonstrated both in vitro and using a transgenic mouse model (Moaven et al. 2013). Interestingly, despite a much lower affinity for AP2 than that of the corresponding arrestin-2 element, the arrestin-1/AP2 interaction turned out to be detrimental for photoreceptor survival (Moaven et al. 2013). This can likely be explained by the high concentration of arrestin-1 in rod cells (Song et al. 2011; Strissel et al. 2006). These results indicate that one can recapitulate protein-protein interactions using peptide elements in both in vitro and in vivo analysis.

Arrestin-Effector Interactions

The role of arrestins in signal transduction was first described in 1999 (Luttrell et al. 1999); however, it is becoming increasingly apparent that arrestins play a critical role in a variety of signal transduction pathways. A global proteomics analysis was used to determine the scope of functions performed by arrestin-2 and -3, and identified 337 non-redundant interacting proteins, 102 of which interacted with both non-visual arrestins (Xiao et al. 2007). This study found a wide distribution for subcellular localization of the interacting proteins, which displayed a range of functions in signal transduction, cellular organization, metabolism, and nucleic acid binding. The screen also identified well-known binding partners, including the extracellular signal-regulated kinase 1 (ERK1), clathrin and AP2, and phosphodiesterase 4. Of the identified signaling proteins that interact with arrestins, the best studied by far are the kinases.

Src Family Kinases

The Src family kinases play a key role in signal transduction from a number of cell surface receptors, most notably the receptor tyrosine kinases. Src family kinases are therefore critical for a variety of cellular processes, including proliferation, differentiation, and migration, which explains why many members of this kinase family are implicated in cancer (Parsons and Parsons 2004). The Src family of kinases includes nine members—Src, Yes, Fyn, Fgr, Lck, Hck, Blk, Lyn, and Frk—that share structural features, including SH2 and SH3 domains, which are necessary for phospho-tyrosine recognition of activated receptor tyrosine kinases. Of these members, Src, Fyn, and Yes are the most widely expressed. Src family kinases are also known to play a role in GPCR signaling, both through direct interaction with GPCRs and through receptor-associated transactivation of receptor tyrosine kinases (Luttrell and Luttrell 2004). Arrestins 1, 2, and 3 have been shown to bind Src family kinases and recruit them to activated GPCRs.

The prototypic non-receptor tyrosine kinase, *c*-Src (short for sarcoma, this kinase was discovered for its role as an oncogene), was the first identified signaling interaction partner of arrestin (Luttrell et al. 1999). Upon activation of receptor tyrosine kinases, *c*-Src is recruited to the plasma membrane where it catalyzes the phosphorylation of tyrosine residues. Confocal microscopy demonstrated that this redistribution could also be induced by agonist activation of the β_2-adrenergic receptor through the assembly of a protein complex that contained receptor, arrestin, and c-Src (Luttrell et al. 1999). This complex could also be isolated using co-immunoprecipitation after exposure to a β_2-adrenergic receptor agonist and covalent crosslinking. The stoichiometry of *c*-Src interaction with arrestin in the receptor immunoprecipitates yielded a ratio of 0.84 ($\pm$0.15):1 (Luttrell et al. 1999)

These results suggest that the arrestin/*c*-Src interaction is important for the localization of *c*-Src to the membrane and trans-activation (Strungs and Luttrell 2014).

The interaction site for *c*-Src on arrestin was identified by co-expressing Flag-tagged deletion or truncation arrestin mutants in COS7 cells with wild-type *c*-Src. Co-immunoprecipitation revealed that deleting amino acids 1–185 resulted in a complete loss of binding to *c*-Src, and that a fragment of arrestin containing amino acids 1–163 was able to bind *c*-Src similar to the full-length arrestin (Luttrell et al. 1999). Site-directed mutagenesis of hydrophobic residues in the P-X-X-P motif in arrestin, which is the minimal consensus sequence for binding to the SH3 domain (Cohen et al. 1995), reduced *c*-Src binding, indicating that *c*-Src interacts with the N-terminal domain of arrestin. Although the *c*-Src binding sites have not been precisely mapped on arrestin, the use of mutational studies to distinguish a specific domain suggest that the binding site can be localized to specific arrestin elements.

In addition to *c*-Src, notable Src family kinases Fyn and Yes interact with arrestins. Activation of luteinizing hormone receptor (LHR) involves several signaling cascades that can lead to Fyn activation, including the formation of a ternary complex between LHR, a non-visual arrestin, and Fyn (Galet and Ascoli 2008). In a similar manner, Yes is recruited to the endothelin type A (ETA) receptor via an interaction with arrestin-2 (Imamura et al. 2001). The exact location of the interaction sites for these proteins on arrestins is unknown; however, it is probable that the kinases bind in a similar manner to *c*-Src, with single element interactions.

MAP Kinases

Arrestin has also been shown to act as a scaffold protein for mitogen activated protein kinase (MAPK) cascades, in particular the subfamilies c-Jun N-terminal kinase 3 (JNK3), extracellular signal-regulated kinases (ERK1/2), and p38, which involve sequential action of three core kinases (MAPK kinase kinase, MAPK kinase, and MAPK) (McDonald et al. 2000; Luttrell et al. 2001; Bruchas et al. 2006) (Fig. 18.3). It is widely accepted that scaffold proteins are important elements in regulating signaling specificity of MAPK cascades (Dhanasekaran et al. 2007; Good et al. 2011). Identification of specific binding sites for MAPKs and their upstream activators on arrestin would offer insight into the signaling process and enable the development of designer arrestins with controlled signaling outputs (Gurevich and Gurevich 2010).

The mechanism for arrestin-dependent assembly of MAPK signaling complexes is not completely understood; however, it has been shown that effector interaction with arrestin is not exclusively dependent on the arrestin-receptor complex formation (Song et al. 2009). For example, receptor-independent arrestin-3 is able to effectively scaffold the ASK1-MKK4/7-JNK3 kinase cascade (Song et al. 2009; Seo et al. 2011; Breitman et al. 2012), while ERK2 binding to arrestin-2/3 occurs independent of receptor binding, but is significantly enhanced in the receptor-bound state (Song et al. 2009; Coffa et al. 2011). It has also been established that effector

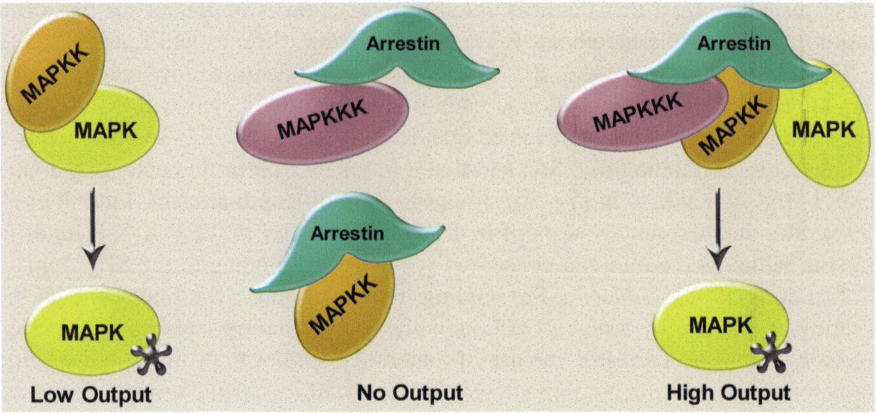

Fig. 18.3 Arrestin acts as a molecular scaffold for mitogen-activated protein kinase cascades. Scaffold proteins promote signaling by localizing molecular components to a specific area of the cell. Arrestin is known to scaffold numerous effectors, most notably the mitogen-activated protein kinase (MAPK) cascades. Cartoon representation of effector binding to the non-receptor bound surface of arrestin is shown. In the absence of arrestin, components of the MAPK cascade (MAPKKK–MAPKK–MAPK) are diffusion limited and exhibit a low output of activated MAPK (phosphorylation indicated by *black star*). When only one component of a MAPK cascade is bound to arrestin, there is no signaling output. In contrast, when arrestin scaffolds the complete MAPK cascade there is robust activation of the final effector kinase

recruitment to arrestin is not necessarily indicative of effector activation (Song et al. 2009; Seo et al. 2011). This is evidenced most clearly by the activation of JNK3 only via arrestin-3 recruitment (McDonald et al. 2000; Song et al. 2009; Seo et al. 2011), despite JNK3 binding to all isoforms of arrestin (Song et al. 2006, 2007). It is therefore necessary to elucidate which regions of arrestin are able to both bind and activate effectors when designing peptides for therapeutics.

Binding regions for several MAPK cascades on arrestin have been determined using in vitro analysis, such as scanning peptide arrays and binding assays with purified proteins, as well as in cell analysis with mutant arrestin constructs (Coffa et al. 2011; Li et al. 2009; Meng et al. 2009; Zhan et al. 2016). Binding does not necessarily cause activation: to promote signaling, arrestins must assemble the appropriate MAPK components in an arrangement that is conducive to activation (Zhan et al. 2014). Next, investigators have moved toward more targeted identification of binding sites for particular kinases.

Miscellaneous Effectors

Several binding sites on arrestin for non-kinase effectors have been mapped to a similar extent. For most effectors, binding occurs on both the N- and C-domains of arrestin (Song et al. 2009; Ahmed et al. 2011). Spot-immobilized peptide array analysis was used to map the interaction site for cAMP-specific phosphodiesterases (PDE4D5, PDE4D3), which revealed a difference in localization of the binding site between the isoforms (Baillie et al. 2007). Calmodulin, an intracellular calcium regulator, was identified as an arrestin-interacting partner in the presence of calcium with a binding site on the concave side of the C-domain and the finger loop of arrestin (Wu et al. 2006). This study was conducted using co-immunoprecipitation and site-directed spin labeling (SDSL) electron paramagnetic resonance (EPR). The N-domain of arrestin-4 was shown to relocalize Mdm2 from the nucleus to the cytoplasm using a nuclear exclusion assay, indicating that this was the site of high affinity binding of this E3 ubiquitin ligase (Song et al. 2007). This list of studies, although not exhaustive for the number of identified interacting partners of arrestin, supports the notion that different regions of arrestin are specialized for binding specific effectors.

Manipulating Signal Transduction Using Monofunctional Elements of Scaffold Proteins

The c-Jun NH_2-terminal protein kinase 3 (JNK3) is the final effector in the ASK1-MKK4/7-JNK3 signaling cascade, and is known for its role in neuronal apoptosis (Yang et al. 1997). Previous studies indicate that all components of the JNK3 signaling cascade interact with non-visual arrestins (McDonald et al. 2000; Song et al. 2009), and that binding of JNK3 is not dependent on the conformational state of arrestin (Song et al. 2006). These data suggested that the members of this MAPK cascade might interact with arrestin via peptide-like surface segments. This hypothesis was tested using an in vitro pull-down assay with MBP-fusion of intact wild-type arrestin-3 as a positive, and empty MBP as a negative control. MBP fusions of the peptides representing the entire non-receptor binding surface of arrestin-3 (Fig. 18.4) were tested for binding to the JNK3α2 isoform (Zhan et al. 2014). This experiment successfully identified three arrestin-derived peptides that bind to JNK3α2. One of the peptides, termed T1 (amino acids M1-R52) demonstrated the greatest binding to JNK3α2. Therefore, it was further divided into three peptide regions T1A (M1-K25), T1B (R26-R52), and T1C (N16-D45), which identified the first twenty-five amino acids of arrestin-3 as a critical element for JNK3α2 binding (Fig. 18.4). Interestingly, this region of arrestin-3 had previously been implicated in binding to the kinase domain of ASK1 (Li et al. 2009). These data suggested that the T1A peptide could be used to recapitulate arrestin-3 scaffolding of the ASK1-MKK4/7-JNK3 signaling cascade.

To assess the functional capabilities of the T1A peptide, binding experiments with purified upstream kinases MKK4 and MKK7 were performed (Zhan et al. 2016). The peptide B1A, which represents the first 25 amino acids of arrestin-2, was used as a "natural" negative control, because arrestin-2 does not promote JNK3 activation despite sharing 78% sequence identity to arrestin-3 (Sterne-Marr et al. 1993). While both peptides bind MKK4 comparably, T1A exhibited more robust

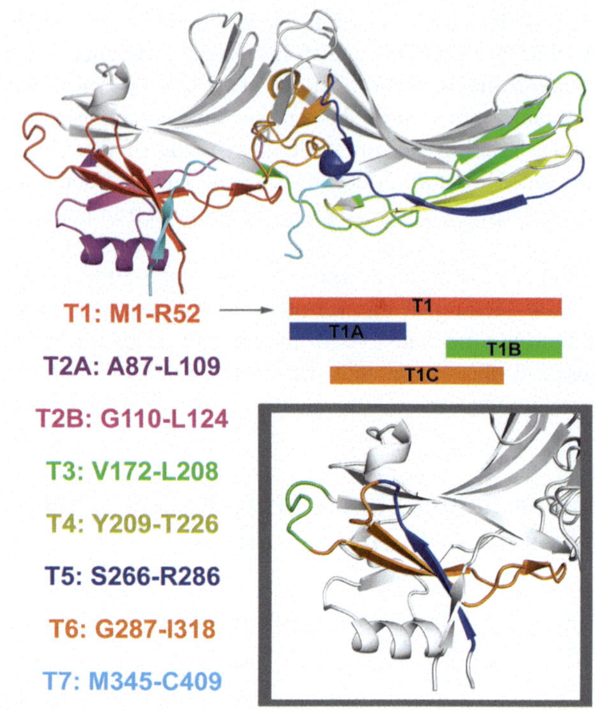

Fig. 18.4 Peptides of the non-receptor-binding surface of arrestin-3. The non-receptor-binding surface of arrestin-3 was divided into peptide fragments to assess the binding of JNK3 to different regions of arrestin-3 (PDB 3P2D) (Zhan et al. 2014). Each segment is color-coded, with the corresponding residues listed. T1 bound with high affinity to arrestin-3, and was further divided into smaller segments to isolate a twenty-five residue peptide, T1A, that not only binds JNK3 with high affinity (Zhan et al. 2014), but can scaffold the complete ASK1-MKK4/7-JNK3 cascade in cells, facilitating JNK3 activation (Zhan et al. 2016)

binding to JNK3 and MKK7. Both MKKs interacted with T1A in a comparable manner to full-length arrestin-3. Since purified ASK1 was not available for in vitro pull downs, HA-ASK1 binding was tested in COS7 cells. It effectively co-immunoprecipitated with YFP-T1A. This demonstrated that the T1A peptide was able to bind all upstream kinases of the JNK3 activation cascade: ASK1, MKK4, and MKK7.

To test whether T1A facilitates JNK3 phosphorylation like full-length arrestin-3, it was co-expressed with HA-ASK1 and HA-JNK3 in COS7 cells and found to be active (Zhan et al. 2016). In contrast to T1A, B1A did not facilitate the phosphorylation of JNK3 in cells, which suggests that the specific sequence of the peptide determines its functional capabilities. Thus, the first twenty-five residues of arrestin-3 are largely responsible for scaffolding the JNK3 cascade. This makes T1A the smallest known MAPK scaffold reported.

A reciprocal study used 25-mer peptides spanning the complete sequence of JNK3 to probe the interaction between the non-visual arrestins and JNK3 (Li et al. 2009). This analysis isolated the N-terminus of JNK3 (amino acids 6–45) as critical for arrestin binding, which was in accordance with previous site-directed mutagenesis studies (Guo and Whitmarsh 2008). Alanine substitution further localized the interaction site to residues F^5 and C^{21}-F^{24}. Similar analysis was conducted with ASK1 and MKK4, identifying two regions on the non-visual arrestins that support ASK1 binding, and one region that supports MKK4 binding.

Conclusions

The use of mono-functional elements has largely been successful when isolating regions of arrestin necessary to interact with receptors and certain downstream effectors. Multiple studies of arrestins have used isolated peptides or domains to localize protein interaction sites, and have been vital for our current understanding of these multifunctional proteins. As the field moves toward the interactions of arrestins with effectors, tools that allow investigators the ability to assess single functions of arrestin are critical for furthering our understanding of arrestin-mediated signaling.

Acknowledgements This work was supported by NIH grants to VVG (GM077561, GM109955, EY011500), EVG (NS065868, DA030103) and TMI (GM095633, GM120569). NAP was supported by T32 GM007628 and the American Heart Association (16PRE30180007).

References

Ahmed MR, Zhan X, Song X, Kook S, Gurevich VV, Gurevich EV (2011) Ubiquitin ligase parkin promotes Mdm2-arrestin interaction but inhibits arrestin ubiquitination. Biochemistry 50:3749–3763

Baillie George S, Adams David R, Bhari N, Houslay Thomas M, Vadrevu S, Meng D, Li X, Dunlop A, Milligan G, Bolger Graeme B, Klussmann E, Houslay Miles D (2007) Mapping binding sites for the PDE4D5 cAMP-specific phosphodiesterase to the N- and C-domains of β-arrestin using spot-immobilized peptide arrays. Biochem J 404:71–80

Breitman M, Kook S, Gimenez LE, Lizama BN, Palazzo MC, Gurevich EV, Gurevich VV (2012) Silent scaffolds: inhibition of c-Jun N-terminal kinase 3 activity in the cell by a dominant-negative arrestin-3 mutant. J Biol Chem 287:19653–19664

Bruchas MR, Macey TA, Lowe JD, Chavkin C (2006) Kappa opioid receptor activation of p38 MAPK Is GRK3- and arrestin-dependent in neurons and astrocytes. J Biol Chem 281:18081–18089

Carpenter B, Nehmé R, Warne T, Leslie AG, Tate CG (2016) Structure of the adenosine A(2A) receptor bound to an engineered G protein. Nature 536:104–107

Coffa S, Breitman M, Hanson SM, Callaway K, Kook S, Dalby KN, Gurevich VV (2011) The effect of arrestin conformation on the recruitment of c-Raf1, MEK1, and ERK1/2 activation. PLoS ONE 6:e28723

Cohen GB, Ren R, Baltimore D (1995) Modular binding domains in signal transduction proteins. Cell 80:237–248

Dhanasekaran DN, Kashef K, Lee CM, Xu H, Reddy EP (2007) Scaffold proteins of MAP-kinase modules. Oncogene 26:3185–3202

Farrens DL, Altenbach C, Yang K, Hubbell WL, Khorana HG (1996) Requirement of rigid-body motion of transmembrane helices for light activation of rhodopsin. Science 274:768–770

Franke RR, König B, Sakmar TP, Khorana HG, Hofmann KP (1990) Rhodopsin mutants that bind but fail to activate transducin. Science 250:123–125

Galet C, Ascoli M (2008) Arrestin-3 is essential for the activation of Fyn by the luteinizing hormone receptor (LHR) in MA-10 cells. Cell Signal 20:1822–1829

Good MC, Tang G, Singleton J, Remenyi A, Lim WA (2009) The Ste5 scaffold directs mating signaling by catalytically unlocking the Fus3 MAP kinase for activation. Cell 136:1085–1097

Good MC, Zalatan JG, Lim WA (2011) Scaffold proteins: hubs for controlling the flow of cellular information. Science 332:680–686

Goodman OB, Krupnick JG, Santini F, Gurevich VV, Penn RB, Gagnon AW, Keen JH, Benovic JL (1996) [beta]-Arrestin acts as a clathrin adaptor in endocytosis of the [beta] 2-adrenergic receptor. Nature 383:447–450

Goodman OB, Krupnick JG, Gurevich VV, Benovic JL, Keen JH (1997) Arrestin/clathrin interaction. Localization of the arrestin binding locus to the clathrin terminal domain. J Biol Chem 272:15017–15022

Guo C, Whitmarsh AJ (2008) The beta-arrestin-2 scaffold protein promotes c-Jun N-terminal kinase-3 activation by binding to its nonconserved N terminus. J Biol Chem 283:15903–15911

Gurevich VV, Benovic JL (1993) Visual arrestin interaction with rhodopsin. Sequential multisite binding ensures strict selectivity toward light-activated phosphorylated rhodopsin. J Biol Chem 268:11628–11638

Gurevich VV, Gurevich EV (2004) The molecular acrobatics of arrestin activation. Trends Pharmacol Sci 25:105–111

Gurevich VV, Gurevich EV (2006) The structural basis of arrestin-mediated regulation of G-protein-coupled receptors. Pharmacol Ther 110:465–502

Gurevich VV, Gurevich EV (2010) Custom-designed proteins as novel therapeutic tools? The case of arrestins. Expert Rev Mol Med 12:e13

Gurevich VV, Gurevich EV (2015) Analyzing the roles of multi-functional proteins in cells: the case of arrestins and GRKs. Crit Rev Biochem Mol Biol 50:440–452

Gurevich EV, Tesmer JJ, Mushegian A, Gurevich VV (2012) G protein-coupled receptor kinases: more than just kinases and not only for GPCRs. Pharmacol Ther 133:40–69

Hanson SM, Gurevich VV (2006) The differential engagement of arrestin surface charges by the various functional forms of the receptor. J Biol Chem 281:3458–3462

Hanson SM, Francis DJ, Vishnivetskiy SA, Kolobova EA, Hubbell WL, Klug CS, Gurevich VV (2006) Differential interaction of spin-labeled arrestin with inactive and active phosphorhodopsin. Proc Natl Acad Sci USA 103:4900–4905

Hirsch JA, Schubert C, Gurevich VV, Sigler PB (1999) The 2.8 A crystal structure of visual arrestin: a model for arrestin's regulation. Cell 97:257–269

Imamura T, Huang J, Dalle S, Ugi S, Usui I, Luttrell LM, Miller WE, Lefkowitz RJ, Olefsky JM (2001) β-Arrestin-mediated recruitment of the Src family kinase yes mediates endothelin-1-stimulated glucose transport. J Biol Chem 276:43663–43667

Kang DS, Kern RC, Puthenveedu MA, von Zastrow M, Williams JC, Benovic JL (2009) Structure of an arrestin2-clathrin complex reveals a novel clathrin binding domain that modulates receptor trafficking. J Biol Chem 284:29860–29872

Kang Y, Zhou XE, Gao X, He Y, Liu W, Ishchenko A, Barty A, White TA, Yefanov O, Han GW, Xu Q, de Waal PW, Ke J, Tan MH, Zhang C, Moeller A, West GM, Pascal BD, Van Eps N, Caro LN, Vishnivetskiy SA, Lee RJ, Suino-Powell KM, Gu X, Pal K, Ma J, Zhi X, Boutet S, Williams GJ, Messerschmidt M, Gati C, Zatsepin NA, Wang D, James D, Basu S, Roy-Chowdhury S, Conrad CE, Coe J, Liu H, Lisova S, Kupitz C, Grotjohann I, Fromme R, Jiang Y, Tan M, Yang H, Li J, Wang M, Zheng Z, Li D, Howe N, Zhao Y, Standfuss J, Diederichs K, Dong Y, Potter CS, Carragher B, Caffrey M, Jiang H, Chapman HN, Spence JC, Fromme P, Weierstall U, Ernst OP, Katritch V, Gurevich VV, Griffin PR, Hubbell WL, Stevens RC, Cherezov V, Melcher K, Xu HE (2015) Crystal structure of rhodopsin bound to arrestin by femtosecond X-ray laser. Nature 523:561–567

Kim M, Vishnivetskiy SA, Van Eps N, Alexander NS, Cleghorn WM, Zhan X, Hanson SM, Morizumi T, Ernst OP, Meiler J, Gurevich VV, Hubbell WL (2012) Conformation of receptor-bound visual arrestin. Proc Nat Acad Sci USA 109:18407–18412

Kim YJ, Hofmann KP, Ernst OP, Scheerer P, Choe HW, Sommer ME (2013) Crystal structure of pre-activated arrestin p44. Nature 497:142–146

König B, Arendt A, McDowell JH, Kahlert M, Hargrave PA, Hofmann KP (1989) Three cytoplasmic loops of rhodopsin interact with transducin. Proc Natl Acad Sci USA 86:6878–6882

Krupnick JG, Gurevich VV, Schepers T, Hamm HE, Benovic JL (1994) Arrestin-rhodopsin interaction. Multi-site binding delineated by peptide inhibition. J Biol Chem 269:3226–3232

Krupnick JG, Gurevich VV, Benovic JL (1997) Mechanism of quenching of phototransduction. Binding competition between arrestin and transducin for phosphorhodopsin. J Biol Chem 272:18125–18131

Laporte SA, Oakley RH, Zhang J, Holt JA, Ferguson SG, Caron MG, Barak LS (1999) The 2-adrenergic receptor/arrestin complex recruits the clathrin adaptor AP-2 during endocytosis. Proc Nat Acad Sci USA 96:3712–3717

Li X, MacLeod R, Dunlop AJ, Edwards HV, Advant N, Gibson LC, Devine NM, Brown KM, Adams DR, Houslay MD, Baillie GS (2009) A scanning peptide array approach uncovers association sites within the JNK/beta arrestin signalling complex. FEBS Lett 583:3310–3316

Luttrell DK, Luttrell LM (2004) Not so strange bedfellows: G-protein-coupled receptors and Src family kinases. Oncogene 23:7969–7978

Luttrell LM, Ferguson SS, Daaka Y, Miller WE, Maudsley S, Della Rocca GJ, Lin F, Kawakatsu H, Owada K, Luttrell DK, Caron MG, Lefkowitz RJ (1999) Beta-arrestin-dependent formation of beta2 adrenergic receptor-Src protein kinase complexes. Science 283:655–661

Luttrell LM, Roudabush FL, Choy EW, Miller WE, Field ME, Pierce KL, Lefkowitz RJ (2001) Activation and targeting of extracellular signal-regulated kinases by β-arrestin scaffolds. Proc Natl Acad Sci 98:2449–2454

McDonald PH, Chow CW, Miller WE, Laporte SA, Field ME, Lin FT, Davis RJ, Lefkowitz RJ (2000) Beta-arrestin 2: a receptor-regulated MAPK scaffold for the activation of JNK3. Science 290:1574–1577

Meng D, Lynch MJ, Huston E, Beyermann M, Eichhorst J, Adams DR, Klussmann E, Houslay MD, Baillie GS (2009) MEK1 binds directly to βArrestin1, influencing both its phosphorylation by ERK and the timing of its isoprenaline-stimulated internalization. J Biol Chem 284:11425–11435

Moaven H, Koike Y, Jao CC, Gurevich VV, Langen R, Chen J (2013) Visual arrestin interaction with clathrin adaptor AP-2 regulates photoreceptor survival in the vertebrate retina. Proc Natl Acad Sci USA 110:9463–9468

Orsini MJ, Benovic JL (1998) Characterization of dominant negative arrestins that inhibit beta2-adrenergic receptor internalization by distinct mechanisms. J Biol Chem 273:34616–34622

Parsons SJ, Parsons JT (2004) Src family kinases, key regulators of signal transduction. Oncogene 23:7906–7909

Peeler JS, Donzell WC, Anderson RG (1993) The appendage domain of the AP-2 subunit is not required for assembly or invagination of clathrin-coated pits. J Cell Biol 120:47–54

Phillips WJ, Wong SC, Cerione RA (1992) Rhodopsin/transducin interactions. II. Influence of the transducin-beta gamma subunit complex on the coupling of the transducin-alpha subunit to rhodopsin. J Biol Chem 267:17040–17046

Pin J-P, Galvez T, Prézeau L (2003) Evolution, structure, and activation mechanism of family 3/C G-protein-coupled receptors. Pharmacol Ther 98:325–354

Rasmussen SG, DeVree BT, Zou Y, Kruse AC, Chung KY, Kobilka TS, Thian FS, Chae PS, Pardon E, Calinski D, Mathiesen JM, Shah ST, Lyons JA, Caffrey M, Gellman SH, Steyaert J, Skiniotis G, Weis WI, Sunahara RK, Kobilka BK (2011) Crystal structure of the β2 adrenergic receptor-Gs protein complex. Nature 477:549–555

Seo J, Tsakem EL, Breitman M, Gurevich VV (2011) Identification of arrestin-3-specific residues necessary for JNK3 kinase activation. J Biol Chem 286:27894–27901

Shenoy SK, Lefkowitz RJ (2011) β-arrestin-mediated receptor trafficking and signal transduction. Trends Pharmacol Sci 32:521–533

Shukla AK, Xiao K, Lefkowitz RJ (2011) Emerging paradigms of β-arrestin-dependent seven transmembrane receptor signaling. Trends Biochem Sci 36:457–469

Shukla AK, Manglik A, Kruse AC, Xiao K, Reis RI, Tseng WC, Staus DP, Hilger D, Uysal S, Huang LY, Paduch M, Tripathi-Shukla P, Koide A, Koide S, Weis WI, Kossiakoff AA, Kobilka BK, Lefkowitz RJ (2013) Structure of active β-arrestin-1 bound to a G-protein-coupled receptor phosphopeptide. Nature 497:137–141

Song X, Raman D, Gurevich EV, Vishnivetskiy SA, Gurevich VV (2006) Visual and both non-visual arrestins in their "inactive" conformation bind JNK3 and Mdm2 and relocalize them from the nucleus to the cytoplasm. J Biol Chem 281:21491–21499

Song X, Gurevich EV, Gurevich VV (2007) Cone arrestin binding to JNK3 and Mdm2: conformational preference and localization of interaction sites. J Neurochem 103:1053–1062

Song X, Coffa S, Fu H, Gurevich VV (2009) How does arrestin assemble MAPKs into a signaling complex? J Biol Chem 284:685–695

Song X, Vishnivetskiy SA, Seo J, Chen J, Gurevich EV, Gurevich VV (2011) Arrestin-1 expression in rods: balancing functional performance and photoreceptor health. Neuroscience 174:37–49

Sterne-Marr R, Gurevich VV, Goldsmith P, Bodine RC, Sanders C, Donoso LA, Benovic JL (1993) Polypeptide variants of beta-arrestin and arrestin3. J Biol Chem 268:15640–15648

Strissel KJ, Sokolov M, Trieu LH, Arshavsky VY (2006) Arrestin translocation is induced at a critical threshold of visual signaling and is superstoichiometric to bleached rhodopsin. J Neurosci 26:1146–1153

Strungs EG, Luttrell LM (2014) Arrestin-dependent activation of ERK and Src family kinases. In: Gurevich VV (ed) Arrestins—pharmacology and therapeutic potential. Springer, Berlin, pp 225–257

Szczepek M, Beyrière F, Hofmann KP, Elgeti M, Kazmin R, Rose A, Bartl FJ, von Stetten D, Heck M, Sommer ME, Hildebrand PW, Scheerer P (2014) Crystal structure of a common GPCR-binding interface for G protein and arrestin. Nat Commun 5

ter Haar E, Harrison SC, Kirchhausen T (2000) Peptide-in-groove interactions link target proteins to the beta-propeller of clathrin. Proc Natl Acad Sci USA 97:1096–1100

Vishnivetskiy SA, Paz CL, Schubert C, Hirsch JA, Sigler PB, Gurevich VV (1999) How does arrestin respond to the phosphorylated state of rhodopsin? J Biol Chem 274:11451–11454

Vishnivetskiy SA, Hosey MM, Benovic JL, Gurevich VV (2004) Mapping the arrestin-receptor interface: structural elements responsible for receptor specificity of arrestin proteins. J Biol Chem 279:1262–1268

Vishnivetskiy SA, Gimenez LE, Francis DJ, Hanson SM, Hubbell WL, Klug CS, Gurevich VV (2011) Few residues within an extensive binding interface drive receptor interaction and determine the specificity of arrestin proteins. J Biol Chem 286:24288–24299

Vishnivetskiy SA, Baameur F, Findley KR, Gurevich VV (2013) Critical role of the central 139-loop in stability and binding selectivity of arrestin-1. J Biol Chem 288:11741–11750

Wilden U (1995) Duration and amplitude of the light-induced cGMP hydrolysis in vertebrate photoreceptors are regulated by multiple phosphorylation of rhodopsin and by arrestin binding. Biochemistry 34:1446–1454

Wilden U, Hall SW, Kuhn H (1986) Phosphodiesterase activation by photoexcited rhodopsin is quenched when rhodopsin is phosphorylated and binds the intrinsic 48-kDa protein of rod outer segments. Proc Natl Acad Sci USA 83:1174–1178

Wu N, Hanson SM, Francis DJ, Vishnivetskiy SA, Thibonnier M, Klug CS, Shoham M, Gurevich VV (2006) Arrestin binding to calmodulin: a direct interaction between two ubiquitous signaling proteins. J Mol Biol 364:955–963

Xiao K, McClatchy DB, Shukla AK, Zhao Y, Chen M, Shenoy SK, Yates JR, Lefkowitz RJ (2007) Functional specialization of β-arrestin interactions revealed by proteomic analysis. Proc Natl Acad Sci 104:12011–12016

Yang DD, Kuan C-Y, Whitmarsh AJ, Rinocn M, Zheng TS, Davis RJ, Rakic P, Flavell RA (1997) Absence of excitotoxicity-induced apoptosis in the hippocampus of mice lacking the Jnk3 gene. Nature 389:865–870

Zhan X, Kook S, Gurevich EV, Gurevich VV (2014a) Arrestin-dependent activation of JNK family kinases. In: Gurevich VV (ed) Arrestins—pharmacology and therapeutic potential. Springer, Berlin, pp 259–280

Zhan X, Perez A, Gimenez LE, Vishnivetskiy SA, Gurevich VV (2014b) Arrestin-3 binds the MAP kinase JNK3α2 via multiple sites on both domains. Cell Signal 26:766–776

Zhan X, Stoy H, Kaoud TS, Perry NA, Chen Q, Perez A, Els-Heindl S, Slagis JV, Iverson TM, Beck-Sickinger AG, Gurevich EV, Dalby KN, Gurevich VV (2016) Peptide mini-scaffold facilitates JNK3 activation in cells. Sci Rep 6:21025

Zhuo Y, Vishnivetskiy SA, Zhan X, Gurevich VV, Klug CS (2014) Identification of receptor binding-induced conformational changes in non-visual arrestins. J Biol Chem 289:20991–21002

Chapter 19
Arrestins in Cell Death

Sandra Berndt, Vsevolod V. Gurevich and Eugenia V. Gurevich

Abstract Programmed cell death (apoptosis) allows multi-cellular organisms to regulate cell number and to react to surrounding environment. Apoptosis can be induced by different stimuli; these can be intrinsic factors (like heat shock and genotoxic stress) or extrinsic factors (like the binding of the death ligand to the death receptor). Depending on the kind of stimulation, two different apoptotic pathways (extrinsic and intrinsic) can be induced. Both multi-step pathways result in the activation of the executor caspases-3, -6 and -7. These proteases trigger endonuclease activation, degradation of chromosomal DNA, protease activation and finally cause the formation of apoptotic bodies. In the last two decades, a variety of regulatory functions of arrestins in cell signaling were elucidated. Arrestins play a role in numerous signaling cascades. They were recently found to play an important role in the regulation of programmed cell death. Visual arrestins are key players in rod cell apoptosis. A dysfunction of arrestin-1 and rhodopsin interaction could lead to retinitis pigmentosa, which results in blindness. Non-visual arrestins also modulate, indirectly or directly, different apoptotic pathways. Recent findings show that arrestin-2 is more likely to act as a pro-survival and arrestin-3 as a pro-apoptosis regulator. Even caspase cleavage products of the non-visual arrestins have a regulatory impact on cell survival. For caspase cleaved arrestin-2, pro-apoptotic characteristics were shown. The cleavage product can induce the intrinsic apoptotic pathway by assisting caspase-cleaved Bid in the release of cytochrome C from mitochondria.

Keywords Arrestin · Apoptosis · Programmed cell death · Intrinsic pathway · Extrinsic pathway · Caspases · Cytochrome C · JNK · ERK · p38 · NFκB · Mdm2 · p53 · cSrc

S. Berndt · V.V. Gurevich (✉) · E.V. Gurevich
Department of Pharmacology, Vanderbilt University, Nashville, TN 46795, USA
e-mail: vsevolod.gurevich@vanderbilt.edu

V.V. Gurevich (ed.), *The Structural Basis of Arrestin Functions*,
DOI 10.1007/978-3-319-57553-7_19

Abbreviations

AIF	Apoptosis inducing factor
Apo1	Apoptosis antigen 1
Apaf-1	Apoptosis protease activating factor
BAD	BCL2 antagonist of cell death
Bcl-2	B-cell lymphoma 2
BID	BH3 interacting-domain death agonist
BAX	BCL2 associated X protein
BAK	BCL2 antagonist killer 1
CD95	Cluster of differentiation 95
Caspase	Cysteinyl aspartic acid protease
DD	Death domain
DED	Death effector domain
DR	Death receptor
DISC	Death induced signaling complex
FADD	Fas-assicated death domain
FasL	Fatty acid synthetase ligand
FasR	Fatty acid synthetase receptor
JNK	c-Jun N-terminal kinase
MOMP	Mitochondrial outer membrane permeabilization
NF-κB	Nuclear factor κ-light-chain-enhancer of activated B cells
PUMA	p53 upregulated modulator of apoptosis
TNFα	Tumor necrosis factor alpha
TNFR	Tumor necrosis factor receptor
TRAIL	TNF-related apoptosis inducing ligand
TRAILR	TNF-related apoptosis inducing ligand receptor
TWEAK	TNF-like weak inducer of apoptosis

Programmed Cell Death

Programmed cell death allows the efficient removal of damaged cells and the preservation of a constant number of cells by multi-cellular organisms. There are different mechanisms of cell death, including apoptosis, necrosis, and autophagy. Here we focus on apoptosis, which is mostly regulated by executor proteases, caspases (Degterev et al. 2003). Cells could be damaged by UV or X-ray irradiation, or other environmental stresses, leading to DNA damage (Norbury and Zhivotovsky 2004). Apoptosis can be induced by intrinsic factors (such as genotoxic stress, heat shock, accumulation of unfolded protein in the ER or nutrient deficiency) or external signals, such as the binding of ligands to death receptors. The apoptosis machinery is very complex and includes many signaling pathways. There are two major apoptotic pathways, the extrinsic and the intrinsic pathway (Fulda and Debatin 2004).

Malfunction in the apoptosis pathway can cause cancer development, ischemic damage, autoimmune diseases or drug resistance. Excessive activity of the apoptosis machinery can result in different degenerative diseases, such as Parkinson's, Alzheimer's, or diabetic retinopathy. On the other hand, the inactivation of apoptosis allows cancer cells to develop and form tumors (Sever and Brugge 2015). The detailed understanding of the apoptosis machinery gives an enormous pharmacological opportunity for an optimized treatment of degenerative diseases and cancer (Fig. 19.1).

Key players regulating this process are the caspases (Thornberry 1998; Thornberry and Lazebnik 1998; Degterev et al. 2003). These cysteinyl, aspartate-specific proteases are divided in two groups, initiator and executioner caspases. Caspase-2, -8, -9 and -10 are initiators and these regulate the beginning of the apoptotic process by forming multi-protein complexes (Hengartner 2000). Initiator Caspases are expressed as zymogens (pro-caspases) and are activated through dimerization followed by cleavage. This activation is induced via different adapter proteins or apoptotic protein complexes like the DISC (death-inducing signaling complex) or the TNFR induced complex-2 (Fuchs and Steller 2015). The initiator caspases cleave and activate executioner caspases, such as caspase-3, -6 and -7. Activated executioners cleave defined cellular components, which results in the destruction of the cell. Caspase-1, -4 and -5 are inflammatory caspases, which are regulators of the immune response to microbial pathogens (McIlwain et al. 2013; Galluzzi et al. 2016). Another important regulatory feature is that activated caspases can activate (cleave) each other, which results in the amplification of the signaling (McIlwain et al. 2013).

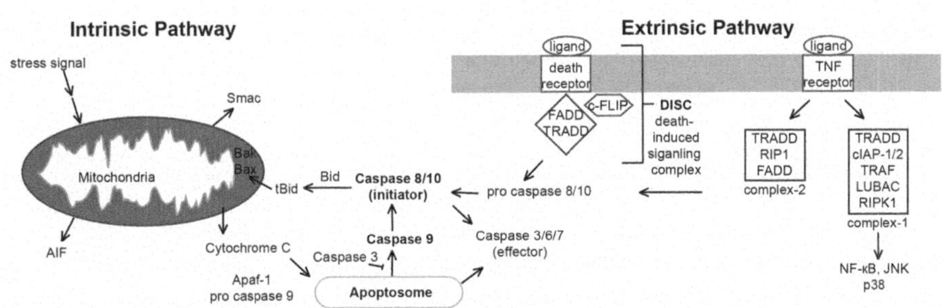

Fig. 19.1 Overview of the intrinsic and extrinsic apoptotic signaling pathway. The intrinsic pathway can be activated by stress factors and induces the release of Smac, AIF and Cytochrome C. Cytochrome C forms together with Apaf-1 and pro-caspase 9 the apoptosome, which is able to activate different caspases (caspase 3, 6, 7 and 9). Activated caspase 9 is stimulating caspase 8 and 10, which lead to the truncation and activation of Bid. TBid is then transferred to the mitochondria and activates Bak and Bax, which regulate the mitochondrial membrane integrity. The extrinsic pathway on the other hand is stimulated by different death receptor ligands. The activated death receptors regulate an intracellular assembly of different regulatory proteins. These can activate pro-caspase 8 and 10 or influence the signaling cascades of NF-κB, JNK or p28

Extrinsic Apoptotic Pathway

This pathway is death receptor-dependent. Death receptors belong to the superfamily of tumor necrosis factor receptors (TNFR) and are activated by various ligands. TNF-α, FasL, TRAIL and TWEAK are ligands, which can induce apoptosis within hours (Smith et al. 1993; Wiley et al. 1995; Pitti et al. 1996; Nagata 1997; Marsters et al. 1998). Death receptor ligands are produced by CD4 + lymphocytes, NK cells, neutrophils, mast cells and neurons (Warren and Smyth 1999; Piliponsky and Levi-Schaffer 2000). LiCl can activate the production of TNF-α and FasL (Kaufmann et al. 2011).

The superfamily of death receptors includes 8 family members, including CD95 (also known as APO-1/Fas) and TRAIL-R1/TRAIL-R2 (Walczak 2013). All receptors have an intracellular death domain (DD), which is the binding site for different adaptor proteins such as FADD (Fas-associated DD) and initiator caspases. These proteins form the death-induced signaling complex (DISC). The complex formation is induced by the activation of CD95 or TRAIL-R. The DISC contains oligomerized death receptors, FADD (Fas-associated Death Domain), pro-caspase-8 and -10 and regulatory protein c-FLIP (cellular FLICE inhibitory protein) (Guicciardi and Gores 2009). Formed DISC allows the activation of initiator caspsase-8 and -10. The dimerization and autocatalytic cleavage of these caspases occurs at the DISC (Salvesen and Dixit 1999). The initiator caspases activate cell type specific effector caspases-3, -6 and -7, which results in cell death (Fig. 19.2).

Multiple molecules of pro-caspase-8 are bound to one death receptor, via a caspase activating chain formation (DED-chains) (Dickens et al. 2012). This feature allows an amplification of the death signaling, so that a limited number of death receptors can activate multiple molecules of caspases.

Initiator caspases-8 and -10 cleave pro-apoptotic Bid (BH3-interacting domain death agonist) into tBid. Bid is a member of the Bcl-2 (B-cell lymphoma 2) family (Li et al. 1998; Schug et al. 2011). The truncated Bid translocates to the mitochondria and activates Bak and Bax, which are also Bcl-2 family members (Martinou and Youle 2011). Bcl-2 proteins act as "apoptotic switch" (Adams and Cory 2007), through their regulation of cytochrome C release. At a certain concentration of the two pro-apoptotic proteins Bak (integral membrane protein from mitochrondria and ER) and Bax (cytosolic protein) pore formation (MOMP—mitochondrial outer membrane permeabilization) is induced (Wei et al. 2001; Dewson and Kluck 2009). This results in the release of cytochrome C and Smac (second mitochondria-derived activator of caspase) from the mitochondrial inner compartment to the cytosol, followed by the assembly of the apoptosome (Rodriguez and Lazebnik 1999). The apoptosome formation can also be induced by cellular stress, which leads to an increase in ROS (reactive oxygen species) or cytosolic calcium levels. Due to the high calcium concentration, non-specific pores are formed in the mitochondrial outer membrane. The pores are formed by three different calcium

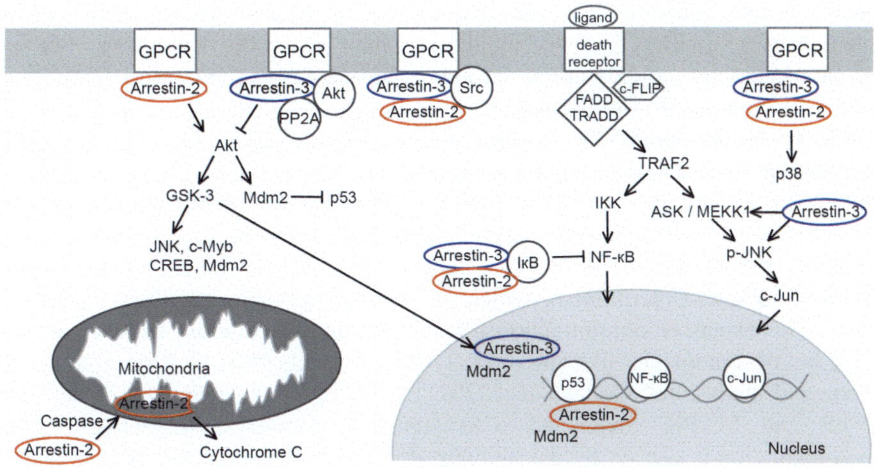

Fig. 19.2 Overview of apoptotic signaling pathways, where arrestins show regulatory functions. Both non-visual arrestins interfere with a lot of apoptotic signaling pathways. GPCR stimulated arrestin-2 promotes Akt activation, which activates GSK-3 or Mdm2. In contrast, arrestin-3 inhibits Akt activation and therefore the activation of the downstream regulators. GSK-3 increases the localization of arrestin-3 in the nucleus. Furthermore, both arrestins are also able to modulate Src function, whereas Src is an important apoptotic regulator. Also, complex-1 of the extrinsic apoptotic pathway, is modulated by arrestin-2 and arrestin-3. The activation of NF-κB via complex-1 can be further enhanced, through the binding of either arrestin-2 or 3 to the inhibitor of NF-κB (IκB). Additionally, arrestin-3 is able to activate ASK/MEKK1 and JNK3. This increases the activity of c-Jun. The transcription factor p53 is also influenced by arrestins, direct and indirect. Mdm2 and arrestin-2 can inhibit p53. In addition, arrestin-2 gets cleaved by caspase-8. This truncation allows a mitochondrial localization which enhances the cytochrome C release and therefore apoptosis

dependent proteins: VDAC (voltage-dependent anion channel), ANT (adenine nucleotide translocator) and cyclophilin D (Pinton et al. 2008).

The apoptosome formation is induced by cytochrome C, which binds to Apaf-1 in an ATP-dependent manner (Kim et al. 2005). This causes Apaf-1 to build a ring structure, which allows pro-caspase-9 binding. Cytochrome C can dissociate freely from the complex. The high local concentration of pro-caspase-9 in the apoptosome makes an activation possible (Malladi et al. 2009). This allows the binding of caspase-3 to the apoptosome, followed by an inhibition of caspase-9. This inhibition is due to an overlapping binding site of both caspases to Apaf-1 (Yuan et al. 2011). Caspase-3 is also capable to cleave Apaf-1, which results in the loss of caspase-9. Another regulatory protein of the apoptosome is XIAP (X-linked inhibitor of apoptosis protein). This is an inhibitor of caspase-3 and -9 (Bratton et al. 2002; Zou et al. 2003).

The extrinsic apoptotic pathway for TNF receptors (TNFRs) does not involve typical DISC formation (Barnhart and Peter 2003). Here the TNFR forms two signaling complexes: complex-1 and complex-2. Activated TNFR-1 recruits the following proteins for complex-1 formation: TRADD (TNFR associated death domain protein), cIAP-1/2 (cellular inhibition of apoptosis), TRAF (TNFR associated factor), LUBAC (linear ubiquitin chain assembly complex), RIPK1 (receptor-interacting serine/threonine protein kinase 1). In contrast to DISC, the complex-1 contains no pro-caspase-8.

TNFR-1 trimers bind TNF (Sedger and McDermott 2014). TRADD and RIP1 bind after each other to the TNFR-1 complex. This is followed by the recruitment of TRAF2 and TRAF5 (Hsu et al. 1995, 1996). After that the ubiquitin ligases clAP1 and clAP2 are able to interact with the complex-1. These ligases cause the ubiquitination of RIPK1 and clAP1. This allows the LUBAC (linear ubiquitin chain assembly complex) recruitment, which induces the formation of linear ubiquitin chains. The extensive ubiquitination patterns provide the binding surface for different kinase complexes like the TAK/TAB and the IκB (IκK) kinase complex (Wertz and Dixit 2008; Gerlach et al. 2011). These complexes are able to activate NF-κB and MAPK (JNK and p38) signaling pathways, respectively. The down-regulation is caused by the binding of deubiquitinase (DUB) to complex-1, which results in the disassembly of this signaling complex. A low activation of complex-1 allows complex-2a to induce apoptosis (Micheau and Tschopp 2003).

The more important complex in terms of cell death regulation is complex-2. This complex can cause apoptosis or necrosis (Christofferson and Yuan 2010; Nikoletopoulou et al. 2013). It is formed after TNFR stimulation. TRADD and RIP1 bind to the activated TNFR-1 via the DD. The complex can dissociate from the receptor, which allows the binding of FADD to the TRADD-RIP1 signaling complex in the cytoplasm (Park et al. 2014). In Complex-2a, TRADD and FADD recruit pro-caspase-8 and activate it to caspase-8. This activates the downstream effectors, caspase-3 and -7, which leads to apoptosis (Wang et al. 2008a).

FLIP is a caspase-8 like protein and can also be recruited to complex-2. It acts as an inhibitor of caspase-8 activation, suppressing apoptosis. The expression of FLIP, Bcl-2, and clAP2 is upregulated by complex-1. The heterodimer of FLIP and caspase-8 does not facilitate apoptosis (Pop et al. 2011).

If the complex-2 has RIP3 instead of RIP1 bound, it is called the necrosome and leads to necroptosis (Moriwaki and Chan 2013).

LUBAC is building an E3 ligase complex, together with HOIP (RNF31), HOIL-1 (RBCK1) and SHARPIN (SIPL1) (Kirisako et al. 2006). This complex is involved in the regulation of cell death, through the interaction with the TNFR complex. It was found that the NFκB pathway is activated by LUBAC (Niu et al. 2011; Dubois et al. 2014).

In complex-2b, TRADD is exchanged by RIP1 (receptor-interacting protein 1). For this interaction RIP1 must be deubiquitinated. The RIP1 kinase represents an important regulator of apoptosis under specific conditions (Wang et al. 2008b). The regulatory functions of this protein include phosphorylation, ubiquitination and deubiquitination.

In contrast to the TNFR1, the TNFR2 is mostly involved in mediating inflammation, instead of cell death.

Intrinsic Apoptotic Pathway

The intrinsic apoptotic pathway is induced by intracellular signals and is independent from receptor signaling. The induction occurs in mitochondria. The signaling is activated by different stress factors, such as radiation, viral infections, or chemotherapy treatment. These conditions can cause DNA damage, hypoxia or an increase of calcium concentration in the cytosol. Another way of induction is through different cytokines, loss of growth factors, or certain hormones. This would lead to an indirect activation of apoptosis, by down-regulation of the inhibitory anti-apoptotic machinery (Kiraz et al. 2016).

Different pro-apoptotic proteins are located in the mitochondria. These are cytochrome C, SMAD (homologues of *C. elegans* SMA protein and Drosophila MAD (mothers against decapentaplegic) proteins) and HtrA2 (high-temperature requirement A2 serine protease) (Saelens et al. 2004). The Bcl-2 family proteins (primarily Bax and Bak) control the intrinsic stimulation of apoptosis. The tumor suppressor p53 is activated by DNA damage and can activate Bcl-2 family proteins. These proteins can function as anti-apoptotic or pro-apoptotic regulators and determine the mitochondrial membrane integrity (Martinou and Youle 2011).

The disruption of the mitochondrial membrane leads to the release of pro-apoptotic proteins to the cytosol, particularly cytochrome C (Babbitt et al. 2015). Cytochrome C is a key player in the intrinsic apoptotic pathway in vertebrates. This protein is encoded in the nucleus and translated as apo-cytochrome C. It is transported to the inter-membrane space of the mitochondria. Here the heme group links covalently to two cysteine residues and forms the holo-cytochrome C. This enzyme works in the respiration chain. When apoptosis is induced, the function and localization of cytochrome C is changed rapidly. The mitochondrial permeability transition is caused by an increasing calcium concentration, which allows the release of cytochrome C (Bossy-Wetzel et al. 1998). In the cytosol cytochrome C binds apoptotic protease activation factor 1 (Apaf1) in an ATP depended manner (Hu et al. 1999). This is followed by the oligomerization of Apaf1, which causes the binding of pro-caspase-9 and leads to an activation of caspase-9. Additionally, the apoptosome results in an extensive activation of executioner caspase-3, -6 and -7, which cleave cellular proteins, mediating apoptotic cell death (Chinnaiyan 1999; Hill et al. 2004).

Executioner caspases cleave different cytoskeletal proteins, including actin, tubulin and fodrin. This results in the loss of the cell shape. Caspase-mediated cleavage of the lamin leads to nuclear shrinking (Luthi and Martin 2007).

The nomenclature committee on cell death (NCCD) summarized the morphological characteristics of apoptotic cells (Kroemer et al. 2009).

Arrestin in Apoptosis

Mammals have only four arrestins that can be divided into two sub-families, visual and non-visual arrestins. Arrestin-1 (S-antigen) and arrestin-4 (cone arrestin) are the visual arrestins and are exclusively expressed in photoreceptor cells in the retina. The non-visual arrestins are arrestin-2 (β-arrestin1) and arrestin-3 (β-arrestin2), which are ubiquitous expressed. The closest relatives of arrestins are arrestin-domain containing proteins and will not be discussed here.

Arrestins were first discovered to desensitize GPCRs (Kuhn et al. 1984; Wilden et al. 1986; Lohse et al. 1990; Attramadal et al. 1992). Arrestins specifically bind activated and phosphorylated GPCRs and terminate G-protein activation by steric hindrance (Wilden 1995; Krupnick et al. 1997). Since there are just two non-visual arrestins and more than 800 GPCRs in primates (many more in other mammals), arrestin-2 and arrestin-3 can bind hundreds of different GPCRs and facilitate their desensitization (Gurevich and Gurevich 2006). Non-visual arrestins also bind dozents of different non-receptor proteins, which are involved in the regulation of a variety of signaling pathways (Lefkowitz and Shenoy 2005; Gurevich and Gurevich 2006; DeWire et al. 2007; Xiao et al. 2007). Hence, arrestins are able to regulate multiple signaling pathways, which include different apoptotic pathways.

Visual Arrestins

The visual arrestins (arrestin-1 and arrestin-4) are solely expressed in retinal photoreceptors. There are two types of photoreceptor cells in mammalian eyes, rods and cones. There are also light-sensitive ganglion cells expressing melanopsin, which is more similar to invertebrate rhodopsins and couples to Gq/11(Hattar et al. 2002). Rods and cones are the main photoreceptor cells. In the outer segments of both cell types are localized photo pigments (rhodopsin or cone opsins) and all the proteins of the visual signaling cascade. Rhodopsin consists of opsin (apo-protein) and a covalently bound ligand 11-*cis*-retinal (Bownds 1967; Matsumoto and Yoshizawa 1975). The ligand is attached to the K296 in transmembrane helix VII of rhodopsin via Schiff-base linkage. The photon activation causes the chromophore to convert from 11-cis to all-trans, which results in the conformational change of the opsin protein. This conformational change allows rhodopsin to interact with the G-protein (transducin), which activates cyclic GMP phosphodiesterase and induces the response of the rod cells (Stryer 1986; Spiegel 1987; Hurley et al. 1998).

The shut-off of light-activated rhodopsin involves two proteins. The rhodopsin kinase phosphorylates serine and threonine residues in the C-terminus. This allows the interaction with visual arrestin-1 (Kuhn and Wilden 1987). The arrestin complex with hyper-phosphorylated rhodopsn initiates the apoptotic pathway (Alloway et al. 2000; Kiselev et al. 2000). Arrestin-1 is observable in rods and cones, whereas arrestin-4 is expressed only in cones, where it constitutes just 2% of the total arrestin complement (Deming et al. 2015).

Rod cells contain around 3 mM rhodopsin and ~2 mM arrestin (Strissel et al. 2006; Song et al. 2011). Rhodopsin represents 30% of the total protein content in rod cells (Filipek et al. 2003). Rods are exceptionally sensitive. The high concentrations of the signaling proteins allow rods to detect a single photon of light (Baylor et al. 1979). This high protein concentration has a disadvantage, because a change in the expression levels and therefore the ratio of rhodopsin to arrestin can result in rod cell death (Tan et al. 2001).

Apoptosis of the photoreceptor cells can be induced through bright light (Noell and Albrecht 1971). This causes permanent activation of rhodopsin and an increase of signaling (Lisman and Fain 1995; Fain and Lisman 1999).

The lack of arrestin-1 in rod cells causes a permanent activation of rhodopsin. This results in excessive signaling and a light-driven rod cell death (Xu et al. 1997; Song et al. 2009b, 2013). An arrestin-1 concentration of at least 5% compared to the wild type, is necessary to maintain visual system functionality (Cleghorn et al. 2011; Song et al. 2011). Apoptotic death of photoreceptor cells can cause retinal dystrophy, age-related macular degeneration and retinitis pigmentosa.

Retinitis pigmentosa (RP) is a form of retinal degeneration that starts with the loss of rod photoreceptors (Mendes et al. 2005). Most forms of RP are caused by different mutations in rhodopsin, some of which result in phosphorylation and arrestin binding (Li et al. 1995). This retinal degeneration can be caused by chronic stimulation of rhodopsin signaling in a transducin dependent or arrestin dependent manner. Transducin-dependent apoptosis is induced through prolonged rhodopsin activation. On the other hand, arrestin-mediated apoptosis is induced through activated rhodopsin. Surprisingly there is a difference in the mechanism of vertebrate and invertebrate phototransduction. The invertebrates show an internalization of rhodopsin. Alloway et al. (2000) showed that in Drosophila a stable rhodopsin-arrestin complex causes apoptosis of photoreceptor cells (Alloway et al. 2000; Kiselev et al. 2000). The adapter protein (AP-2) binds to rhodopsin-associated arrestin-1, apparently via released arrestin-1 C-tail, and allows the accumulation in late endosomes (Orem et al. 2006; Chinchore et al. 2009). The arrestin-1 mediated rhodopsin internalization is not observable in vertebrates; here rhodopsin is transported to the outer segment disc and stays there (Young 1967). Therefore, the concentration of rhodopsin in outer segment discs does not change.

The best studied RP-causing mutant of rhodopsin is K296E. This mutation removes the retinal attachment site and prevents the formation of the salt bridge to E113 (Sakmar et al. 1989; Zhukovsky and Oprian 1989). Therefore, rhodopsin K296E is locked in the active conformation and is deactivated by phosphorylation and arrestin binding (Li et al. 1995). This rhodopsin mutation leads to death of vertebrate rods, but the mechanism is not well understood. The rhodopsin K296E with bound arrestin-1 is localized to the plasma membrane, instead to the outer segment discs (Chen et al. 2006). This transport is triggered via AP-2, which binds to rhodopsin-bound arrestin-1 (Laporte et al. 1999, 2000). The interaction takes place within the last 25 amino acids of arrestin-1. The localization at the plasma membrane allows the accessibility to other signaling proteins and the induction of cell death.

The splice variant p44 (lacks the last 35 amino acids in the C-terminus) of arrestin-1 binds activated and phosphorylated rhodopsin, but has no AP-2 binding site. This prevents retinal degeneration and rescues the malfunction of rhodopsin mutant K296E (Moaven et al. 2013).

Recently, a study showed that dominant and recessive RP have a common mechanism of cell death—the calpain-mediated pathway (Comitato et al. 2016). Dominant RP is caused by point mutations in rhodopsin and recessive RP by loss of function mutations in rhodopsin. Calpain is a calcium-dependent cysteine protease that plays an important role in apoptosis and necrosis. The mechanism of calpain-mediated apoptosis is not well understood. There is evidence that calcium increase in the cytosol caused by the endoplasmatic reticulum (ER) or the mitochondria activates calpain. Calpain can also be activated via the ERK signaling pathway (Glading et al. 2000, 2001). Calpains are known to interfere within caspase cascade by cleavage of regulatory proteins of the apoptosis pathway. It is believed that calpains can modulate the intrinsic apoptotic pathway through Bcl-2 deactivation (Gil-Parrado et al. 2002). Additionally, Bid could be activated by calpain-mediated cleavage (Mandic et al. 2002). During retinal degeneration, the induction of calpain activity is triggered in photoreceptors. In different RP animal models activated calpains were found (Paquet-Durand et al. 2006; Sancho-Pelluz et al. 2008; Sothilingam et al. 2015).

A study form 1995 showed that calpain cleaves the visual arrestin bound to the activated and phosphorylated rhodopsin (Azarian et al. 1995). It cleaves off selectively the last 26 amino acids of the C-terminus. C-terminally truncated versions of arrestin-1 have a higher affinity for phosphorylated rhodopsin, like p44. This cleavage of arrestin-1 bound to the activated phosphorylated receptor, allows the rod cells to adapt to bright light exposure. If this exposure is chronic, the amount of truncated arrestin-1 would increase. This represents another regulation mechanism whereby arrestin-1 can prevent rod cell death.

Another binding partner of arrestin-1 is NSF (N-ethylmaleimide-sensitive factor) (Huang et al. 2010). This protein is localized in photoreceptor synapses and is a regulator of the SNARE (soluble NSF attachment protein receptor) complex disassembly. It appears that arrestin-1 can modulate the ATPase activity of NSF and therefore increases the disassembly of the SNARE complex. In consequence, arrestin-1 regulates the neurotransmitter exocytosis. The NSF attachment protein α (αSNAP) is an important regulator of epithelial cell apoptosis (Naydenov et al. 2012).

The arrestin-1 concentration in rod cells is highly significant for proper function of the signaling mechanism. This makes the capability of arrestin-1 to oligomerize an important regulatory feature (Hanson et al. 2007b; Chen et al. 2014). Only a small fraction of arrestin-1 is present as monomer (Schubert et al. 1999; Imamoto et al. 2003; Kim et al. 2011). The oligomers of arrestin-1 are believed to be storage forms, since a high concentration of monomeric arrestin-1 seems to be toxic for photoreceptor cells (Song et al. 2013). Interestingly, arrestin-4 is a constitutive monomer (Hanson et al. 2008), which might explain its relatively low levels in cones (Chan et al. 2007; Nikonov et al. 2008).

Non-visual Arrestins in Apoptosis

The non-visual arrestins, arrestin-2 and arrestin-3, are ubiquitously expressed and bind hundreds of different receptors and non-receptor proteins. Therefore, both arrestins have an immense regulatory impact on a variety of signaling pathways. This functional diversity of both homologues is further expanded by different posttranslational modifications, including phosphorylation, acetylation, nitrosylation, ubiquitination, and SUMOylation.

Both non-visual arrestins have an N-terminal nuclear localization signal, but only arrestin-3 has a nuclear export signal (Wang et al. 2003b). This means that subcellular localization is another distinguishing feature for the two non-visual arrestins. Also, IP6-mediated oligomerization of arrestins affects their localization and functions (Milano et al. 2006).

The regulation of apoptosis by non-visual arrestins can be initiated via GPCR or non-GPCR activation. Several stress-regulated proteins, such as JNK, ERK, and NFkB are important regulators for cell survival and apoptosis (Boulton et al. 1991; Davis 2000; Karin and Lin 2002). Via their interactions with key players of different MAPK cascades and AKT pathway, non-visual arrestins have a significant impact on cell survival and death.

Apoptosis can be induced by GPCR activation. GPCRs can regulate the extrinsic or intrinsic apoptotic pathway, through the activation of different G-proteins. It was shown that all heterotrimeric G-protein families can modulate apoptosis signaling, acting as pro-apoptotic or pro-survival factors. This regulation is mediated by their impact on proteins of the Bcl-2 family, NF-κB, PI3 kinase, small GTPases and MAP kinases (Yanamadala et al. 2009). Due to their role in GPCR desensitization, arrestins are in this case indirect regulators of apoptosis.

It was shown that the activation of certain GPCRs results in apoptosis of arrestin-2 and arrestin-3 double knockout MEFs (Revankar et al. 2004; Lefkowitz et al. 2006).

One of the earlier discovered example of GPCR mediated cell death is the neurokinin-1 receptor (NK1R). This receptor is activated by a neuropeptide substance P (SP) and binds arrestin-2. This interaction allows the internalization, which is followed by the formation of a scaffolding complex of NK1R, arrestin, Src and ERK1/2. This explains the pro-survival effect of SP via ERK1/2 signaling (Eistetter et al. 1992; DeFea et al. 2000). Another study showed that the SP/NK1R system could also induce programmed cell death. This is activated via MAPK signaling pathway including Raf-1, MEK2 and ERK2 (Castro-Obregon et al. 2004). The MAP kinases phosphorylated the orphan nuclear receptor Nur77, which leads to non-apoptotic programmed cell death. Here the pro- and anti-survival effect for the same ligand-receptor system is regulated via arrestin interaction.

The PTHR (parathyroid hormone receptor) can act as a pro- or anti-apoptotic stimulus, by regulating the activation of Bcl-2 proteins (Amling et al. 1997; Turner et al. 1998). The PTH1R seems to regulate apoptosis through the activation of G_S protein, which activates the PKA pathway (Chen et al. 2002). This results in an increase of calcium level and the induction of the intrinsic apoptotic pathway. It was also reported that the activation of PTHR can stimulate death receptor signaling, initiating the extrinsic apoptotic pathway (Gagiannis et al. 2009).

Another pathway involves GPCR-mediated regulation of apoptosis through arrestin activation. The N-formyl peptide receptor (FPR) mediates apoptosis though Src kinase and ERK1/2, which is regulated via arrestin-Src binding. In this case, without arrestin, the cells would go into apoptosis. How the FPR stimulation activates these signaling proteins is not well understood (Wagener et al. 2016). GPCR-arrestin-mediated signaling can be either pro-survival or pro-apoptotic, depending on the signaling pathway it activates.

Arrestins act as scaffold proteins for different signaling pathways. This can be GPCR-dependent or GPCR-independent. A typical signaling cascade is the JNK activation pathway. Here arrestin-3 is required to facilitate the ASK1-mediated JNK3 activation. Arrestin-3 interacts directly with JNK3 and ASK1. This allows arrestin-3-dependent formation of a signalosome for the MAPK signaling cascade (McDonald et al. 2000). Similar mechanisms were shown for the ERK signaling pathway. Here is also a docking model of the receptor, arrestin, Src, Raf, MEK, ERK complex available (Bourquard et al. 2015). All these signaling pathways are involved in the regulation of cell survival and apoptosis.

An over-stimulation of apoptosis-initiating signaling can cause neurodegenerative diseases, such as Alzheimer's or Parkinson's disease. Alzheimer's disease (AD) ends in the neuronal cell death triggered by Aβ, the amyloid precursor peptide. Two GPCRs were shown to play a role in Alzheimer's disease, the GPR3 and the b2AR (Ni et al. 2006; Thathiah and De Strooper 2009). Aβ was found to interact with the N-terminus of the b2AR. This interaction induces internalization of the receptor, and this internalization is mediated by arrestin-3, not arrestin-2 (Wang et al. 2011). Recently, it was shown that with an overexpression of arrestin-3 there is also an increase in Aβ concentration (Thathiah et al. 2013). On the other hand, the silencing of the arrestin-3 gene results in a decrease of Aβ accumulation. This impact is possible due to the arrestin-3 interaction with the Aph-1a subunit of γ-secretase complex. This complex is known to regulate Aβ degradation (Thathiah and De Strooper 2011). These new findings suggest that arrestin-3 is a potential new target for the treatment of Alzheimer's disease (Jiang et al. 2013). Here the development of new biased ligands, which favor the arrestin-3-mediated pathways for the GPR3 or the b2AR could become a novel approach for AD treatment.

Recently discovered caspase-mediated cleavage of non-visual arrestins is yet another mechanism of arrestin impact on cell death. In the extrinsic and the intrinsic apoptotic pathway arrestin-2 is cleaved by caspases (Kook et al. 2014).

The cleavage fragment of arrestin-2 contains amino acids 1–380. This fragment translocates to the mitochondria and increases cytochrome C release induced by tBid, thereby accelerating apoptotic cell death.

JNK Signaling Cascade in Apoptosis

The MAP kinase JNK is an essential player in cell proliferation, cell differentiation, and stress-induced apoptosis (Nishitoh et al. 2002). JNK interacts with pro-survival and pro-apoptotic proteins (Lin and Dibling 2002; Liu and Lin 2005). JNKs are encoded by three different genes, which divide the JNK family in JNK1, JNK2 and JNK3. Each sub-family has several isoforms generated by alternative splicing. JNK1 and JNK2 are ubiquitous, whereas JNK3 is mainly found in the brain, testis, and heart. The JNK signaling pathway is activated by environmental stress, growth factors, inflammatory cytokines and GPCRs. The stress response is triggered by small GTPases Rho and Rac. JNKs are the last module of the MAPK cascade consisting of MAP3K (MEKK1-4), MAP2K (MKK4 or MKK7) and a MAPK (JNK). ASK1 (apoptosis signal-regulating kinase 1) is a serine-threonine protein kinase, which is one or the MAP3K of JNK pathways. It activates MKK 4 and 7. The activation of these MAP2Ks results in the activation of JNK. Activated JNK translocates in the nucleus and phosphorylates c-Jun, which regulates the transcription of a variety of different pro-apoptotic proteins (Derijard et al. 1994).

This process plays a role in the extrinsic and intrinsic apoptotic pathways (Fan et al. 2001; Schroeter et al. 2003). JNK activity regulates the balance of pro- and anti-apoptotic proteins (Aoki et al. 2002). For example, JNKs interact with BAX and Bcl-2 and modulate their function. JNK regulation of BAX is essential for the release of cytochrome C during apoptosis (Papadakis et al. 2006).

Two other Bcl family members, BID and BIM, are regulated by JNK. BID is the main connection between the intrinsic and the mitochondrial apoptotic pathway (Li et al. 1998; Luo et al. 1998). JNK phosphorylates BID at Thr59, which prevents the caspase-8 cleavage (Prakasam et al. 2014). This results in the accumulation of the full length BID, which has some pro-apoptotic functions (Billen et al. 2008).

JNK dependent BID cleavage at T59 (Deng et al. 2003) generates jBid. The translocation of this Bid fragment results in the release of Smac/DIABLO, but not cytochrome C. This results in the disruption of the TRAF2-cIAP1 complex and allows further caspase 8 activation. Thus, jBid is the connection of extrinsic and intrinsic apoptotic pathway.

GABA is well known to inhibit tumor growth of different cancer cells (Joseph et al. 2002). GABAb receptor inhibits Akt signaling pathway via arrestins (Lu et al. 2012). Recently it was shown in different cell lines that the inhibitory neurotransmitter GABA triggers the interaction of GABAbR1, arrestin-2/3 and JNK (Tian et al. 2015). Here arrestin functions as an adaptor protein, which regulates the GABA-mediated tumor cells apoptosis through the JNK cascade.

Only JNK3 interacts with arrestin-3 and relocalizes it to the cytoplasm (McDonald et al. 2000; Song et al. 2009a; Zhan et al. 2011, 2013). Additionally, the three upstream kinases of this cascade (ASK1, MKK7, and MKK4) also bind arrestins (McDonald et al. 2000; Song et al. 2009a; Seo et al. 2011). Arrestin-3 acts as a scaffolding protein for the whole JNK signaling cascade and makes it more efficient, due to the local increase in protein concentration. With an increasing H_2O_2 concentration, arrestin-2 and -3 suppress JNK activation through interaction with

ASK1 (Zhang et al. 2009). This binding enhances ASK1 ubiquitination by the E3 ligase CHIP, whereupon ASK1 undergoes proteosomal degradation. In consequence the MAPK cascade-mediated JNK activation is down-regulated, which has a pro-survival effect on the cell.

NF-κB Signaling Mediated Apoptosis

NF-κB is a transcription factor and a key regulator of cell survival, cell migration and inflammation. Several genes of regulatory proteins of the apoptosis machinery are under the control of this transcription factor, including some Bcl-2 family members, c-Myc, p53 and FasL (Wu and Lozano 1994; Bellas and Sonenshein 1999; Chen et al. 1999; Grumont et al. 1999; Kasibhatla et al. 1999; Lee et al. 1999). The NF-κB expression is up-regulated by different stimuli, such as stress, cytokines, and UV irradiation. There are 5 family members of the NF-κB/Rel family. These are p50 (NF-κB1), p52 and p100 (NF-κB2), p65 (RelA), c-Rel and RelB. The inactive NF-κB is bound to inhibitory proteins (IκBs) (Baeuerle and Baltimore 1988). The IκB kinases phosphorylate IκBs after stimulation by cytokines, oxidative stress or GPCR activation (Ye 2001). The phosphorylation is followed by ubiquitination and subsequent degradation of IκB (Chen et al. 1995). This results in the accessibility of the nuclear translocation sequence in NF-κB and in consequence the translocation to the nucleus. Therefore, proteins that bind IκB are potentially important regulators of the NF-κB signaling pathway.

Both non-visual arrestins bind the NF-κB inhibitor IκBα (Witherow et al. 2004). Additionally, some upstream kinases of this signaling pathway interact with arrestins (e.g. NF-κB-inducing kinase, IκB kinase α and β). Arrestin-3 stabilizes IκBα by direct binding (Gao et al. 2004). Additionally, arrestin-3 inhibits the UV induced NF-κB activation (Luan et al. 2005). Therefore, arrestin-3 has anti-apoptotic characteristics. In contrast, arrestin-2 can act as positive regulator for NF-κB activation, as in the endothelin-induced NF-κB activation (Cianfrocca et al. 2014; Rosano and Bagnato 2016). This is a pro-survival effect. The suppression of arrestin-2 by RNA interference resulted in an increase of TNFα-stimulated NF-κB activity (Witherow et al. 2004).

These opposing regulatory functions could be a consequence of the difference in subcellular localization of arrestins. In general, the NF-κB activation has an anti-apoptotic effect.

MAPK p38 Signaling Cascade and Apoptosis

The p38 MAPK cascade plays an important role in the regulation of stress response and inflammation (Tibbles and Woodgett 1999). Different signaling cascades, involving MEKK1-4, ASK1/2 or MLK2/3, can activate this MAPK. The phosphorylation of p38 allows the downstream activation of a number of different transcription factors, such as NFAT, STAT1, Myc, ELK1 or p53.

This signaling pathway is arrestin-regulated (Sun et al. 2002; Miller et al. 2003; Bruchas et al. 2006). The down regulation of arrestin-2 inhibits Rac-1 membrane translocation. This localization is obligatory for early p38 activation, which is induced in an arrestin-2, Rac-1, NADPH oxidase pathway (Gong et al. 2008). The late p38 activation is regulated by GPCR dependent G_s signaling.

A direct regulatory function of arrestins was shown for the chemokine receptor and κ-opioid receptor (Sun et al. 2002; Bruchas et al. 2006). The activation of p38 by the kappa opioid receptor is GRK3- and arrestin-dependent (Bruchas et al. 2006). The cytokine receptor US28 is also able to activate p38; this process is GRK- and arrestin-dependent, as well (Miller et al. 2003). The activation of PAF receptors results in the formation of arrestin-2 scaffolding of ASK-1, MKK-3 and p38 (McLaughlin et al. 2006).

Another example for an indirect arrestin-mediated regulation of apoptosis is based on interleukin 10 (IL-10). The TLR4 (toll-like receptor)-triggered activation of p38 by arrestin-3 causes an increase in IL-10 expression (Li et al. 2014). IL-10 can be a positive or negative regulator of apoptosis through regulation of the NF-κB signaling.

In MEFs the arrestin-mediated p38 MAPK activation regulates apoptosis (Yang et al. 2012). Here it was shown that both non-visual arrestins inhibit cell apoptosis via p38 MAPKs and pro-survival Akt signaling.

Mdm2 and p53 Mediated Apoptosis

In 2001 the interaction of Mdm2 ("murine double minute 2") and arrestin-3 was elucidated (Strous and Schantl 2001; Wang et al. 2003b; Shenoy et al. 2009). It was shown that agonist activation of the β2AR results in an ubiquitination of arrestin-3, which is regulated by Mdm2 (Shenoy et al. 2009). Mdm2 is an E3 ubiquitin ligase. It has a zinc finger-type catalytic RING domain (Deshaies and Joazeiro 2009). Ubiquitination and other posttranslational modifications (phosphorylation, S-nitrosylation and acetylation) modulate arrestin functions. These structural changes could be an important feature for the versatile functions of arrestins. For example, the ubiquitination of arrestin-3 enhances the activation of ERK (Shenoy et al. 2007). Deubiquitination of arrestins is triggered by the ubiquitin-specific protease 33 (USP33). The N-domain (Wang et al. 2003a; Boularan et al. 2007) or both N- and C-domains (Song et al. 2007) of arrestins bind Mdm2; USP33 binds both N- and C-domains. Class A and class B receptors differ in the level of ubiquitination. Class A GPCRs bind arrestins more transiently, and that appears to promote USP33 association. This results in the deubiquitination and the dissociation of arrestin from the receptor, which reduces arrestin-mediated ERK activation. In case of class B GPCRs (such as the angiotensin II type 1a receptor and the vasopressin V2 receptor) tighter arrestin interaction appears to result in an arrestin conformation, which is not favored for USP33 binding. This results in a higher level of arrestin ubiquitination and therefore an increased ERK activation. Consequently, the GPCR-dependent ubiquitination and deubiquitination of arrestins by Mdm2 and USP33 are important regulatory inputs for the interaction with pro-survival effector ERK.

The E3 ubiquitin ligase Mdm2 is an oncogenic protein and major regulator of the p53 tumor suppressor gene (Ryan et al. 2001). Mdm2 can inhibit p53 by ubiquitination that leads to proteosomal degradation (Brooks and Gu 2006; Lee and Gu 2010). The expression of p53 is stress-dependent. If the concentration exceeds a certain level, p53 is able to induce the intrinsic apoptotic pathway. This is due to the regulation of Bcl-2 and Bax expression by p53 (Miyashita et al. 1994). Bcl-2 and Bax are the main regulators of the apoptosis. The binding of arrestin-3 to Mdm2 results in the inhibition of the p53 ubiquitination. Thus, arrestin-3 reduces the suppression of p53 by Mdm2, thereby facilitating apoptosis. In contrast, arrestin-2 activation by the β2AR results in AKT-mediated activation of Mdm2 (Hara et al. 2011). The activated Mdm2 facilitates the degradation of p53. Therefore, arrestin-2 has an anti-apoptotic effect.

An additional important regulatory aspect is the localization of Mdm2. The Mdm2 concentration in the cytosol is low compared to the nucleus (Wang et al. 2003b; Song et al. 2006). In both compartments Mdm2 has important functions. In the nucleus it is mainly the regulation of p53 activity, whereas in the cytosol it regulates the stability of the GPCR-arrestin-complex. The cytosolic concentration of Mdm2 is dependent on the nuclear interaction with p53, the degradation rate and also the activity of the Akt pathway (Strous and Schantl 2001). Translocation of Mdm2 to the cytosplasm is assisted by arrestin-3. The C-terminal leucine-rich nuclear export signal of arrestin-3 makes the shuttling possible (Scott et al. 2002; Wang et al. 2003b; Song et al. 2006). Nuclear arrestin-3 increases the nuclear localization of Mdm2. Additionally the IP6-regulated oligomerization of arrestin-3 could also affect the Mdm2 regulation (Milano et al. 2006; Boularan et al. 2007). Although both non-visual arrestins bind Mdm2 (Song et al. 2006; Hanson et al. 2007a), only arrestin-3 translocates it out of the nucleus.

These functions make Mdm2 a promising drug target. Mdm2 inhibition is a potential venue for cancer therapy (Vassilev 2007). In particular, the treatment of leukemia could benefit by the use of small-molecule Mdm2 inhibitors (Kojima et al. 2016).

To summarize, overall arrestin-3 appears to have a pro-apoptotic effect, whereas arrestin-2 has a pro-survival effect via the regulation of Mdm2-p53 axis.

Src Family Kinases and Their Role in Apoptosis

The non-receptor tyrosine-protein kinase c-Src (cellular Src kinase, Src stands for sarcoma) is a proto-oncogene and a key player in cancer development. Depending on the cellular environment, c-Src can decide between cell multiplication, senescence or apoptosis. c-Src regulates apoptosis via different pathways. One is the stimulation of c-myc expression (Sovova et al. 1993), which activates p53. c-Src also activates the PI3K/AKT pathway, which has an anti-apoptotic effect. Another way is c-Src-mediated activation of p38. Due to this diversity c-Src has anti- as well as pro-apoptotic characteristics.

The protein kinase c-Src was the first non-receptor signaling protein that was discovered to interact with arrestins (Luttrell et al. 1999). Src family kinases can be activated by different GPCRs (Luttrell and Luttrell 2004). Arrestins can activate these tyrosine kinases (Strungs and Luttrell 2014). Arrestin-2 recruits c-Src to activated GPCRs (Luttrell et al. 1999; DeFea et al. 2000). The phosphorylation of c-Src is important for the downstream signaling for a variety of pathways. The first receptor-arrestin-c-Src complex was described for b2AR (Luttrell et al. 1999).

A similar complex was shown to form with the angiotensin II type 1 receptor (AT1R) (Fessart et al. 2005). This complex contains the agonist-stimulated receptor, arrestin-3, AP-2 and c-Src. Here c-Src regulates the interaction of AP-2 with the complex and receptor internalization.

Neurokinin-1 receptor activated by substance P was shown to have an anti-apoptotic effect via arrestin-mediated Src recruitment (DeFea et al. 2000).

Overall the recruitment of Src to a GPCR via arrestin is important for multiple cellular processes, including internalization and the activation of MAPK pathways. The recruitment is not limited to c-Src, other Src family members can bind to arrestin-GPCR complexes. Recently it was shown that arrestin-3 modulates TRAIL-induced HepG2 cell apoptosis (Qi et al. 2016). The data suggest that in the presence of TRAIL arrestin-3 can directly interact with death receptors 4 and 5. Thus, arrestin-3 appears to be a facilitator of the TRAIL-induced apoptosis via scaffolding DR and Src.

The role of c-Src in indirect regulation of arrestin-mediated apoptosis was shown. In double arrestin-2/3 knockout MEFs, the activation of the FPR (N-formyl peptide receptor) causes G-protein mediated Src activation, which results in apoptosis of the cell. When present, arrestin-2 binds the activated FPR and induces its internalization. Src and AP-2 bind arrestin-2 and AP-2 gets phosphorylated, which makes arrestin-2 effect anti-apoptotic. The arrestin-2 mutant (P91G/P212E) is incapable of binding Src (Luttrell et al. 1999). This mutant leads to FPR-induced apoptosis. Here the FPR-arrestin-2 (P91G/P212E) complex is still internalizing but accumulates in perinuclear recycling endosomes, precluding normal receptor trafficking. In contrast, the FPR complex with WT arrestin-2 is transported to endosomes and does not cause cell death (Wagener et al. 2016). Therefore, Src is involved in receptor trafficking. It seems that the FPR activation, without arrestin regulation, causes Src- and ERK1/2-mediated apoptosis. For FPR signaling, Src plays an anti-apoptotic and pro-apoptotic role. Also here arrestin-2 shows again an anti-apoptotic effect.

Arrestin also binds to calmodulin, which could have a regulatory effect on ubiquitination of regulatory proteins like Src (Wu et al. 2006).

Table 19.1 Pro- and anti-apoptotic function of arrestin-2 and arrestin-3 by signaling pathway

Signaling cascade	Arrestin-2	Arrestin-3
JNK	Anti-apoptotic	Anti-apoptotic
NF-κB	Anti-apoptotic	Anti-apoptotic
p38	Anti-apoptotic	Anti-apoptotic
Mdm2 and p53	Anti-apoptotic	Pro-apoptotic
Src	Anti-apoptotic	–
Akt	Anti-apoptotic	Pro-apoptotic

AKT as Apoptotic Protein

The protein kinase B (PKB or AKT) is a serine/threonine kinase and a pro-survival regulator. Activated AKT phosphorylates a large number of apoptotic regulatory proteins. One example of a downstream effector is glycogen synthase kinase-3β (GSK-3β). GSK-3β is inactivated by AKT phosphorylation. It is a key regulator of cell survival and apoptosis. GSK-3 phosphorylates and activates JNK, C-Myb, CREB and Mdm2. The phosphorylation regulates the Mdm2-mediated degradation of p53 (Kulikov et al. 2005). In H9c2 cells it was found that the inhibition of GSK-3β causes an increase in nuclear arrestin-3 (Chang et al. 2016). This nuclear localization of arrestin-3 could result in an increased nuclear localization of Mdm2. This promotes the pro-apoptotic transcription factor p53. In this case arrestin-3 would show pro-apoptotic characteristics.

Additionally, arrestin-2 or arrestin-3 KO mice showed differences in terms of survival of liver injury due to BDL. For arrestin-3 KO an increase in survival rate was observed, due to enhanced AKT activity and reduced GSK-3β and p38 MAPK signaling. It was shown that the N-terminus of arrestin-3 is required for the AKT interaction (Yin et al. 2016).

Overall AKT has an anti-apoptotic function (Yin et al. 2006; Xie et al. 2010). The AKT-signaling is arrestin-2-dependent (Povsic et al. 2003). Arrestin-2 binds to the agonist activated insulin-like growth factor 1 (IGF-1) receptor and mediates the activation of phosphatidylinositol 3 kinase. This kinase activates AKT. Thus, arrestin-2 shows a pro-survival function.

Arrestin-3 scaffolds AKT and PP2A in the cytosol. By interacting with both proteins, arrestin-3 decreases the phosphorylation level of AKT. This results in a pro-apoptotic role of arrestin-3 (Table 19.1).

Conclusion

Cell death, and apoptosis in particular, is essential for the development of multi-cellular organisms and an enormous variety of pathological processes. A malfunction of the apoptosis regulation can have dramatic consequences for human health. In general, excessive apoptosis can cause neurodegenerative diseases, whereas its suppression can cause the development of tumors and cancer.

The improvement of living standards in the last century increased average life expectancy. Longer lifespan increases the chances that the complex signaling machinery makes mistakes, which could cause neurodegenerative diseases or cancer. Additionally, malfunction of the apoptotic signaling system, can also be caused by environmental circumstances (extrinsic apoptotic factors). Even in ancient Egyptian mummies researchers found signs of cancer (David and Zimmerman 2010).

Complex signaling machinery involving a variety of pathways, regulates apoptosis. Because arrestins are involved in the regulation of many of these signaling pathways, they play a major regulatory role in apoptosis. It is still not well understood at the structural level how arrestins fulfill numerous regulatory functions. Arrestins oligomerize, are post-translationally modified, assume different conformations, and are cleaved by caspases. All of these processes increase conformational variation, helping arrestins to interact with numerous partners with distinct functional consequences.

Arrestins are intimately involved in the regulation of apoptosis. They can influence the anti- or pro-apoptotic decision of the cell. In the majority of cases, arrestin-2 has an anti-apoptotic and arrestin-3 a pro-apoptotic functions. This makes arrestins attractive as, molecular tools and pharmaceutical drug targets (Gurevich and Gurevich 2012).

Cancer is the second (after heart disease) most common cause of death in the US. That is why a lot of research focuses on cancer and treatment optimization. One conceivable strategy of promoting cell death is the use of pro-apoptotic arrestin-2-(1–380) fragment. This protein can likely cause a natural apoptotic death of malignant cells. The key problem is precise tumor targeting, to deliver the ORF of this protein solely to tumor cells. In the last decades, there was a lot of improvement in the area, but further optimization is necessary for clinical use (Bae and Park 2011). Another option could be mono-functional N-terminal fragment of arrestin-3, which activates usually pro-apoptotic JNK family kinases (Zhan et al. 2016). In the case of degenerative diseases, caspase-resistant arrestin-2 mutant can be useful, as it is cyto-protective (Kook et al. 2014).

Considering the complexity of the functions of non-visual arrestins, it is hard to predict the effect of any particular mutant on apoptosis. For example, the effects on cell death and survival of different signaling-biased mutants of arrestins (Celver et al. 2002; Breitman et al. 2012) should be tested. Furthermore, the caspase cleaved arrestins and post-translationally modified and their impact on other signaling cascades should be tested.

Acknowledgements This work was supported by NIH grants to VVG (GM077561, GM109955, EY011500), EVG (NS065868, DA030103).

References

Adams JM, Cory S (2007) Bcl-2-regulated apoptosis: mechanism and therapeutic potential. Curr Opin Immunol 19(5):488–496

Alloway PG, Howard L, Dolph PJ (2000) The formation of stable rhodopsin-arrestin complexes induces apoptosis and photoreceptor cell degeneration. Neuron 28(1):129–138

Amling M, Neff L, Tanaka S, Inoue D, Kuida K, Weir E, Philbrick WM, Broadus AE, Baron R (1997) Bcl-2 lies downstream of parathyroid hormone-related peptide in a signaling pathway that regulates chondrocyte maturation during skeletal development. J Cell Biol 136(1):205–213

Aoki H, Kang PM, Hampe J, Yoshimura K, Noma T, Matsuzaki M, Izumo S (2002) Direct activation of mitochondrial apoptosis machinery by c-Jun N-terminal kinase in adult cardiac myocytes. J Biol Chem 277(12):10244–10250

Attramadal H, Arriza JL, Aoki C, Dawson TM, Codina J, Kwatra MM, Snyder SH, Caron MG, Lefkowitz RJ (1992) Beta-arrestin2, a novel member of the arrestin/beta-arrestin gene family. J Biol Chem 267(25):17882–17890

Azarian SM, King AJ, Hallett MA, Williams DS (1995) Selective proteolysis of arrestin by calpain. Molecular characteristics and its effect on rhodopsin dephosphorylation. J Biol Chem 270(41):24375–24384

Babbitt SE, Sutherland MC, San Francisco B, Mendez DL, Kranz RG (2015) Mitochondrial cytochrome c biogenesis: no longer an enigma. Trends Biochem Sci 40(8):446–455

Bae YH, Park K (2011) Targeted drug delivery to tumors: myths, reality and possibility. J Control Release 153(3):198–205

Baeuerle PA, Baltimore D (1988) I kappa B: a specific inhibitor of the NF-kappa B transcription factor. Science 242(4878):540–546

Barnhart BC, Peter ME (2003) The TNF receptor 1: a split personality complex. Cell 114(2): 148–150

Baylor DA, Lamb TD, Yau KW (1979) Responses of retinal rods to single photons. J Physiol 288:613–634

Bellas RE, Sonenshein GE (1999) Nuclear factor kappaB cooperates with c-Myc in promoting murine hepatocyte survival in a manner independent of p53 tumor suppressor function. Cell Growth Differ 10(5):287–294

Billen LP, Shamas-Din A, Andrews DW (2008) Bid: a Bax-like BH3 protein. Oncogene 27(Suppl 1):S93–104

Bossy-Wetzel E, Newmeyer DD, Green DR (1998) Mitochondrial cytochrome c release in apoptosis occurs upstream of DEVD-specific caspase activation and independently of mitochondrial transmembrane depolarization. EMBO J 17(1):37–49

Boularan C, Scott MG, Bourougaa K, Bellal M, Esteve E, Thuret A, Benmerah A, Tramier M, Coppey-Moisan M, Labbe-Jullie C, Fahraeus R, Marullo S (2007) beta-arrestin 2 oligomerization controls the Mdm2-dependent inhibition of p53. Proc Natl Acad Sci U S A 104 (46):18061–18066

Boulton TG, Nye SH, Robbins DJ, Ip NY, Radziejewska E, Morgenbesser SD, DePinho RA, Panayotatos N, Cobb MH, Yancopoulos GD (1991) ERKs: a family of protein-serine/threonine kinases that are activated and tyrosine phosphorylated in response to insulin and NGF. Cell 65 (4):663–675

Bourquard T, Landomiel F, Reiter E, Crepieux P, Ritchie DW, Aze J, Poupon A (2015) Unraveling the molecular architecture of a G protein-coupled receptor/beta-arrestin/Erk module complex. Sci Rep 5:10760

Bownds D (1967) Site of attachment of retinal in rhodopsin. Nature 216(5121):1178–1181

Bratton SB, Lewis J, Butterworth M, Duckett CS, Cohen GM (2002) XIAP inhibition of caspase-3 preserves its association with the Apaf-1 apoptosome and prevents CD95- and Bax-induced apoptosis. Cell Death Differ 9(9):881–892

Breitman M, Kook S, Gimenez LE, Lizama BN, Palazzo MC, Gurevich EV, Gurevich VV (2012) Silent scaffolds: inhibition OF c-Jun N-terminal kinase 3 activity in cell by dominant-negative arrestin-3 mutant. J Biol Chem 287(23):19653–19664

Brooks CL, Gu W (2006) p53 ubiquitination: Mdm2 and beyond. Mol Cell 21(3):307–315

Bruchas MR, Macey TA, Lowe JD, Chavkin C (2006) Kappa opioid receptor activation of p38 MAPK is GRK3- and arrestin-dependent in neurons and astrocytes. J Biol Chem 281 (26):18081–18089

Castro-Obregon S, Rao RV, del Rio G, Chen SF, Poksay KS, Rabizadeh S, Vesce S, Zhang XK, Swanson RA, Bredesen DE (2004) Alternative, nonapoptotic programmed cell death: mediation by arrestin 2, ERK2, and Nur77. J Biol Chem 279(17):17543–17553

Celver J, Vishnivetskiy SA, Chavkin C, Gurevich VV (2002) Conservation of the phosphate-sensitive elements in the arrestin family of proteins. J Biol Chem 277(11):9043–9048

Chan S, Rubin WW, Mendez A, Liu X, Song X, Hanson SM, Craft CM, Gurevich VV, Burns ME, Chen J (2007) Functional comparisons of visual arrestins in rod photoreceptors of transgenic mice. Invest Ophthalmol Vis Sci 48(5):1968–1975

Chang F, Liu J, Fu H, Wang J, Li F, Yue H, Li W, Zhao J, Yin D (2016) GSK-3beta promotes PA-induced apoptosis through changing beta-arrestin 2 nucleus location in H9c2 cardiomyocytes. Apoptosis 21(9):1045–1055

Chen Z, Hagler J, Palombella VJ, Melandri F, Scherer D, Ballard D, Maniatis T (1995) Signal-induced site-specific phosphorylation targets I kappa B alpha to the ubiquitin-proteasome pathway. Genes Dev 9(13):1586–1597

Chen F, Castranova V, Shi X, Demers LM (1999) New insights into the role of nuclear factor-kappaB, a ubiquitous transcription factor in the initiation of diseases. Clin Chem 45 (1):7–17

Chen HL, Demiralp B, Schneider A, Koh AJ, Silve C, Wang CY, McCauley LK (2002) Parathyroid hormone and parathyroid hormone-related protein exert both pro- and anti-apoptotic effects in mesenchymal cells. J Biol Chem 277(22):19374–19381

Chen J, Shi G, Concepcion FA, Xie G, Oprian D, Chen J (2006) Stable rhodopsin/arrestin complex leads to retinal degeneration in a transgenic mouse model of autosomal dominant retinitis pigmentosa. J Neurosci 26(46):11929–11937

Chen Q, Zhuo Y, Kim M, Hanson SM, Francis DJ, Vishnivetskiy SA, Altenbach C, Klug CS, Hubbell WL, Gurevich VV (2014) Self-association of arrestin family members. Handb Exp Pharmacol 219:205–223

Chinchore Y, Mitra A, Dolph PJ (2009) Accumulation of rhodopsin in late endosomes triggers photoreceptor cell degeneration. PLoS Genet 5(2):e1000377

Chinnaiyan AM (1999) The apoptosome: heart and soul of the cell death machine. Neoplasia 1 (1):5–15

Christofferson DE, Yuan J (2010) Necroptosis as an alternative form of programmed cell death. Curr Opin Cell Biol 22(2):263–268

Cianfrocca R, Tocci P, Semprucci E, Spinella F, Di Castro V, Bagnato A, Rosano L (2014) beta-Arrestin 1 is required for endothelin-1-induced NF-kappaB activation in ovarian cancer cells. Life Sci 118(2):179–184

Cleghorn WM, Tsakem EL, Song X, Vishnivetskiy SA, Seo J, Chen J, Gurevich EV, Gurevich VV (2011) Progressive reduction of its expression in rods reveals two pools of arrestin-1 in the outer segment with different roles in photoresponse recovery. PLoS One 6(7): e22797

Comitato A, Di Salvo MT, Turchiano G, Montanari M, Sakami S, Palczewski K, Marigo V (2016) Dominant and recessive mutations in rhodopsin activate different cell death pathways. Hum Mol Genet 25(13):2801–2812

David AR, Zimmerman MR (2010) Cancer: an old disease, a new disease or something in between? Nat Rev Cancer 10(10):728–733

Davis RJ (2000) Signal transduction by the JNK group of MAP kinases. Cell 103(2):239–252

DeFea KA, Vaughn ZD, O'Bryan EM, Nishijima D, Dery O, Bunnett NW (2000) The proliferative and antiapoptotic effects of substance P are facilitated by formation of a beta-arrestin-dependent scaffolding complex. Proc Natl Acad Sci U S A 97(20):11086–11091

Degterev A, Boyce M, Yuan J (2003) A decade of caspases. Oncogene 22(53):8543–8567

Deming JD, Pak JS, Shin JA, Brown BM, Kim MK, Aung MH, Lee EJ, Pardue MT, Craft CM (2015) Arrestin 1 and cone arrestin 4 have unique roles in visual function in an All-cone mouse retina. Invest Ophthalmol Vis Sci 56(13):7618–7628

Deng Y, Ren X, Yang L, Lin Y, Wu X (2003) A JNK-dependent pathway is required for TNFalpha-induced apoptosis. Cell 115(1):61–70

Derijard B, Hibi M, Wu IH, Barrett T, Su B, Deng T, Karin M, Davis RJ (1994) JNK1: a protein kinase stimulated by UV light and Ha-Ras that binds and phosphorylates the c-Jun activation domain. Cell 76(6):1025–1037

Deshaies RJ, Joazeiro CA (2009) RING domain E3 ubiquitin ligases. Annu Rev Biochem 78: 399–434

DeWire SM, Ahn S, Lefkowitz RJ, Shenoy SK (2007) Beta-arrestins and cell signaling. Annu Rev Physiol 69:483–510

Dewson G, Kluck RM (2009) Mechanisms by which Bak and Bax permeabilise mitochondria during apoptosis. J Cell Sci 122(Pt 16):2801–2808

Dickens LS, Boyd RS, Jukes-Jones R, Hughes MA, Robinson GL, Fairall L, Schwabe JW, Cain K, Macfarlane M (2012) A death effector domain chain DISC model reveals a crucial role for caspase-8 chain assembly in mediating apoptotic cell death. Mol Cell 47(2):291–305

Dubois SM, Alexia C, Wu Y, Leclair HM, Leveau C, Schol E, Fest T, Tarte K, Chen ZJ, Gavard J, Bidere N (2014) A catalytic-independent role for the LUBAC in NF-kappaB activation upon antigen receptor engagement and in lymphoma cells. Blood 123(14):2199–2203

Eistetter HR, Mills A, Brewster R, Alouani S, Rambosson C, Kawashima E (1992) Functional characterization of neurokinin-1 receptors on human U373MG astrocytoma cells. Glia 6(2): 89–95

Fain GL, Lisman JE (1999) Light, Ca^{2+}, and photoreceptor death: new evidence for the equivalent-light hypothesis from arrestin knockout mice. Invest Ophthalmol Vis Sci 40 (12):2770–2772

Fan M, Goodwin ME, Birrer MJ, Chambers TC (2001) The c-Jun NH(2)-terminal protein kinase/AP-1 pathway is required for efficient apoptosis induced by vinblastine. Cancer Res 61 (11):4450–4458

Fessart D, Simaan M, Laporte SA (2005) c-Src regulates clathrin adapter protein 2 interaction with beta-arrestin and the angiotensin II type 1 receptor during clathrin- mediated internalization. Mol Endocrinol 19(2):491–503

Filipek S, Stenkamp RE, Teller DC, Palczewski K (2003) G protein-coupled receptor rhodopsin: a prospectus. Annu Rev Physiol 65:851–879

Fuchs Y, Steller H (2015) Live to die another way: modes of programmed cell death and the signals emanating from dying cells. Nat Rev Mol Cell Biol 16(6):329–344

Fulda S, Debatin KM (2004) Targeting apoptosis pathways in cancer therapy. Curr Cancer Drug Targets 4(7):569–576

Gagiannis S, Muller M, Uhlemann S, Koch A, Melino G, Krammer PH, Nawroth PP, Brune M, Schilling T (2009) Parathyroid hormone-related protein confers chemoresistance by blocking apoptosis signaling via death receptors and mitochondria. Int J Cancer 125(7):1551–1557

Galluzzi L, Lopez-Soto A, Kumar S, Kroemer G (2016) Caspases connect cell-death signaling to organismal homeostasis. Immunity 44(2):221–231

Gao H, Sun Y, Wu Y, Luan B, Wang Y, Qu B, Pei G (2004) Identification of beta-arrestin2 as a G protein-coupled receptor-stimulated regulator of NF-kappaB pathways. Mol Cell 14(3): 303–317

Gerlach B, Cordier SM, Schmukle AC, Emmerich CH, Rieser E, Haas TL, Webb AI, Rickard JA, Anderton H, Wong WW, Nachbur U, Gangoda L, Warnken U, Purcell AW, Silke J, Walczak H (2011) Linear ubiquitination prevents inflammation and regulates immune signalling. Nature 471(7340):591–596

Gil-Parrado S, Fernandez-Montalvan A, Assfalg-Machleidt I, Popp O, Bestvater F, Holloschi A, Knoch TA, Auerswald EA, Welsh K, Reed JC, Fritz H, Fuentes-Prior P, Spiess E, Salvesen GS, Machleidt W (2002) Ionomycin-activated calpain triggers apoptosis. A probable role for Bcl-2 family members. J Biol Chem 277(30):27217–27226
Glading A, Chang P, Lauffenburger DA, Wells A (2000) Epidermal growth factor receptor activation of calpain is required for fibroblast motility and occurs via an ERK/MAP kinase signaling pathway. J Biol Chem 275(4):2390–2398
Glading A, Uberall F, Keyse SM, Lauffenburger DA, Wells A (2001) Membrane proximal ERK signaling is required for M-calpain activation downstream of epidermal growth factor receptor signaling. J Biol Chem 276(26):23341–23348
Gong K, Li Z, Xu M, Du J, Lv Z, Zhang Y (2008) A novel protein kinase A-independent, beta-arrestin-1-dependent signaling pathway for p38 mitogen-activated protein kinase activation by beta2-adrenergic receptors. J Biol Chem 283(43):29028–29036
Grumont RJ, Rourke IJ, Gerondakis S (1999) Rel-dependent induction of A1 transcription is required to protect B cells from antigen receptor ligation-induced apoptosis. Genes Dev 13 (4):400–411
Guicciardi ME, Gores GJ (2009) Life and death by death receptors. FASEB J 23(6):1625–1637
Gurevich VV, Gurevich EV (2006) The structural basis of arrestin-mediated regulation of G-protein-coupled receptors. Pharmacol Ther 110(3):465–502
Gurevich VV, Gurevich EV (2012) Synthetic biology with surgical precision: targeted reengineering of signaling proteins. Cell Signal 24(10):1899–1908
Hanson SM, Cleghorn WM, Francis DJ, Vishnivetskiy SA, Raman D, Song X, Nair KS, Slepak VZ, Klug CS, Gurevich VV (2007a) Arrestin mobilizes signaling proteins to the cytoskeleton and redirects their activity. J Mol Biol 368(2):375–387
Hanson SM, Van Eps N, Francis DJ, Altenbach C, Vishnivetskiy SA, Arshavsky VY, Klug CS, Hubbell WL, Gurevich VV (2007b) Structure and function of the visual arrestin oligomer. EMBO J 26(6):1726–1736
Hanson SM, Vishnivetskiy SA, Hubbell WL, Gurevich VV (2008) Opposing effects of inositol hexakisphosphate on rod arrestin and arrestin2 self-association. Biochemistry 47(3):1070–1075
Hara MR, Kovacs JJ, Whalen EJ, Rajagopal S, Strachan RT, Grant W, Towers AJ, Williams B, Lam CM, Xiao K, Shenoy SK, Gregory SG, Ahn S, Duckett DR, Lefkowitz RJ (2011) A stress response pathway regulates DNA damage through beta2-adrenoreceptors and beta-arrestin-1. Nature 477(7364):349–353
Hattar S, Liao HW, Takao M, Berson DM, Yau KW (2002) Melanopsin-containing retinal ganglion cells: architecture, projections, and intrinsic photosensitivity. Science 295 (5557):1065–1070
Hengartner MO (2000) The biochemistry of apoptosis. Nature 407(6805):770–776
Hill MM, Adrain C, Duriez PJ, Creagh EM, Martin SJ (2004) Analysis of the composition, assembly kinetics and activity of native Apaf-1 apoptosomes. EMBO J 23(10):2134–2145
Hsu H, Xiong J, Goeddel DV (1995) The TNF receptor 1-associated protein TRADD signals cell death and NF-kappa B activation. Cell 81(4):495–504
Hsu H, Shu HB, Pan MG, Goeddel DV (1996) TRADD-TRAF2 and TRADD-FADD interactions define two distinct TNF receptor 1 signal transduction pathways. Cell 84(2):299–308
Hu Y, Benedict MA, Ding L, Nunez G (1999) Role of cytochrome c and dATP/ATP hydrolysis in Apaf-1-mediated caspase-9 activation and apoptosis. EMBO J 18(13):3586–3595
Huang SP, Brown BM, Craft CM (2010) Visual Arrestin 1 acts as a modulator for N-ethylmaleimide-sensitive factor in the photoreceptor synapse. J Neurosci 30(28):9381–9391
Hurley JB, Spencer M, Niemi GA (1998) Rhodopsin phosphorylation and its role in photoreceptor function. Vis Res 38(10):1341–1352
Imamoto Y, Tamura C, Kamikubo H, Kataoka M (2003) Concentration-dependent tetramerization of bovine visual arrestin. Biophys J 85(2):1186–1195
Jiang T, Yu JT, Tan MS, Zhu XC, Tan L (2013) beta-Arrestins as potential therapeutic targets for Alzheimer's disease. Mol Neurobiol 48(3):812–818

Joseph J, Niggemann B, Zaenker KS, Entschladen F (2002) The neurotransmitter gamma-aminobutyric acid is an inhibitory regulator for the migration of SW 480 colon carcinoma cells. Cancer Res 62(22):6467–6469

Karin M, Lin A (2002) NF-kappaB at the crossroads of life and death. Nat Immunol 3(3):221–227

Kasibhatla S, Genestier L, Green DR (1999) Regulation of fas-ligand expression during activation-induced cell death in T lymphocytes via nuclear factor kappaB. J Biol Chem 274 (2):987–992

Kaufmann L, Marinescu G, Nazarenko I, Thiele W, Oberle C, Sleeman J, Blattner C (2011) LiCl induces TNF-alpha and FasL production, thereby stimulating apoptosis in cancer cells. Cell Commun Signal 9:15

Kim HE, Du F, Fang M, Wang X (2005) Formation of apoptosome is initiated by cytochrome c-induced dATP hydrolysis and subsequent nucleotide exchange on Apaf-1. Proc Natl Acad Sci U S A 102(49):17545–17550

Kim M, Hanson SM, Vishnivetskiy SA, Song X, Cleghorn WM, Hubbell WL, Gurevich VV (2011) Robust self-association is a common feature of mammalian visual arrestin-1. Biochemistry 50(12):2235–2242

Kiraz Y, Adan A, Kartal Yandim M, Baran Y (2016) Major apoptotic mechanisms and genes involved in apoptosis. Tumour Biol 37(7):8471–8486

Kirisako T, Kamei K, Murata S, Kato M, Fukumoto H, Kanie M, Sano S, Tokunaga F, Tanaka K, Iwai K (2006) A ubiquitin ligase complex assembles linear polyubiquitin chains. EMBO J 25 (20):4877–4887

Kiselev A, Socolich M, Vinos J, Hardy RW, Zuker CS, Ranganathan R (2000) A molecular pathway for light-dependent photoreceptor apoptosis in Drosophila. Neuron 28(1):139–152

Kojima K, Ishizawa J, Andreeff M (2016) Pharmacological activation of wild-type p53 in the therapy of leukemia. Exp Hematol 44(9):791–798

Kook S, Zhan X, Cleghorn WM, Benovic JL, Gurevich VV, Gurevich EV (2014) Caspase-cleaved arrestin-2 and BID cooperatively facilitate cytochrome C release and cell death. Cell Death Differ 21(1):172–184

Kroemer G, Galluzzi L, Vandenabeele P, Abrams J, Alnemri ES, Baehrecke EH, Blagosklonny MV, El-Deiry WS, Golstein P, Green DR, Hengartner M, Knight RA, Kumar S, Lipton SA, Malorni W, Nunez G, Peter ME, Tschopp J, Yuan J, Piacentini M, Zhivotovsky B, Melino G (2009) Classification of cell death: recommendations of the Nomenclature Committee on Cell Death 2009. Cell Death Differ 16(1):3–11

Krupnick JG, Gurevich VV, Benovic JL (1997) Mechanism of quenching of phototransduction. Binding competition between arrestin and transducin for phosphorhodopsin. J Biol Chem 272 (29):18125–18131

Kuhn H, Wilden U (1987) Deactivation of photoactivated rhodopsin by rhodopsin-kinase and arrestin. J Recept Res 7(1–4):283–298

Kuhn H, Hall SW, Wilden U (1984) Light-induced binding of 48-kDa protein to photoreceptor membranes is highly enhanced by phosphorylation of rhodopsin. FEBS Lett 176(2):473–478

Kulikov R, Boehme KA, Blattner C (2005) Glycogen synthase kinase 3-dependent phosphorylation of Mdm2 regulates p53 abundance. Mol Cell Biol 25(16):7170–7180

Laporte SA, Oakley RH, Zhang J, Holt JA, Ferguson SS, Caron MG, Barak LS (1999) The beta2-adrenergic receptor/betaarrestin complex recruits the clathrin adaptor AP-2 during endocytosis. Proc Natl Acad Sci U S A 96(7):3712–3717

Laporte SA, Oakley RH, Holt JA, Barak LS, Caron MG (2000) The interaction of beta-arrestin with the AP-2 adaptor is required for the clustering of beta 2-adrenergic receptor into clathrin-coated pits. J Biol Chem 275(30):23120–23126

Lee JT, Gu W (2010) The multiple levels of regulation by p53 ubiquitination. Cell Death Differ 17 (1):86–92

Lee HH, Dadgostar H, Cheng Q, Shu J, Cheng G (1999) NF-kappaB-mediated up-regulation of Bcl-x and Bfl-1/A1 is required for CD40 survival signaling in B lymphocytes. Proc Natl Acad Sci U S A 96(16):9136–9141

Lefkowitz RJ, Shenoy SK (2005) Transduction of receptor signals by beta-arrestins. Science 308 (5721):512–517

Lefkowitz RJ, Rajagopal K, Whalen EJ (2006) New roles for beta-arrestins in cell signaling: not just for seven-transmembrane receptors. Mol Cell 24(5):643–652

Li T, Franson WK, Gordon JW, Berson EL, Dryja TP (1995) Constitutive activation of phototransduction by K296E opsin is not a cause of photoreceptor degeneration. Proc Natl Acad Sci U S A 92(8):3551–3555

Li H, Zhu H, Xu CJ, Yuan J (1998) Cleavage of BID by caspase 8 mediates the mitochondrial damage in the Fas pathway of apoptosis. Cell 94(4):491–501

Li H, Hu D, Fan H, Zhang Y, LeSage GD, Caudle Y, Stuart C, Liu Z, Yin D (2014) beta-Arrestin 2 negatively regulates Toll-like receptor 4 (TLR4)-triggered inflammatory signaling via targeting p38 MAPK and interleukin 10. J Biol Chem 289(33):23075–23085

Lin A, Dibling B (2002) The true face of JNK activation in apoptosis. Aging Cell 1(2):112–116

Lisman J, Fain G (1995) Support for the equivalent light hypothesis for RP. Nat Med 1(12):1254–1255

Liu J, Lin A (2005) Role of JNK activation in apoptosis: a double-edged sword. Cell Res 15 (1):36–42

Lohse MJ, Benovic JL, Codina J, Caron MG, Lefkowitz RJ (1990) beta-Arrestin: a protein that regulates beta-adrenergic receptor function. Science 248(4962):1547–1550

Lu FF, Su P, Liu F, Daskalakis ZJ (2012) Activation of GABA(B) receptors inhibits protein kinase B/glycogen synthase kinase 3 signaling. Mol Brain 5:41

Luan B, Zhang Z, Wu Y, Kang J, Pei G (2005) Beta-arrestin2 functions as a phosphorylation-regulated suppressor of UV-induced NF-kappaB activation. EMBO J 24 (24):4237–4246

Luo X, Budihardjo I, Zou H, Slaughter C, Wang X (1998) Bid, a Bcl2 interacting protein, mediates cytochrome c release from mitochondria in response to activation of cell surface death receptors. Cell 94(4):481–490

Luthi AU, Martin SJ (2007) The CASBAH: a searchable database of caspase substrates. Cell Death Differ 14(4):641–650

Luttrell DK, Luttrell LM (2004) Not so strange bedfellows: G-protein-coupled receptors and Src family kinases. Oncogene 23(48):7969–7978

Luttrell LM, Ferguson SS, Daaka Y, Miller WE, Maudsley S, Della Rocca GJ, Lin F, Kawakatsu H, Owada K, Luttrell DK, Caron MG, Lefkowitz RJ (1999) Beta-arrestin-dependent formation of beta2 adrenergic receptor-Src protein kinase complexes. Science 283(5402):655–661

Malladi S, Challa-Malladi M, Fearnhead HO, Bratton SB (2009) The Apaf-1*procaspase-9 apoptosome complex functions as a proteolytic-based molecular timer. EMBO J 28(13):1916–1925

Mandic A, Viktorsson K, Strandberg L, Heiden T, Hansson J, Linder S, Shoshan MC (2002) Calpain-mediated Bid cleavage and calpain-independent Bak modulation: two separate pathways in cisplatin-induced apoptosis. Mol Cell Biol 22(9):3003–3013

Marsters SA, Sheridan JP, Pitti RM, Brush J, Goddard A, Ashkenazi A (1998) Identification of a ligand for the death-domain-containing receptor Apo3. Curr Biol 8(9):525–528

Martinou JC, Youle RJ (2011) Mitochondria in apoptosis: Bcl-2 family members and mitochondrial dynamics. Dev Cell 21(1):92–101

Matsumoto H, Yoshizawa T (1975) Existence of a beta-ionone ring-binding site in the rhodopsin molecule. Nature 258(5535):523–526

McDonald PH, Chow CW, Miller WE, Laporte SA, Field ME, Lin FT, Davis RJ, Lefkowitz RJ (2000) Beta-arrestin 2: a receptor-regulated MAPK scaffold for the activation of JNK3. Science 290(5496):1574–1577

McIlwain DR, Berger T, Mak TW (2013) Caspase functions in cell death and disease. Cold Spring Harb Perspect Biol 5(4):a008656

McLaughlin NJ, Banerjee A, Kelher MR, Gamboni-Robertson F, Hamiel C, Sheppard FR, Moore EE, Silliman CC (2006) Platelet-activating factor-induced clathrin-mediated endocytosis requires beta-arrestin-1 recruitment and activation of the p38 MAPK signalosome at the plasma membrane for actin bundle formation. J Immunol 176(11):7039–7050

Mendes HF, van der Spuy J, Chapple JP, Cheetham ME (2005) Mechanisms of cell death in rhodopsin retinitis pigmentosa: implications for therapy. Trends Mol Med 11(4):177–185

Micheau O, Tschopp J (2003) Induction of TNF receptor I-mediated apoptosis via two sequential signaling complexes. Cell 114(2):181–190

Milano SK, Kim YM, Stefano FP, Benovic JL, Brenner C (2006) Nonvisual arrestin oligomerization and cellular localization are regulated by inositol hexakisphosphate binding. J Biol Chem 281(14):9812–9823

Miller WE, Houtz DA, Nelson CD, Kolattukudy PE, Lefkowitz RJ (2003) G-protein-coupled receptor (GPCR) kinase phosphorylation and beta-arrestin recruitment regulate the constitutive signaling activity of the human cytomegalovirus US28 GPCR. J Biol Chem 278(24):21663–21671

Miyashita T, Krajewski S, Krajewska M, Wang HG, Lin HK, Liebermann DA, Hoffman B, Reed JC (1994) Tumor suppressor p53 is a regulator of bcl-2 and bax gene expression in vitro and in vivo. Oncogene 9(6):1799–1805

Moaven H, Koike Y, Jao CC, Gurevich VV, Langen R, Chen J (2013) Visual arrestin interaction with clathrin adaptor AP-2 regulates photoreceptor survival in the vertebrate retina. Proc Natl Acad Sci U S A 110(23):9463–9468

Moriwaki K, Chan FK (2013) RIP3: a molecular switch for necrosis and inflammation. Genes Dev 27(15):1640–1649

Nagata S (1997) Apoptosis by death factor. Cell 88(3):355–365

Naydenov NG, Harris G, Brown B, Schaefer KL, Das SK, Fisher PB, Ivanov AI (2012) Loss of soluble N-ethylmaleimide-sensitive factor attachment protein alpha (alphaSNAP) induces epithelial cell apoptosis via down-regulation of Bcl-2 expression and disruption of the Golgi. J Biol Chem 287(8):5928–5941

Ni Y, Zhao X, Bao G, Zou L, Teng L, Wang Z, Song M, Xiong J, Bai Y, Pei G (2006) Activation of beta2-adrenergic receptor stimulates gamma-secretase activity and accelerates amyloid plaque formation. Nat Med 12(12):1390–1396

Nikoletopoulou V, Markaki M, Palikaras K, Tavernarakis N (2013) Crosstalk between apoptosis, necrosis and autophagy. Biochim Biophys Acta 1833(12):3448–3459

Nikonov SS, Brown BM, Davis JA, Zuniga FI, Bragin A, Pugh EN Jr, Craft CM (2008) Mouse cones require an arrestin for normal inactivation of phototransduction. Neuron 59(3):462–474

Nishitoh H, Matsuzawa A, Tobiume K, Saegusa K, Takeda K, Inoue K, Hori S, Kakizuka A, Ichijo H (2002) ASK1 is essential for endoplasmic reticulum stress-induced neuronal cell death triggered by expanded polyglutamine repeats. Genes Dev 16(11):1345–1355

Niu J, Shi Y, Iwai K, Wu ZH (2011) LUBAC regulates NF-kappaB activation upon genotoxic stress by promoting linear ubiquitination of NEMO. EMBO J 30(18):3741–3753

Noell WK, Albrecht R (1971) Irreversible effects on visible light on the retina: role of vitamin A. Science 172(3978):76–79

Norbury CJ, Zhivotovsky B (2004) DNA damage-induced apoptosis. Oncogene 23(16):2797–2808

Orem NR, Xia L, Dolph PJ (2006) An essential role for endocytosis of rhodopsin through interaction of visual arrestin with the AP-2 adaptor. J Cell Sci 119(Pt 15):3141–3148

Papadakis ES, Finegan KG, Wang X, Robinson AC, Guo C, Kayahara M, Tournier C (2006) The regulation of Bax by c-Jun N-terminal protein kinase (JNK) is a prerequisite to the mitochondrial-induced apoptotic pathway. FEBS Lett 580(5):1320–1326

Paquet-Durand F, Azadi S, Hauck SM, Ueffing M, van Veen T, Ekstrom P (2006) Calpain is activated in degenerating photoreceptors in the rd1 mouse. J Neurochem 96(3):802–814

Park YH, Jeong MS, Jang SB (2014) Death domain complex of the TNFR-1, TRADD, and RIP1 proteins for death-inducing signaling. Biochem Biophys Res Commun 443(4):1155–1161

Piliponsky AM, Levi-Schaffer F (2000) Regulation of apoptosis in mast cells. Apoptosis 5(5): 435–441

Pinton P, Giorgi C, Siviero R, Zecchini E, Rizzuto R (2008) Calcium and apoptosis: ER-mitochondria Ca^{2+} transfer in the control of apoptosis. Oncogene 27(50):6407–6418
Pitti RM, Marsters SA, Ruppert S, Donahue CJ, Moore A, Ashkenazi A (1996) Induction of apoptosis by Apo-2 ligand, a new member of the tumor necrosis factor cytokine family. J Biol Chem 271(22):12687–12690
Pop C, Oberst A, Drag M, Van Raam BJ, Riedl SJ, Green DR, Salvesen GS (2011) FLIP(L) induces caspase 8 activity in the absence of interdomain caspase 8 cleavage and alters substrate specificity. Biochem J 433(3):447–457
Povsic TJ, Kohout TA, Lefkowitz RJ (2003) Beta-arrestin1 mediates insulin-like growth factor 1 (IGF-1) activation of phosphatidylinositol 3-kinase (PI3K) and anti-apoptosis. J Biol Chem 278(51):51334–51339
Prakasam A, Ghose S, Oleinik NV, Bethard JR, Peterson YK, Krupenko NI, Krupenko SA (2014) JNK1/2 regulate Bid by direct phosphorylation at Thr59 in response to ALDH1L1. Cell Death Dis 5:e1358
Qi Z, Qi S, Gui L, Shen L (2016) beta-arrestin2 regulates TRAIL-induced HepG2 cell apoptosis via the Src-extracellular signal-regulated signaling pathway. Mol Med Rep 14(1):263–270
Revankar CM, Vines CM, Cimino DF, Prossnitz ER (2004) Arrestins block G protein-coupled receptor-mediated apoptosis. J Biol Chem 279(23):24578–24584
Rodriguez J, Lazebnik Y (1999) Caspase-9 and APAF-1 form an active holoenzyme. Genes Dev 13(24):3179–3184
Rosano L, Bagnato A (2016) beta-arrestin1 at the cross-road of endothelin-1 signaling in cancer. J Exp Clin Cancer Res 35(1):121
Ryan KM, Phillips AC, Vousden KH (2001) Regulation and function of the p53 tumor suppressor protein. Curr Opin Cell Biol 13(3):332–337
Saelens X, Festjens N, Vande Walle L, van Gurp M, van Loo G, Vandenabeele P (2004) Toxic proteins released from mitochondria in cell death. Oncogene 23(16):2861–2874
Sakmar TP, Franke RR, Khorana HG (1989) Glutamic acid-113 serves as the retinylidene Schiff base counterion in bovine rhodopsin. Proc Natl Acad Sci U S A 86(21):8309–8313
Salvesen GS, Dixit VM (1999) Caspase activation: the induced-proximity model. Proc Natl Acad Sci U S A 96(20):10964–10967
Sancho-Pelluz J, Arango-Gonzalez B, Kustermann S, Romero FJ, van Veen T, Zrenner E, Ekstrom P, Paquet-Durand F (2008) Photoreceptor cell death mechanisms in inherited retinal degeneration. Mol Neurobiol 38(3):253–269
Schroeter H, Boyd CS, Ahmed R, Spencer JP, Duncan RF, Rice-Evans C, Cadenas E (2003) c-Jun N-terminal kinase (JNK)-mediated modulation of brain mitochondria function: new target proteins for JNK signalling in mitochondrion-dependent apoptosis. Biochem J 372(Pt 2): 359–369
Schubert C, Hirsch JA, Gurevich VV, Engelman DM, Sigler PB, Fleming KG (1999) Visual arrestin activity may be regulated by self-association. J Biol Chem 274(30):21186–21190
Schug ZT, Gonzalvez F, Houtkooper RH, Vaz FM, Gottlieb E (2011) BID is cleaved by caspase-8 within a native complex on the mitochondrial membrane. Cell Death Differ 18(3):538–548
Scott MG, Le Rouzic E, Perianin A, Pierotti V, Enslen H, Benichou S, Marullo S, Benmerah A (2002) Differential nucleocytoplasmic shuttling of beta-arrestins. Characterization of a leucine-rich nuclear export signal in beta-arrestin2. J Biol Chem 277(40):37693–37701
Sedger LM, McDermott MF (2014) TNF and TNF-receptors: From mediators of cell death and inflammation to therapeutic giants - past, present and future. Cytokine Growth Factor Rev 25 (4):453–472
Seo J, Tsakem EL, Breitman M, Gurevich VV (2011) Identification of arrestin-3-specific residues necessary for JNK3 kinase activation. J Biol Chem 286(32):27894–27901
Sever R, Brugge JS (2015) Signal transduction in cancer. Cold Spring Harb Perspect Med 5(4)
Shenoy SK, Barak LS, Xiao K, Ahn S, Berthouze M, Shukla AK, Luttrell LM, Lefkowitz RJ (2007) Ubiquitination of beta-arrestin links seven-transmembrane receptor endocytosis and ERK activation. J Biol Chem 282(40):29549–29562

Shenoy SK, Modi AS, Shukla AK, Xiao K, Berthouze M, Ahn S, Wilkinson KD, Miller WE, Lefkowitz RJ (2009) Beta-arrestin-dependent signaling and trafficking of 7-transmembrane receptors is reciprocally regulated by the deubiquitinase USP33 and the E3 ligase Mdm2. Proc Natl Acad Sci U S A 106(16):6650–6655

Smith CA, Gruss HJ, Davis T, Anderson D, Farrah T, Baker E, Sutherland GR, Brannan CI, Copeland NG, Jenkins NA et al (1993) CD30 antigen, a marker for Hodgkin's lymphoma, is a receptor whose ligand defines an emerging family of cytokines with homology to TNF. Cell 73 (7):1349–1360

Song X, Raman D, Gurevich EV, Vishnivetskiy SA, Gurevich VV (2006) Visual and both non-visual arrestins in their "inactive" conformation bind JNK3 and Mdm2 and relocalize them from the nucleus to the cytoplasm. J Biol Chem 281(30):21491–21499

Song X, Gurevich EV, Gurevich VV (2007) Cone arrestin binding to JNK3 and Mdm2: conformational preference and localization of interaction sites. J Neurochem 103(3):1053–1062

Song X, Coffa S, Fu H, Gurevich VV (2009a) How does arrestin assemble MAPKs into a signaling complex? J Biol Chem 284(1):685–695

Song X, Vishnivetskiy SA, Gross OP, Emelianoff K, Mendez A, Chen J, Gurevich EV, Burns ME, Gurevich VV (2009b) Enhanced arrestin facilitates recovery and protects rods lacking rhodopsin phosphorylation. Curr Biol 19(8):700–705

Song X, Vishnivetskiy SA, Seo J, Chen J, Gurevich EV, Gurevich VV (2011) Arrestin-1 expression level in rods: balancing functional performance and photoreceptor health. Neuroscience 174:37–49

Song X, Seo J, Baameur F, Vishnivetskiy SA, Chen Q, Kook S, Kim M, Brooks EK, Altenbach C, Hong Y, Hanson SM, Palazzo MC, Chen J, Hubbell WL, Gurevich EV, Gurevich VV (2013) Rapid degeneration of rod photoreceptors expressing self-association-deficient arrestin-1 mutant. Cell Signal 25(12):2613–2624

Sothilingam V, Garcia Garrido M, Jiao K, Buena-Atienza E, Sahaboglu A, Trifunovic D, Balendran S, Koepfli T, Muhlfriedel R, Schon C, Biel M, Heckmann A, Beck SC, Michalakis S, Wissinger B, Seeliger MW, Paquet-Durand F (2015) Retinitis pigmentosa: impact of different Pde6a point mutations on the disease phenotype. Hum Mol Genet 24 (19):5486–5499

Sovova V, Friis R, Fidlerova H, Hlozanek I (1993) c-myc gene activation as a permanent trait of RSV-infected quail cells. Int J Cancer 53(6):983–987

Spiegel AM (1987) Signal transduction by guanine nucleotide binding proteins. Mol Cell Endocrinol 49(1):1–16

Strissel KJ, Sokolov M, Trieu LH, Arshavsky VY (2006) Arrestin translocation is induced at a critical threshold of visual signaling and is superstoichiometric to bleached rhodopsin. J Neurosci 26(4):1146–1153

Strous GJ, Schantl JA (2001) Beta-arrestin and Mdm2, unsuspected partners in signaling from the cell surface. Sci STKE 2001(110):pe41

Strungs EG, Luttrell LM (2014) Arrestin-dependent activation of ERK and Src family kinases. Handb Exp Pharmacol 219:225–257

Stryer L (1986) Cyclic GMP cascade of vision. Annu Rev Neurosci 9:87–119

Sun Y, Cheng Z, Ma L, Pei G (2002) Beta-arrestin2 is critically involved in CXCR4-mediated chemotaxis, and this is mediated by its enhancement of p38 MAPK activation. J Biol Chem 277(51):49212–49219

Tan E, Wang Q, Quiambao AB, Xu X, Qtaishat NM, Peachey NS, Lem J, Fliesler SJ, Pepperberg DR, Naash MI, Al-Ubaidi MR (2001) The relationship between opsin overexpression and photoreceptor degeneration. Invest Ophthalmol Vis Sci 42(3):589–600

Thathiah A, De Strooper B (2009) G protein-coupled receptors, cholinergic dysfunction, and Abeta toxicity in Alzheimer's disease. Sci Signal 2(93):re8

Thathiah A, De Strooper B (2011) The role of G protein-coupled receptors in the pathology of Alzheimer's disease. Nat Rev Neurosci 12(2):73–87

Thathiah A, Horre K, Snellinx A, Vandewyer E, Huang Y, Ciesielska M, De Kloe G, Munck S, De Strooper B (2013) beta-arrestin 2 regulates Abeta generation and gamma-secretase activity in Alzheimer's disease. Nat Med 19(1):43–49

Thornberry NA (1998) Caspases: key mediators of apoptosis. Chem Biol 5(5):R97–103

Thornberry NA, Lazebnik Y (1998) Caspases: enemies within. Science 281(5381):1312–1316

Tian H, Wu JX, Shan FX, Zhang SN, Cheng Q, Zheng JN, Pei DS (2015) Gamma-aminobutyric acid induces tumor cells apoptosis via GABABR1.beta-arrestins.JNKs signaling module. Cell Biochem Biophys 71(2):679–688

Tibbles LA, Woodgett JR (1999) The stress-activated protein kinase pathways. Cell Mol Life Sci 55(10):1230–1254

Turner PR, Mefford S, Bambino T, Nissenson RA (1998) Transmembrane residues together with the amino terminus limit the response of the parathyroid hormone (PTH) 2 receptor to PTH-related peptide. J Biol Chem 273(7):3830–3837

Vassilev LT (2007) MDM2 inhibitors for cancer therapy. Trends Mol Med 13(1):23–31

Wagener BM, Marjon NA, Prossnitz ER (2016) Regulation of N-formyl peptide receptor signaling and trafficking by arrestin-Src kinase interaction. PLoS One 11(1):e0147442

Walczak H (2013) Death receptor-ligand systems in cancer, cell death, and inflammation. Cold Spring Harb Perspect Biol 5(5):a008698

Wang P, Gao H, Ni Y, Wang B, Wu Y, Ji L, Qin L, Ma L, Pei G (2003a) Beta-arrestin 2 functions as a G-protein-coupled receptor-activated regulator of oncoprotein Mdm2. J Biol Chem 278 (8):6363–6370

Wang P, Wu Y, Ge X, Ma L, Pei G (2003b) Subcellular localization of beta-arrestins is determined by their intact N domain and the nuclear export signal at the C terminus. J Biol Chem 278 (13):11648–11653

Wang L, Du F, Wang X (2008a) TNF-alpha induces two distinct caspase-8 activation pathways. Cell 133(4):693–703

Wang Y, Sun X, Wu J, Xu BE, Gu C, Wang H, Wang X, Tan F, Peng X, Qiang B, Yuan J, Luo Y (2008b) Casein kinase 1alpha interacts with RIP1 and regulates NF-kappaB activation. Biochemistry 47(1):441–448

Wang D, Yuen EY, Zhou Y, Yan Z, Xiang YK (2011) Amyloid beta peptide-(1-42) induces internalization and degradation of beta2 adrenergic receptors in prefrontal cortical neurons. J Biol Chem 286(36):31852–31863

Warren HS, Smyth MJ (1999) NK cells and apoptosis. Immunol Cell Biol 77(1):64–75

Wei MC, Zong WX, Cheng EH, Lindsten T, Panoutsakopoulou V, Ross AJ, Roth KA, MacGregor GR, Thompson CB, Korsmeyer SJ (2001) Proapoptotic BAX and BAK: a requisite gateway to mitochondrial dysfunction and death. Science 292(5517):727–730

Wertz IE, Dixit VM (2008) Ubiquitin-mediated regulation of TNFR1 signaling. Cytokine Growth Factor Rev 19(3–4):313–324

Wilden U (1995) Duration and amplitude of the light-induced cGMP hydrolysis in vertebrate photoreceptors are regulated by multiple phosphorylation of rhodopsin and by arrestin binding. Biochemistry 34(4):1446–1454

Wilden U, Hall SW, Kuhn H (1986) Phosphodiesterase activation by photoexcited rhodopsin is quenched when rhodopsin is phosphorylated and binds the intrinsic 48-kDa protein of rod outer segments. Proc Natl Acad Sci U S A 83(5):1174–1178

Wiley SR, Schooley K, Smolak PJ, Din WS, Huang CP, Nicholl JK, Sutherland GR, Smith TD, Rauch C, Smith CA et al (1995) Identification and characterization of a new member of the TNF family that induces apoptosis. Immunity 3(6):673–682

Witherow DS, Garrison TR, Miller WE, Lefkowitz RJ (2004) beta-Arrestin inhibits NF-kappaB activity by means of its interaction with the NF-kappaB inhibitor IkappaBalpha. Proc Natl Acad Sci U S A 101(23):8603–8607

Wu H, Lozano G (1994) NF-kappa B activation of p53. A potential mechanism for suppressing cell growth in response to stress. J Biol Chem 269(31):20067–20074

Wu N, Hanson SM, Francis DJ, Vishnivetskiy SA, Thibonnier M, Klug CS, Shoham M, Gurevich VV (2006) Arrestin binding to calmodulin: a direct interaction between two ubiquitous signaling proteins. J Mol Biol 364(5):955–963

Xiao K, McClatchy DB, Shukla AK, Zhao Y, Chen M, Shenoy SK, Yates JR 3rd, Lefkowitz RJ (2007) Functional specialization of beta-arrestin interactions revealed by proteomic analysis. Proc Natl Acad Sci U S A 104(29):12011–12016

Xie N, Li H, Wei D, LeSage G, Chen L, Wang S, Zhang Y, Chi L, Ferslew K, He L, Chi Z, Yin D (2010) Glycogen synthase kinase-3 and p38 MAPK are required for opioid-induced microglia apoptosis. Neuropharmacology 59(6):444–451

Xu J, Dodd RL, Makino CL, Simon MI, Baylor DA, Chen J (1997) Prolonged photoresponses in transgenic mouse rods lacking arrestin. Nature 389(6650):505–509

Yanamadala V, Negoro H, Denker BM (2009) Heterotrimeric G proteins and apoptosis: intersecting signaling pathways leading to context dependent phenotypes. Curr Mol Med 9 (5):527–545

Yang X, Zhou G, Ren T, Li H, Zhang Y, Yin D, Qian H, Li Q (2012) beta-Arrestin prevents cell apoptosis through pro-apoptotic ERK1/2 and p38 MAPKs and anti-apoptotic Akt pathways. Apoptosis 17(9):1019–1026

Ye RD (2001) Regulation of nuclear factor kappaB activation by G-protein-coupled receptors. J Leukoc Biol 70(6):839–848

Yin D, Woodruff M, Zhang Y, Whaley S, Miao J, Ferslew K, Zhao J, Stuart C (2006) Morphine promotes Jurkat cell apoptosis through pro-apoptotic FADD/P53 and anti-apoptotic PI3K/Akt/NF-kappaB pathways. J Neuroimmunol 174(1–2):101–107

Yin D, Yang X, Li H, Fan H, Zhang X, Feng Y, Stuart C, Hu D, Caudle Y, Xie N, Liu Z, LeSage G (2016) beta-Arrestin 2 Promotes Hepatocyte Apoptosis by Inhibiting Akt Protein. J Biol Chem 291(2):605–612

Young RW (1967) The renewal of photoreceptor cell outer segments. J Cell Biol 33(1):61–72

Yuan S, Yu X, Asara JM, Heuser JE, Ludtke SJ, Akey CW (2011) The holo-apoptosome: activation of procaspase-9 and interactions with caspase-3. Structure 19(8):1084–1096

Zhan X, Kaoud TS, Dalby KN, Gurevich VV (2011) Nonvisual arrestins function as simple scaffolds assembling the MKK4-JNK3alpha2 signaling complex. Biochemistry 50(48):10520–10529

Zhan X, Kaoud TS, Kook S, Dalby KN, Gurevich VV (2013) JNK3 enzyme binding to arrestin-3 differentially affects the recruitment of upstream mitogen-activated protein (MAP) kinase kinases. J Biol Chem 288(40):28535–28547

Zhan X, Stoy H, Kaoud TS, Perry NA, Chen Q, Perez A, Els-Heindl S, Slagis JV, Iverson TM, Beck-Sickinger AG, Gurevich EV, Dalby KN, Gurevich VV (2016) Peptide mini-scaffold facilitates JNK3 activation in cells. Sci Rep 6(1)

Zhang Z, Hao J, Zhao Z, Ben P, Fang F, Shi L, Gao Y, Liu J, Wen C, Luo L, Yin Z (2009) beta-Arrestins facilitate ubiquitin-dependent degradation of apoptosis signal-regulating kinase 1 (ASK1) and attenuate H_2O_2-induced apoptosis. Cell Signal 21(7):1195–1206

Zhukovsky EA, Oprian DD (1989) Effect of carboxylic acid side chains on the absorption maximum of visual pigments. Science 246(4932):928–930

Zou H, Yang R, Hao J, Wang J, Sun C, Fesik SW, Wu JC, Tomaselli KJ, Armstrong RC (2003) Regulation of the Apaf-1/caspase-9 apoptosome by caspase-3 and XIAP. J Biol Chem 278 (10):8091–8098

Index

V.V. Gurevich (ed.), *The Structural Basis of Arrestin Functions*,
DOI 10.1007/978-3-319-57553-7

The manufacturer's authorised representative in the EU is Springer Nature Customer Service Centre GmbH, Europaplatz 3, 69115 Heidelberg, Germany. If you have any concerns regarding our products, please contact ProductSafety@springernature.com

Printed and bound by CPI Group (UK) Ltd, Croydon, CR0 4YY
05/10/2026
02228133-0003